Worldwide Trends in Agronomy Research: Bibliometric Studies

Worldwide Trends in Agronomy Research: Bibliometric Studies

Editors

Francisco Manzano Agugliaro
Esther Salmerón-Manzano

MDPI • Basel • Beijing • Wuhan • Barcelona • Belgrade • Manchester • Tokyo • Cluj • Tianjin

Editors
Francisco Manzano Agugliaro
Department of Engineering
Universidad de Almería
Almeria
Spain

Esther Salmerón-Manzano
Faculty of Law
Universidad Internacional
de la Rioja (UNIR)
La Rioja
Spain

Editorial Office
MDPI
St. Alban-Anlage 66
4052 Basel, Switzerland

This is a reprint of articles from the Special Issue published online in the open access journal *Agronomy* (ISSN 2073-4395) (available at: www.mdpi.com/journal/agronomy/special_issues/bibliometric_studies).

For citation purposes, cite each article independently as indicated on the article page online and as indicated below:

LastName, A.A.; LastName, B.B.; LastName, C.C. Article Title. *Journal Name* **Year**, *Volume Number*, Page Range.

ISBN 978-3-0365-2190-9 (Hbk)
ISBN 978-3-0365-2189-3 (PDF)

© 2021 by the authors. Articles in this book are Open Access and distributed under the Creative Commons Attribution (CC BY) license, which allows users to download, copy and build upon published articles, as long as the author and publisher are properly credited, which ensures maximum dissemination and a wider impact of our publications.

The book as a whole is distributed by MDPI under the terms and conditions of the Creative Commons license CC BY-NC-ND.

Contents

About the Editors . . . . **vii**

Preface to "Worldwide Trends in Agronomy Research: Bibliometric Studies" . . . . **ix**

Esther Salmerón-Manzano and Francisco Manzano-Agugliaro
Worldwide Trends in Agronomy Research: Bibliometric Studies
Reprinted from: *Agronomy* **2021**, *11*, 1993, doi:10.3390/agronomy11101993 . . . . **1**

Claudia A. Ochoa-Noriega, José A. Aznar-Sánchez, Juan F. Velasco-Muñoz and Alejandro Álvarez-Bejar
The Use of Water in Agriculture in Mexico and Its Sustainable Management: A Bibliometric Review
Reprinted from: *Agronomy* **2020**, *10*, 1957, doi:10.3390/agronomy10121957 . . . . **5**

Ernesto Mesa-Vázquez, Juan F. Velasco-Muñoz, José A. Aznar-Sánchez and Belén López-Felices
Experimental Economics in Agriculture: A Review of Worldwide Research
Reprinted from: *Agronomy* **2021**, *11*, 1566, doi:10.3390/agronomy11081566 . . . . **25**

José Luis Ruiz-Real, Juan Uribe-Toril, José Antonio Torres Arriaza and Jaime de Pablo Valenciano
A Look at the Past, Present and Future Research Trends of Artificial Intelligence in Agriculture
Reprinted from: *Agronomy* **2020**, *10*, 1839, doi:10.3390/agronomy10111839 . . . . **41**

Mila Cascajares, Alfredo Alcayde, Esther Salmerón-Manzano and Francisco Manzano-Agugliaro
Transfer of Agricultural and Biological Sciences Research to Patents: The Case of EU-27
Reprinted from: *Agronomy* **2021**, *11*, 252, doi:10.3390/agronomy11020252 . . . . **57**

Juan D. Borrero and Alberto Zabalo
Identification and Analysis of Strawberries' Consumer Opinions on Twitter for Marketing Purposes
Reprinted from: *Agronomy* **2021**, *11*, 809, doi:10.3390/agronomy11040809 . . . . **83**

Héctor Madrid-Casaca, Guido Salazar-Sepúlveda, Nicolás Contreras-Barraza, Miseldra Gil-Marín and Alejandro Vega-Muñoz
Global Trends in Coffee Agronomy Research
Reprinted from: *Agronomy* **2021**, *11*, 1471, doi:10.3390/agronomy11081471 . . . . **103**

Lucas Santos Santana, Gabriel Araújo e Silva Ferraz, Alberdan José da Silva Teodoro, Mozarte Santos Santana, Giuseppe Rossi and Enrico Palchetti
Advances in Precision Coffee Growing Research: A Bibliometric Review
Reprinted from: *Agronomy* **2021**, *11*, 1557, doi:10.3390/agronomy11081557 . . . . **123**

Lucia Jimenez-Montenegro, Matilde Lopez-Fernandez and Estela Gimenez
Worldwide Research on the Ozone Influence in Plants
Reprinted from: *Agronomy* **2021**, *11*, 1504, doi:10.3390/agronomy11081504 . . . . **139**

Chris Lytridis, Vassilis G. Kaburlasos, Theodore Pachidis, Michalis Manios, Eleni Vrochidou, Theofanis Kalampokas and Stamatis Chatzistamatis
An Overview of Cooperative Robotics in Agriculture
Reprinted from: *Agronomy* **2021**, *11*, 1818, doi:10.3390/agronomy11091818 . . . . **157**

About the Editors

Francisco Manzano Agugliaro

Francisco Manzano-Agugliaro, full professor at the Engineering Department in the University of Almeria (Spain), received his M.Sc. in Agricultural Engineering and completed his Ph.D. in Geomatics at the University of Cordoba (Spain). He has published over 180 papers in JCR journals, H-index 38. His main interests are energy, sustainability, agronomy, water, and engineering. He has been the supervisor of 29 Ph.D. theses. He has also been the Vice Dean of the Engineering Faculty (2001–2004); the Director of Central Research Services (2016–2019); Ph.D. Program Coordinator for Environmental Engineering (2000 to 2012), Greenhouse Technology, Industrial and Environmental Engineering (from 2010); and General Manager of Infrastructures (from 2019) at University of Almeria. He has won the following awards: Top Reviewer in Cross-Field—September 2019 (Web of Science), 2019 Outstanding Reviewer Award (*Energies*), 2019 Winner of Sustainability Best Paper Awards.

Esther Salmerón-Manzano

Esther Salmeron-Manzano is a lecturer at the Faculty of Law in the Universidad Internacional de la Rioja (Spain) and Lecturer at Law Department of University of Almeria (Spain). She received her M.Sc. degree in Law and completed her Ph.D. in Law at the University of Almeria (Spain). She has published over 16 papers in JCR journals, H-index 10. Her main interests are laws and emerging technologies, and contract law. She has been the supervisor of 25 bachelor's and master's Final Reports. She is Academic Director of the master's degree in legal consultancy for companies and the master's degree in family law at Universidad Internacional de la Rioja (UNIR).

Preface to "Worldwide Trends in Agronomy Research: Bibliometric Studies"

Agriculture is the world's most pressing and responsible sector, given that seven billion people must eat every day. To achieve this, there are the following three priority issues: health, variety, and quantity. Agriculture, therefore, is the cultivation of land or the production of crops from the soil, but its main science of study is agronomy. Agronomy can be understood as the field of science that oversees organizing the knowledge of various applied sciences, focused on enhancing the quality of production processes and the transformation of agricultural products.

Globally, food security is at risk, and for this reason, agronomy must achieve agricultural sustainability on Earth. In summary, agronomy should contribute to improve the efficiency in the use of resources for food production.

One of the main objectives of this book is to contribute studies that help to identify the global research trends in agronomy, especially if they have an approach related to sustainability. Therefore, articles reviewing this state of the art in any of these issues, bibliometric or scientometric studies, and research chapters with a global perspective are welcome. These studies are recommended to identify the research trends in each scientific field related to agronomy and, if possible, identify the open challenges in that particular field of study.

Francisco Manzano Agugliaro, Esther Salmerón-Manzano
Editors

Editorial

Worldwide Trends in Agronomy Research: Bibliometric Studies

Esther Salmerón-Manzano [1] and Francisco Manzano-Agugliaro [2,*]

[1] Faculty of Law, Universidad Internacional de La Rioja (UNIR), Av. de la Paz, 137, 26006 Logroño, Spain; esther.salmeron@unir.net
[2] Department of Engineering, CEIA3, University of Almeria, 04120 Almeria, Spain
* Correspondence: fmanzano@ual.es; Tel.: +34-950-015-346; Fax: +34-950-015-491

Abstract: Agriculture has the large challenge of providing food for a continuously growing world population, while natural resources remain the same. This great challenge is certainly supported in the future by *Agronomy*, which brings together practical knowledge and scientifically based techniques and applies them to agricultural productivity. Research in agronomy at a global level must reflect global interests, while considering the particular conditions of each country or region. One of the main objectives of this Special Issue is to contribute studies that help to identify the global research trends in agronomy, especially if they have an approach related to sustainability.

Keywords: agronomy; patents; scopus; sustainability; precision agriculture; coffee; ozone; environment; health; agroforestry; bibliometrics; berry growers; artificial intelligence; agriculture; robots; farming automation; economy; irrigation

Citation: Salmerón-Manzano, E.; Manzano-Agugliaro, F. Worldwide Trends in Agronomy Research: Bibliometric Studies. *Agronomy* 2021, 11, 1993. https://doi.org/10.3390/agronomy11101993

Received: 10 September 2021
Accepted: 24 September 2021
Published: 1 October 2021

Publisher's Note: MDPI stays neutral with regard to jurisdictional claims in published maps and institutional affiliations.

Copyright: © 2021 by the authors. Licensee MDPI, Basel, Switzerland. This article is an open access article distributed under the terms and conditions of the Creative Commons Attribution (CC BY) license (https://creativecommons.org/licenses/by/4.0/).

1. Introduction

Agriculture is the world's most pressing and responsible sector, given that seven billion people must eat every day. To achieve this, there are the following three priority issues: health, variety, and quantity. Agriculture, therefore, is the cultivation of land or the production of crops from the soil, but its main science of study is agronomy. Agronomy can be understood as the field of science that oversees organizing the knowledge of various applied sciences, focused on enhancing the quality of production processes and the transformation of agricultural products.

Globally, food security is at risk, and for this reason, agronomy must achieve agricultural sustainability on Earth. In summary, *Agronomy* should contribute to improve the efficiency in the use of resources for food production.

One of the main objectives of this Special Issue is to contribute studies that help to identify the global research trends in agronomy, especially if they have an approach related to sustainability. Therefore, articles reviewing this state of the art in any of these issues, bibliometric or scientometric studies, and research articles with a global perspective are welcome. These studies are recommended to identify the research trends in each scientific field related to agronomy and, if possible, identify the open challenges in that particular field of study.

2. Publications Statistics

The summary of the call for papers for this Special Issue on the 12 manuscripts submitted is as follows: three rejected (25%) and nine published (75%).

The submitted manuscripts come from seven countries and are summarized in Table 1. For this statistic, only the first affiliation of the authors has been considered, in which it gives us the opportunity to observe 37 authors from 7 countries. Note that it is common for a manuscript to be signed by more than one author and for authors to belong to different affiliations. The average number of authors per published manuscript in this Special Issue was four authors.

Table 1. Authors' countries: statistics.

Country	Authors
Spain	18
Mexico	1
Honduras	1
Brazil	4
Chile	4
Italy	2
Greece	7
Total	37

3. Authors' Affiliation

There are 12 different affiliations of the authors. Note that only the first affiliation per author has been considered. Table 2 summarizes the authors and their first affiliations.

Table 2. Authors' affiliation: statistics.

Author	First Affiliation	References
Ochoa-Noriega, C.A.	University of Almeria	[1]
Aznar-Sánchez, J.A.	University of Almeria	[1,2]
Velasco-Muñoz, J.F.	University of Almeria	[1,2]
Álvarez-Bejar, A.	National Autonomous University of México	[1]
Mesa-Vázquez, E.	University of Almeria	[2]
López-Felices, B.	University of Almeria	[2]
Ruiz-Real, J.L.	University of Almeria	[3]
Uribe-Toril, J.	University of Almeria	[3]
Torres Arriaza, J.A.	University of Almeria	[3]
de Pablo Valenciano J. A.	University of Almeria	[3]
Cascajares, M.	University of Almeria	[4]
Alcayde, A.	University of Almeria	[4]
Salmerón-Manzano, E.	Universidad Internacional de La Rioja (UNIR)	[4]
Manzano-Agugliaro, F.	University of Almeria	[4]
Borrero, J.D.	University of Huelva	[5]
Zabalo, A.	University of Huelva	[5]
Madrid-Casaca, H.	Universidad Nacional Autónoma de Honduras	[6]
Salazar-Sepúlveda, G.	Universidad Católica de la Santísima Concepción	[6]
Contreras-Barraza, N.	Universidad Andres Bello	[6]
Gil-Marín, M.	Universidad Autónoma de Chile	[6]
Vega-Muñoz, A.	Universidad Autónoma de Chile	[6]
Jimenez-Montenegro, L.;	Universidad Politécnica de Madrid	[7]
Lopez-Fernandez, M.;	Universidad Politécnica de Madrid	[7]
Gimenez, E.	Universidad Politécnica de Madrid	[7]
Santana, L.S.	Federal University of Lavras	[8]
Ferraz, G.A.e.S.	Federal University of Lavras	[8]
Teodoro, A.J.d.S.	Federal University of Lavras	[8]
Santana, M.S.	Federal University of Lavras	[8]
Rossi, G.	University of Florence	[8]
Palchetti, E.	University of Florence	[8]
Lytridis, C.	International Hellenic University (IHU)	[9]
Kaburlasos, V.G.	International Hellenic University (IHU)	[9]
Pachidis, T.	International Hellenic University (IHU)	[9]
Manios, M.	International Hellenic University (IHU)	[9]
Vrochidou, E.	International Hellenic University (IHU)	[9]
Kalampokas, T.	International Hellenic University (IHU)	[9]
Chatzistamatis, S.	International Hellenic University (IHU)	[9]

4. Topics

Table 3 summarizes the research conducted by the authors on this Special Issue, by identifying the areas to which they report. It was noted that they have been grouped into the following five main lines of research: Crops, Technologies, Water and Environment, Plant response, and Bibliometry.

Table 3. Topics for Worldwide Trends in Agronomy Research: Bibliometric Studies.

Bibliometric Studies	Number of Manuscripts	References
Water and Environment	2	[1,2]
Technologies	2	[3,9]
Bibliometry	1	[4]
Crops	3	[5,6,8]
Plant response	1	[7]

Author Contributions: E.S.-M. and F.M.-A. all made equal contributions to this article. All authors have read and agreed to the published version of the manuscript.

Funding: This research received no external funding.

Institutional Review Board Statement: Not applicable.

Informed Consent Statement: Not applicable.

Data Availability Statement: Not applicable.

Acknowledgments: The authors would like to thank to the CIAIMBITAL (University of Almeria, CeiA3) for its support.

Conflicts of Interest: The authors declare no conflict of interest.

References

1. Ochoa-Noriega, C.A.; Aznar-Sánchez, J.A.; Velasco-Muñoz, J.F.; Álvarez-Bejar, A. The Use of Water in Agriculture in Mexico and Its Sustainable Management: A Bibliometric Review. *Agronomy* **2020**, *10*, 1957. [CrossRef]
2. Mesa-Vázquez, E.; Velasco-Muñoz, J.F.; Aznar-Sánchez, J.A.; López-Felices, B. Experimental Economics in Agriculture: A Review of Worldwide Research. *Agronomy* **2021**, *11*, 1566. [CrossRef]
3. Ruiz-Real, J.L.; Uribe-Toril, J.; Torres Arriaza, J.A.; de Pablo Valenciano, J.A. Look at the Past, Present and Future Research Trends of Artificial Intelligence in Agriculture. *Agronomy* **2020**, *10*, 1839. [CrossRef]
4. Cascajares, M.; Alcayde, A.; Salmerón-Manzano, E.; Manzano-Agugliaro, F. Transfer of Agricultural and Biological Sciences Research to Patents: The Case of EU-27. *Agronomy* **2021**, *11*, 252. [CrossRef]
5. Borrero, J.D.; Zabalo, A. Identification and Analysis of Strawberries' Consumer Opinions on Twitter for Marketing Purposes. *Agronomy* **2021**, *11*, 809. [CrossRef]
6. Madrid-Casaca, H.; Salazar-Sepúlveda, G.; Contreras-Barraza, N.; Gil-Marín, M.; Vega-Muñoz, A. Global Trends in Coffee Agronomy Research. *Agronomy* **2021**, *11*, 1471. [CrossRef]
7. Jimenez-Montenegro, L.; Lopez-Fernandez, M.; Gimenez, E. Worldwide Research on the Ozone Influence in Plants. *Agronomy* **2021**, *11*, 1504. [CrossRef]
8. Santana, L.S.; Ferraz, G.A.e.S.; Teodoro, A.J.d.S.; Santana, M.S.; Rossi, G.; Palchetti, E. Advances in Precision Coffee Growing Research: A Bibliometric Review. *Agronomy* **2021**, *11*, 1557. [CrossRef]
9. Lytridis, C.; Kaburlasos, V.G.; Pachidis, T.; Manios, M.; Vrochidou, E.; Kalampokas, T.; Chatzistamatis, S. An Overview of Cooperative Robotics in Agriculture. *Agronomy* **2021**, *11*, 1818. [CrossRef]

Review

The Use of Water in Agriculture in Mexico and Its Sustainable Management: A Bibliometric Review

Claudia A. Ochoa-Noriega [1], José A. Aznar-Sánchez [1,*], Juan F. Velasco-Muñoz [1] and Alejandro Álvarez-Bejar [2]

1 Department of Economy and Business, Research Centre CIAIMBITAL and CAESCG, University of Almería, 04120 Almería, Spain; claudia08a@hotmail.com (C.A.O.-N.); jfvelasco@ual.es (J.F.V.-M.)
2 Department of Economy, National Autonomous University of México, Mexico D.F. 04510, Mexico; abejar@unam.mx
* Correspondence: jaznar@ual.es

Received: 20 November 2020; Accepted: 9 December 2020; Published: 12 December 2020

Abstract: The development of agricultural activity in Mexico is generating environmental externalities that could compromise its future. One of the principal challenges facing the Mexican agricultural sector is to find a way to continue growing without jeopardising the availability and quality of its water resources. The objective of this article is to analyse the dynamics of the research on the use of water in agriculture in Mexico and its sustainable management. To do this, a review and a bibliometric analysis have been carried out on a sample of 1490 articles. The results show that the research has focused on the pollution of water bodies, climate change, the quality of water, the application of technology in order to make water use more efficient, biodiversity, erosion, agronomic practices that reduce water consumption, underground water sources, and conservation agriculture. Although research focusing on sustainability is still in its infancy, it has become a priority field. A gap in the research has been detected in terms of the economic and social dimensions of sustainability. There is also a lack of holistic studies that include all three of the pillars of sustainability (environmental, economic, and social).

Keywords: agriculture; water management; water resources; irrigation; sustainable management; sustainability; bibliometric analysis; Mexico

1. Introduction

Today's society must face a series of challenges in order to guarantee the survival of a constantly growing population, ensuring the same opportunities for future generations based on the principal of sustainability [1,2]. These include the supply of water and food, the eradication of hunger and poverty, and the conservation of a healthy natural environment [3,4]. These challenges are closely related to one another and are particularly relevant in the most disadvantaged regions. Agriculture is an economic activity that connects the different objectives proposed. It is the principal supplier of food on a global level. Therefore, it plays a fundamental role in food provision [5]. In addition, agriculture is one of the principal activities in rural areas. In some cases, it is the only possible activity and, therefore, the only engine of growth for the economies of these areas [6,7]. On the other hand, this sector is the principal consumer of water resources, so it has a direct impact on the availability of water [8,9]. Furthermore, agriculture is a source of environmental pollution and is too large of a degree responsible for the over-exploitation and degradation of water sources [10,11].

Mexican agriculture is a paradigmatic example of the relevance of this sector. According to data of the Food and Agriculture Organization of the United Nations (FAO), Mexico has a national territory of 198 million hectares, of which 145 million are dedicated to agricultural activity [12,13].

This area is divided into 30 million hectares for crops and 115 million for pastures. Although its share of Gross domestic product (GDP) is barely 4%, agriculture is an important element for the country's development, as it constitutes a tool that helps to ensure food security [14,15]. Furthermore, it also forms a base for reinforcing progress and the growth of production, which can lead to improved standards of living and a greater production capacity of the rural sectors. Mexican agriculture is a fundamental activity for the rural environment, where 24 million Mexicans live, which is a quarter of the country's population. It also represents 50% of the revenue of this population [13,15,16].

The extensive area of Mexico encompasses a diverse range of climate areas [17]. In general, two clearly differentiated areas can be distinguished. First, two-thirds of the country's territory has arid and semi-arid climates while the areas in the southern part of the country have a mild, tropical climate [12]. Overall, Mexico has 451,585 million cubic metres of renewable fresh water, taking into account rainfall, evapotranspiration, and the exit and entry flows of water with neighbouring countries [18]. The agricultural sector is the principal consumer of water, representing 76% of total consumption. In total, 63.6% of the water used in agricultural comes from surface sources and 36.4% of the water comes from underground sources. The National Water Plan 2019–2024 identifies the inefficient use of water as one of the problems related to water resources, particularly in the agricultural sector, which generates water losses of more than 40% [19].

In a global context, Mexico's overuse on its water resources is low, at 19.5% [18]. However, two-thirds of its territory is in arid or semi-arid areas (north, centre, and north-east) with annual rainfalls of less than 500 mm [20]. Since the 1920s, large hydraulic infrastructures have benefited the northwest, contributing to the take-off of a modern and capitalist agriculture, but also a great demand for water. This is why, in these regions, there is a high level of overuse, which fluctuates between 40% and 100% [18,19]. Furthermore, 105 of the 653 aquifers in Mexico are over-exploited, 32 have saline soil and brackish water, and 18 are affected by sea intrusion [19,21]. On the other hand, approximately 69 of the country's 757 water basins have deficits, as the flow granted or assigned exceeds that of the renewable water [19,21]. In addition, the possible effects derived from climate change could have a significant impact on water resources in the whole of the Mexican territory with the increase in temperature and the alteration of rainfall patterns. It is estimated, for example, that, by the end of the century, rainfall will have decreased by up to 30% [19]. On the other hand, one of the principal problems highlighted by farmers in relation to the development of the agricultural activity is the loss of crops due to climate causes, particularly droughts [22]. The areas most affected by drought in recent years are Baja California, Sonora, and Sinaloa [18].

In recent years, the country has been boosting its agricultural activity and is now among the leading producers on a global level [12,23]. There has been a strong concentration of exports in fruit and vegetables in only one country (the United States). This is due to the increase of the presence of Mexican products in external markets, driven by the quality and variety of the produce, as well as the tariff advantages arising from the North American Free Trade Agreement (NAFTA). Furthermore, there is a need to feed the growing Mexican population, which is estimated to increase by 17% by 2050 [15]. The struggle to eradicate poverty is another reason to strengthen agriculture, given that almost 20% of the population is living below the national food poverty line, and 5% of the population is classified as undernourished [13]. Even so, the margin to improve the use of natural resources in Mexico is still wide and could increase the levels of agricultural production and productivity [24]. However, this commitment by the sector could put water resources at risk in the medium and long term [20,25]. In this situation, there is an urgent need to develop agricultural water management models aimed at guaranteeing the sustainability of a strategic sector for the Mexican economy, increasing production and ensuring the supply of water resources [26–28].

Within this context, an increasing number of contributions have been published that study the use of water in agriculture in Mexico. However, to date, no studies have analysed these contributions as a whole. Therefore, the objective of this article is to analyse the dynamics of the research on the use of water in agriculture in Mexico and its sustainable management. The methodology selected for achieving this objective is bibliometric analysis. Additionally, the results obtained will enable us to identify the principal driving agents of the knowledge in this field and the most relevant lines of research.

2. Methodology

2.1. Bibliometric Analysis

This methodology was developed in the middle of the last century in order to identify, organise, and evaluate the constituent elements of a specific field of study [29]. Today, bibliometric analysis has become one of the principal tools for reviewing a large amount of existing literature in any scientific discipline [30–32]. Its success is largely due to the availability of cartographic techniques for representing the bibliographic information stored in different databases and statistical and mathematical methods for determining the trends in a research field [33,34]. According to Durieux and Gevenois [35], bibliometric analysis can be based on three different kinds of indicators: quantity indicators, which measure productivity, relevance indicators, which show the impact of the publications, and structural indicators, which identify the connections between the different elements of the same research field. In this study, the three types of indicators are analysed and the traditional approach based on co-occurrence analysis is applied following Robinson et al. [36].

2.2. Sample Selection

The Scopus database has been chosen for selecting the sample of studies to analyse in this review because it is the largest database of abstracts and citations of peer-reviewed literature, it is the most accessible, it offers greater processing capabilities, and it is the most used in bibliometric studies on agriculture and water resources [7,37,38]. In addition, there are other search engines such as Web of Science (WoS), according to Gavel and Iselid [39]. Furthermore, 84% of the WoS titles are also indexed in Scopus, while only 54% of the Scopus titles are indexed in WoS. To carry out this study, two samples of studies were selected including one general sample on the use of water in agriculture in Mexico and another focused on its sustainable management. Both searches had common restrictions. The search was specified for the period of 1990 to 2019. This period is marked by the implementation of NAFTA, which is of great importance in shaping Mexico's export agriculture. Documents published in 2020 have not been included so that complete annual periods can be compared [40]. In order to avoid duplications, only original articles have been included in the sample [41]. The parameters used to select the sample of documents on the use of water in agriculture in Mexico were: TITLE-ABS-KEY (water OR irrigation OR "water management" OR "water resource*" OR "water *use*" OR "hydrological resource*") AND TITLE-ABS-KEY (agricultur* OR crop* OR farm* OR cultivation OR agrosystem* OR agroecosystem*) AND TITLE-ABS-KEY (Mexico OR Aguascalientes OR "Baja California" OR Campeche OR Chiapas OR Chihuahua OR Coahuila OR Colima OR Durango OR Guanajuato OR Guerrero OR Hidalgo OR Jalisco OR Michoacán OR Morelos OR Nayarit OR "Nuevo León" OR Oaxaca OR Puebla OR Querétaro OR "Quintana Roo" OR "San Luis Potosí" OR Sinaloa OR Sonora OR Tabasco OR Tamaulipas OR Tlaxcala OR Veracruz OR Yucatán OR Zacatecas). In order to obtain the second sample, the following was added to the parameters used in the first: TITLE-ABS-KEY (sustainab*). As a result, a final sample of 1490 articles on the use of water in agriculture in Mexico was obtained and 436 articles for the case of sustainable management were obtained. The selection of the sample was carried out in May 2020.

2.3. Data Processing

After selecting the samples of articles, the information was downloaded. The data were prepared before being analysed. To do this, duplications were eliminated, omissions and errors were corrected, and incomplete information was sought [42]. The analysis phase was then carried out. First, the evolution of the number of articles was examined, together with the subject areas in which the documents are classified in the Scopus database. Then the journals, institutions, and authors that had most published on the subject area, which is the object of this study, were identified, as were the principal international collaborations in the articles. The number of studies was used as the indicator of productivity. To evaluate the impact of the publications, the following quality indicators were selected: the counting of citations, the H index, and the impact factor of the Scimago Journal Rank journals (SJR). The H index shows the number h of a total of N documents that include h citations in each of them [43]. The SJR shows a weighting of the number of the citations received, taking into account the material and the prestige of the journal in which the citation is made [44]. Finally, cartographic techniques were used to visualise the co-occurrence network of keywords to determine the research trends [45]. The tools used were Excel (version 2016, Microsoft, Redmond, DC, USA), SciMaT (v1.1.04, Soft Computing and Intelligent Information Systems research group, University of Granada, Granada, Spain), and VOSviewer (version 1.6.5., Leiden University, Leiden, the Netherlands). The methodology described above has been used in other works [28,32,40]. Figure 1 shows an overall view of the methodology applied in this study.

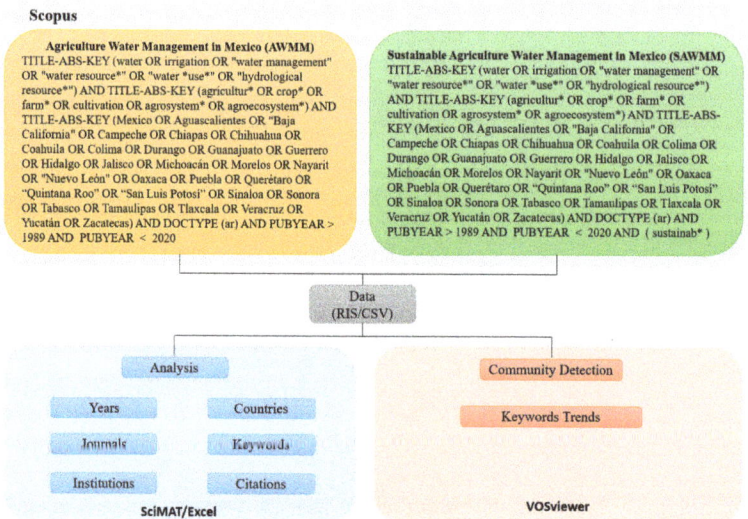

Figure 1. Summary of the methodology.

3. Results and Discussion

3.1. General Evolution on Agricultural Water Management in Mexico Research

Table 1 shows the evolution of the principal variables related to the research on agricultural water management in Mexico (AWMM) and sustainable agricultural water management in Mexico (SAWMM) in the period of 1990 to 2019. The total number of articles published in this period was 1490 in the case of research on AWMM and 436 in the case of SAWMM. The research on SAWMM represented 29.3% of the overall research on AWMM. The number of articles on AWMM increased from three in 1990 to 129 in 2019. In the case of the articles on SAWMM, in 1990, only one article was published, while, in 2019, this figure increased to 55. Both lines of research have gained importance in recent years, as 63.62% of the articles on AWMM and 73.85% on SAWMM have been published in the last 10 years. After the year 2000, we can observe a point of inflection, where the research on SAWMM began to gain more relevance within the research on AWMM. The average annual growth of the articles on SAWMM was 14.8% while that of articles on AWMM was 13.9%. This enables us to affirm that the research line on SAWMM has been gaining relevance within the general research on AWMM in recent years.

Table 1. Major characteristics of agricultural water management research.

Year	A SAWMM	A AWMM	AU SAWMM	AU AWMM	J SAWMM	J AWMM	C SAWMM	C AWMM	TC SAWMM	TC AWMM	TC/CA SAWMM	TC/CA AWMM
1990	1	3	1	7	1	3	1	3	0	0	0.0	0.0
1991	1	6	1	21	1	6	1	3	0	0	0.0	0.0
1992	0	2	0	7	0	2	0	3	1	1	0.5	0.1
1993	1	6	1	13	1	5	1	3	0	1	0.3	0.1
1994	0	6	0	20	0	6	0	4	2	4	1.0	0.3
1995	0	7	0	13	0	6	0	1	0	12	1.0	0.6
1996	2	20	2	60	2	18	3	9	0	19	0.6	0.7
1997	1	12	1	46	1	11	1	3	0	31	0.5	1.1
1998	4	18	4	64	4	17	2	5	7	71	1.0	1.7
1999	4	23	4	81	4	22	4	6	11	78	1.5	2.1
2000	8	27	7	92	7	26	5	6	13	104	1.5	2.5
2001	4	28	4	110	4	21	5	7	14	117	1.8	2.8
2002	9	37	7	136	7	29	5	6	21	164	2.0	3.1
2003	6	26	5	89	5	20	4	12	31	197	2.4	3.6
2004	12	46	10	174	10	33	9	7	43	225	2.7	3.8
2005	8	35	7	145	7	30	4	13	49	300	3.1	4.4
2006	12	67	11	293	11	56	4	10	68	379	3.6	4.6
2007	10	51	9	240	9	37	7	12	106	585	4.4	5.4
2008	17	58	15	259	15	49	4	10	119	652	4.9	6.2
2009	14	64	11	275	11	52	8	18	203	812	6.0	6.9
2010	10	51	9	242	9	41	3	9	178	889	7.0	7.8
2011	28	90	22	380	22	60	9	14	235	1023	7.2	8.3
2012	26	73	24	312	24	60	11	18	271	1210	7.7	9.1
2013	24	96	21	422	21	67	8	20	365	1334	8.6	9.6
2014	24	81	22	413	22	63	7	21	431	1617	9.6	10.5
2015	26	98	24	490	24	75	10	21	453	1789	10.4	11.3
2016	36	87	30	412	30	69	13	15	512	1941	10.9	12.1
2017	50	127	43	601	43	86	13	21	564	2107	10.9	12.6
2018	43	116	34	676	34	73	9	14	699	2430	11.5	13.3
2019	55	129	45	651	45	95	14	25	833	2782	12.2	14.1

A: The annual number of total articles. AU: the annual number of authors. J: the annual number of journals. C: the annual number of countries. TC: the annual number of citations in cumulative articles. TC/CA: annual total citation per cumulative article.

During the whole period analysed, a total of 5314 authors participated in the 1490 articles on AWMM. In the case of research on SAWMM, 1759 authors collaborated on the 436 articles published on this subject matter. In both cases, this variable has grown considerably. Specifically, the number of authors grew from seven in 1990 to 651 in 2019 in the case of research on AWMM and from one to 295 in the case of research on SAWMM. The average number of authors per article increased from 2.33 to 5.04 in the research on AWMM and from one to 5.36 in that on SAWMM. In total, 1490 articles on AWMM were published in 541 different journals, while 436 articles on SAWMM were published in 226 journals. The average number of articles per journal remained practically constant during the whole period at around one in the case of research on SAWMM and 1.22 in the case of research on AWMM. With respect to the countries that participated in carrying out the studies, during the whole period analysed, there were a total of 54 for AWMM and 35 for SAWMM. The number of countries increased from three to 25 for AWMM and from one to 14 for SAWMM.

In the case of citations, as a whole, the studies on AWMM obtained a total of 20,874 citations during the whole period analysed, while, in the case of SAWMM, there were 5229. The citations in the case of SAWMM represent around 25% of the total citations obtained in the general subject area. The number of citations increased from one in 1992 to 2782, and 833 in 2019, for the articles on AWMM and SAWMM, respectively. The average number of citations obtained per article increased from 0.1 to 14.1 in the research on AWMM and from 0.5 to 12.2 in that on SAWMM.

3.2. Evolution of Research by Subject Area

Table 2 shows the number of articles published during the whole period analysed in both lines of research, classified in accordance with the subject categories established by Scopus. It should be taken into account that the same article may be classified in more than one category simultaneously. As we can see, in both lines of research, the categories with the highest number of articles are Environmental Sciences, Agricultural and Biological Sciences, and Earth and Planetary Sciences. In the period of 1990 to 2019, 54.4% of the articles on AWMM were published under the category Environmental Sciences, 49.6% in the category of Agricultural and Biological Sciences, and 16.3% in the category of Earth and Planetary Sciences. In the case of research on SAWMM, these percentages were 62.2%, 47.5%, and 15.1%, respectively. In general, in both lines of research, the categories related to environmental and technical fields predominated.

Table 2. Number of articles published by subject category.

AWMM	Total	%	SAWMM	Total	%
Environmental Sciences	810	54.4	Environmental Sciences	271	62.2
Agricultural and Biological Sciences	724	49.6	Agricultural and Biological Sciences	207	47.5
Earth and Planetary Sciences	243	16.3	Earth and Planetary Sciences	66	15.1
Engineering	144	10.7	Social Sciences	64	14.7
Biochemistry, Genetics, and Molecular Biology	123	8.3	Engineering	45	10.3
Social Sciences	121	8.1	Biochemistry, Genetics, and Molecular Biology	39	9.9
Medicine	79	5.3	Energy	28	6.4
Immunology and Microbiology	62	4.2	Chemical Engineering	11	2.5
Pharmacology, Toxicology, and Pharmaceutics	60	4.0	Medicine	11	2.5
Business, Management, and Accounting	15	1.0	Business, Management, and Accounting	9	2.1
Economics, Econometrics, and Finance	13	0.9	Economics, Econometrics, and Finance	9	2.1

Sustainability spans across three fields: environmental, economic, and social. In the case of research on SAWMM, higher percentages were found in the categories of the social and economic dimensions, showing the greater importance that these areas have in this line of research. Specifically, the Social Sciences category represents 14.7% in the case of SAWMM while it only accounts for 8.1% in the case of AWMM. The economic categories (Economics, Econometrics and Finance, and Business,

Management, and Accounting) represent 4.2% in the case of SAWMM and only 1.9% in the case of research on AWMM. Therefore, although the social and economic fields have a greater relevance in the case of research related to sustainability, the still incipient nature of this line of research means that it still has not reached values similar to those in the environmental field. Hence, it is necessary to broaden the research from the social and economic perspectives and carry out holistic studies that take into account all three dimensions of sustainability.

3.3. Most Relevant Journals

Tables 3 and 4 show the most prolific journals in terms of AWMM and SAWMM research in the period of 1990 to 2019 and the principal characteristics of their articles. If we compare the two tables, we can observe that only five journals have published on both subject areas (*Tecnologia y Ciencias del Agua, Agrociencia, Revista Internacional de Contaminación Ambiental, Science of the Total Environment,* and *Soil and Tillage Research*). Furthermore, in both cases, the journal with the highest number of articles published is *Tecnologia y Ciencias del Agua*. If we analyse Table 3, we can see that the principal journals in the case of research on AWMM are from five different countries, three in Europe (UK, Spain, and Netherlands) and two in America (Mexico and USA). In total, this group of journals has published 336 articles within the sample, which represent 22.6% of the total. These data do not enable us to confirm whether there is a central nucleus of journals that leads this line of publication. *Tecnologia y Ciencias del Agua*, with a total of 102 articles, is the journal that published the most articles on AWMM. This journal has an H index of 6, a total of 165 citations, and its average number of citations per article is 1.6. Moreover, it has a Scimago Journal Rank (SJR) impact factor of 0.195 and has been publishing on AWMM since the year 2000. With almost half the number of articles, it is followed by the journals *Agrociencia* and *Revista Internacional de Contaminación Ambiental*, which have published a total of four articles each. *Agrociencia* has an H index of 7, a total of 155 citations, and 3.4 citations per article and its SJR impact factor is 0.181. *Revista Internacional de Contaminación Ambiental*, meanwhile, has an H index of 8, a total of 211 citations, an average number of citations per article of 4.6, and an SJR impact factor of 0.190. Despite having published only 16 articles on the subject area, the journal *Soil and Tillage Research* has the highest H index in the entire table (12). Furthermore, it has the highest values of the total citations and average number of citations per article with 557 and 34.8, respectively. The journal that has been publishing on the subject for the longest is *Bulletin of Environmental Contamination and Toxicology*, as it published its first articles on the subject in 1993 and continues publishing in this line of research today.

Table 3. Major characteristics of the most active journals related to agricultural water management in Mexico (AWMM) research.

Journal	A	SJR	H index	C	TC	TC/A	1st A	Last A
Tecnologia y Ciencias del Agua *	102	0.195 (Q3)	6	Mexico	165	1.6	2000	2019
Agrociencia	46	0.181 (Q3)	7	Mexico	155	3.4	2004	2019
Revista Internacional de Contaminacion Ambiental	46	0.190 (Q4)	8	Mexico	211	4.6	1998	2019
Bulletin of Environmental Contamination and Toxicology	24	0.515 (Q2)	11	USA	306	12.8	1993	2019
Water Science and Technology	20	0.471 (Q2)	11	UK	341	17.1	1995	2016
Investigaciones Geograficas	18	0.190 (Q3)	4	Spain	48	2.7	2004	2017
Agricultural Water Management	16	1.369 (Q1)	10	Netherlands	289	18.1	1999	2018
Environmental Monitoring and Assessment	16	0.571 (Q2)	9	Netherlands	144	9.0	2000	2019
Science of the Total Environment	16	1.661 (Q1)	9	Netherlands	295	18.4	2006	2019
Soil and Tillage Research	16	1.791 (Q1)	12	Netherlands	557	34.8	2000	2018
Wit Transactions on Ecology and the Environment	16	0.142 (Q4)	2	UK	14	0.9	2006	2019

A: the annual number of total articles. SJR: Scimago Journal Ranking. C: country. TC: the annual number of citations in total articles. TC/A: total citation per article. 1st A: first article of SPMM research by journal. Last A: last article. * Includes *Ingenieria Hidraulica En Mexico*. This journal changed its name in 2009. In 2010, it became *Tecnologia y Ciencias del Agua*.

Table 4. Major characteristics of the most active journals related to sustainable agricultural water management in Mexico (SAWMM) research.

Journal	A	SJR	H index	C	TC	TC/A	1st A	Last A
Tecnologia y Ciencias del Agua *	29	0.195 (Q3)	3	Mexico	48	1.7	2004	2019
Revista Internacional de Contaminacion Ambiental	12	0.190 (Q4)	4	Mexico	38	3.2	2011	2018
Soil and Tillage Research	11	1.791 (Q1)	10	Netherlands	477	43.4	2000	2018
Water	11	0.657 (Q1)	4	Switzerland	39	3.5	2012	2019
Agrociencia	10	0.181 (Q3)	2	Mexico	13	1.3	2007	2019
Sustainability	8	0.581 (Q2)	3	Switzerland	21	2.6	2015	2019
Agriculture Ecosystems and Environment	7	1.719 (Q1)	6	Netherlands	235	33.6	1991	2018
Ecological Engineering	7	1.122 (Q1)	4	Netherlands	37	5.3	2013	2019
Environmental Earth Sciences	7	0.604 (Q2)	5	Germany	50	7.1	2010	2019
Field Crops Research	7	1.767 (Q1)	5	Netherlands	349	49.9	2002	2018
Science of the Total Environment	7	1.661 (Q1)	5	Netherlands	45	6.4	2012	2019

A: the annual number of total articles. SJR: Scimago Journal Ranking. C: country. TC: the annual number of citations in total articles. TC/A: total citation per article. 1st A: first article of SPMM research by journal. Last A: last article. * Includes *Ingenieria Hidraulica En Mexico*. This journal changed its name in 2009. In 2010, it became *Tecnologia y Ciencias del Agua*.

Meanwhile, if we analyse the research on SAWMM, we can see that the principal journals belong to only four countries, including three in Europe (Netherlands, Switzerland, and Germany) and Mexico. In this case, the total articles published by these journals during the period analysed represent 26.6% of the total. *Tecnologia y Ciencias del Agua* is also the journal with the highest number of articles published, with a total of 29. This journal has an H index of 3, a total of 48 citations, and its average number of citations per article is 1.7. This journal began to publish on AWMM in the year 2000 and on SAWMM in 2004. The *Revista Internacional de Contaminación Ambiental* is the journal with the second highest number of articles with a total of 12. It has an H index of 4 and 38 citations in total. This journal obtained 3.2 citations per article. It began to publish on AWMM in 1998 and published its first article on SAWMM in 2011. It is followed by the journals *Soil and Tillage Research* and *Water*, with 11 articles each. *Soil and Tillage Research* has the highest H index of the group (10) and also the highest average number of citations per article (43.4 citations per article). The journal *Water* has an H index of 4, a total of 39 citations, and its average number of citations per article is 3.5. The journal that has been publishing in the research on SAWMM for the longest in the table is *Agriculture, Ecosystems, and Environment*, which published its first article on the subject in 1991, even though it has only published seven articles in total.

3.4. International Collaboration

Table 5 shows the results of the analysis of the collaboration networks established between Mexico and its principal collaborators in the research on AWMM and SAWMM. The average percentage of studies carried out through international collaboration is higher in the research on SAWMM than in the case of AWMM with 41.3% and 35.5%, respectively. This difference can be explained because the research on sustainability is considered as being more multidisciplinary and, therefore, more collaborative. The table also shows the principal international collaborators in both cases, with the majority being common to both. If we analyse the differences, in the case of research on AWMM, we find Italy and China in the group of principal collaborators while, in the case of research on SAWMM, Saudi Arabia is incorporated. Seven of the principal collaborators in the case of the research on SAWMM are from the most prolific countries with respect to research on a global level on Sustainable Water Use in Agriculture (USA, Spain, Germany, France, Australia, UK, and Netherlands) [42]. It is noteworthy that, although China is the most important country on a global level in research on SWUA, it does not appear among the most important collaborators in the case of research on Mexico. In terms of the number of citations, the articles carried out through international collaboration have a higher average number in both lines of research than the articles carried out without an international collaboration.

Table 5. Main characteristics of the international collaboration of Mexico related to AWMM and SAWMM research.

	IC (%)	NC	Main Collaborators	TC/A IC	TC/A NIC
AWMM	35.5	59	USA, Spain, Germany, France, Canada, UK, Australia, Belgium, Italy, Chile, Netherlands	22.7	9.2
SAWMM	41.3	36	USA, Spain, Germany, Canada, France, Belgium, Australia, UK, Netherlands, Saudi Arabia	19.7	6.6

IC: international collaborations. NC: total number of international collaborators. TC/A: total citation per article. NIC: no international collaborations.

3.5. Most Relevant Institutions

Tables 6 and 7 show the most prolific institutions in terms of AWMM and SAWMM research in the period of 1990 to 2019 and the principal characteristics of their articles. In both cases, all of the institutions are in Mexico except for the *University of Arizona* in the USA. The majority of the institutions have published in both lines of research except for the *Centro de Investigaciones Biológicas del Noroeste, Universidad Autónoma de Chapingo, Universidad de Sonora, Universidad Michoacana de San Nicolás de Hidalgo* and *Tecnológico de Monterrey*.

Table 6. Major characteristics of the most active institutions related to AWMM research.

Institution	C	A	TC	TC/A	H Index	IC (%)	TC/A IC	TC/A NIC
Universidad Nacional Autónoma de México	Mexico	338	5723	16.9	40	28.4	23.6	14.3
Colegio de Postgraduados	Mexico	122	1378	11.3	14	26.2	32.7	3.7
Instituto Nacional de Investigaciones Forestales, Agricolas y Pecuarias	Mexico	117	1118	9.6	15	33.3	20.1	4.3
Instituto Politécnico Nacional	Mexico	97	877	9.0	17	18.6	12.9	8.2
Centro Internacional de Mejoramiento de Maiz y Trigo	Mexico	74	4583	61.9	36	82.4	61.3	64.8
Centro de Investigaciones Biologicas Del Noroeste	Mexico	65	732	11.3	16	24.6	8.4	12.2
Instituto Mexicano de Tecnologia del Agua	Mexico	64	363	5.7	13	23.4	3.7	6.3
Universidad Autónoma de Chapingo	Mexico	60	347	5.8	9	25.0	12.8	3.4
Universidad de Sonora	Mexico	49	530	10.8	15	30.6	12.3	10.1
University of Arizona	USA	48	992	20.7	16	100.0	20.7	0.0
Instituto de Ecología, A.C.	Mexico	48	577	12.0	14	37.5	18.9	7.9

C: country. A: the annual number of total articles. TC: the annual number of citations in total articles. TC/A: total citation per article. IC: international collaborations. NIC: no international collaborations.

Table 7. Major characteristics of the most active institutions related to SAWMM research.

Institution	C	A	TC	TC/A	H Index	IC (%)	TC/A IC	TC/A NIC
Universidad Nacional Autónoma de México	Mexico	82	969	11.8	17	29.3	10.9	12.2
Centro Internacional de Mejoramiento de Maiz y Trigo	Mexico	32	1337	41.8	22	81.3	43.8	33.2
Instituto Nacional de Investigaciones Forestales, Agricolas y Pecuarias	Mexico	29	436	15.0	9	48.3	23.4	7.3
Colegio de Postgraduados	Mexico	28	304	10.9	6	35.7	26.2	2.3
Instituto Politécnico Nacional	Mexico	27	184	6.8	9	18.5	4.6	7.3
University of Arizona	USA	23	383	16.7	10	100.0	16.7	0.0
Instituto Mexicano de Tecnologia del Agua	Mexico	22	71	3.2	4	31.8	4.7	2.5
Universidad Michoacana de San Nicolás de Hidalgo	Mexico	19	153	8.1	7	42.1	15.0	3.0
Instituto de Ecología, A.C.	Mexico	18	183	10.2	8	33.3	15.8	7.3
Tecnologico de Monterrey	Mexico	17	155	9.1	7	47.1	16.8	2.3

C: country. A: the annual number of total articles. TC: the annual number of citations in total articles. TC/A: total citation per article. IC: international collaborations. NIC: no international collaborations.

In the research on AWMM, the *Universidad Nacional Autónoma de México* is in first place with 338 articles. It has the highest total number of citations with 5723, an average of 16.9 citations per article, and an H index of 40. This is followed by the *Colegio de Postgraduados* with 122 articles, 1378 citations, an average of 11.3 citations per article, and an H index of 14. Next is the *Instituto Nacional de Investigaciones Forestales, Agrícolas y Pecuarias* with 117 articles, a total of 1118 citations, an average of 9.6 citations per article, and an H index of 15. The *Centro Internacional de Mejoramiento de Maíz y Trigo* holds the fifth position in terms of the number of articles with a total of 74 and it is the institution with the highest average number of citations per article at 61.9. Furthermore, it has a total number of citations of 4583 and an H index of 36. With respect to the international collaboration of the institutions, the average percentage of articles carried out through collaboration is 39.1%. In this respect, The *University of Arizona* reveals 100% of collaboration, given that the whole of the sample has had the participation of a Mexican institution. *The Centro Internacional de Mejoramiento de Maíz y Trigo*, with 82.4%, is the Mexican institution with the highest percentage of an international collaboration. The average number of citations in the articles written through international collaboration was 20.7 while, for the rest of the articles, it was 12.3.

In the case of research on SAWMM, the first position is also held by the *Universidad Nacional Autónoma de México* with 82 articles. Furthermore, it has an H index of 17 and a total of 969 citations. The institution with the second highest number of articles is the *Centro Internacional de Mejoramiento de Maíz y Trigo* with 32 articles. This institution has 1337 citations and the highest H index of the group with 22. It also has the highest average number of citations per article (41.8). Next is the *Instituto Nacional de Investigaciones Forestales, Agrícolas y Pecuarias*, which has 29 articles, 436 citations, and an H index of 9. The average number of citations of the articles written through international collaboration in this group of institutions was 17.8 as opposed to 7.7 in the rest.

3.6. Most Relevant Authors

Tables 8 and 9 include the most productive authors in the research on AWMM and SAWMM and show the most salient characteristics of their articles. In general, these groups of authors are affiliated to nine different institutions in three countries. As we would expect, the majority of the authors are affiliated with a Mexican institution. There are only two authors affiliated with an American entity and another with a Belgian institution. The majority of the authors have published in both lines of research. In the case of research on AWMM, the author with the highest number of articles is Federico Páez-Osuna with a total of 28. Furthermore, this author has been publishing on this subject matter for a long time, as his first article was published in 1993 and he continues to publish today. His articles have received a total of 753 citations, an average number of citations per article of 26.9, and an H index of 15. He is followed by Christina D. Siebe with 26 articles. This author has accumulated a total of 807 citations, has an average of 31.1 citations per article, and an H index of 16. The following author is Bram Govaerts with 19 articles. This author accumulates a total of 675 citations, an average of 35.5 citations per article, and an H index of 13. This is the only author who does not belong to a Mexican institution. Matthew P. Reynolds is the author with the highest number of citations with a total of 1723 and the highest average number of citations per article with 95.7.

In the research on SAWMM, we find that the most prolific author, with 17 articles, is Bram Govaerts. This author also has the highest number of citations with a total of 650 and the highest H index (12). The average number of citations per article of this author is higher in the case of research on SAWMM than in general research (38.2 and 35.5 citations, respectively). The following author is Nele Verhulst with 13 articles. This author has a total of 278 citations and an H index of 8. The average number of citations per article obtained by this author in the case of research on SAWMM is 21.4 citations, as opposed to 20.2 citations of general research. The next author is José María Ponce-Ortega with 11 articles. This author accumulates a total of 116 citations, an average of 10.5 citations per article, and an H index of six. Jozef A. Deckers, affiliated with an institution in Belgium, is the author with the highest average number of citations per article (75.3). The most veteran author is also, in this case, Federico Páez-Osuna, who began to publish on SAWMM in 1998 and still does today.

Table 8. Major characteristics of the most active authors related to AWMM research.

Author	A	TC	TC/A	H Index	C	Affiliation	First Article	Last Article
Páez-Osuna, Federico	28	753	26.9	15	Mexico	Universidad Nacional Autónoma de México	1993	2019
Siebe, Christina D.	26	807	31.1	16	Mexico	Universidad Nacional Autónoma de México	1995	2019
Govaerts, Bram	19	675	35.5	13	USA	Cornell University	2006	2018
Dendooven, Luc	18	529	29.4	13	Mexico	Centro de Investigacion y de Estudios Avanzados	2002	2019
Reynolds, Matthew P.	18	1723	95.7	16	Mexico	Centro Internacional de Mejoramiento de Maiz y Trigo	1996	2016
Verhulst, Nele	15	303	20.2	9	Mexico	Centro Internacional de Mejoramiento de Maiz y Trigo	2011	2019
Ruiz-Fernández, Ana C.	14	496	35.4	11	Mexico	Universidad Nacional Autónoma de México	1997	2016
Sayre, Kenneth D.	14	1195	85.4	14	Mexico	Centro Internacional de Mejoramiento de Maiz y Trigo	1998	2012
López-López, Eugenia	13	227	17.5	8	Mexico	Instituto Politécnico Nacional	1998	2018
Mahlknecht, Jürgen	13	340	26.2	9	Mexico	Tecnologico de Monterrey	2004	2019

A: the annual number of total articles. TC: total number of citations in total articles. TC/A: total citations per article. C: country.

Table 9. Major characteristics of the most active authors related to SAWMM research.

Author	A	TC	TC/A	H Index	C	Affiliation	1st Article	Last Article
Govaerts, Bram	17	650	38.2	12	USA	Cornell University	2006	2018
Verhulst, Nele	13	278	21.4	8	Mexico	Centro Internacional de Mejoramiento de Maiz y Trigo	2011	2019
Ponce-Ortega, José María	11	116	10.5	6	Mexico	Universidad Michoacana de San Nicolás de Hidalgo	2012	2019
Sayre, Kenneth D.	9	535	59.4	9	Mexico	Centro Internacional de Mejoramiento de Maiz y Trigo	2006	2012
Páez-Osuna, Federico	7	189	27.0	6	Mexico	Universidad Nacional Autónoma de México	1998	2019
Siebe, Christina D.	7	149	21.3	5	Mexico	Universidad Nacional Autónoma de México	2012	2019
Deckers, Jozef A.	6	452	75.3	6	Belgium	University of Leuven	2006	2011
Dendooven, Luc	6	221	36.8	5	Mexico	Centro de Investigacion y de Estudios Avanzados	2009	2019
El-Halwagi, Mahmoud M.	6	109	18.2	5	USA	Texas A&M University	2012	2017
Mahlknecht, Jürgen	6	80	13.3	5	Mexico	Tecnologico de Monterrey	2008	2019

A: the annual number of total articles. TC: total number of citations in total articles. TC/A: total citation per article. C: country.

3.7. Keywords Analysis

Figures 2 and 3 show the network maps of keywords in the different lines of research on AWMM and SAWMM. The size of the circle varies depending on the number of times the term has been used, while the colour represents the group in which the keyword is included depending on the number of co-occurrences.

As could be expected, in Figure 2, we can find a large number of different clusters (a total of 9), reflecting the diversity of the topics within the general research. The red cluster refers to the pollution of water bodies. In Mexico, more than half of the waste water is not treated [46]. The uncontrolled discharging of untreated, reused water can generate negative effects derived from the pollution of water bodies and agricultural soils [47]. Pérez-Castresana et al. [48] find, for example, that the quality of the water of the River Atoyacd, on which the agricultural activities greatly depend in the area of Puebla, has been compromised due to the discharging of large amounts of poorly treated waste water.

Meanwhile, the application of fertilizers has also been shown to be a cause of pollution of water and agricultural soils [49]. García-Hernández et al. [24] analysed the research on the effect of the use of pesticides in Mexico, finding that they have had negative impacts on the land and coastal ecosystems and on the health of the agricultural workers and their families.

The green cluster refers to the effects of climate change on the availability and management of water. Hernández-Bedolla et al. [50] developed indices to evaluate the availability of water in different scenarios, concluding that the principal factors that affect its availability are the decrease in rainfall and the high temperatures. A study on the possible effects of climate change on the Guadalupe River basin in the north of Mexico shows that the run-off could decrease by anywhere from 45% to 60% while the recharging of the underground waters could fall by up to 74% [51]. The scarcity of water resources as a consequence of the effects of climate change place the survival of the agricultural sector at risk, and, therefore, jeopardize the capacity to feed the population. For example, in Mexico, it is estimated that wheat production, which currently amounts to around 3.3 million tonnes, will decrease as a result of climate change [52].

The blue cluster refers to the quality of the water since the spread of certain anthropogenic activities causes the pollution of water resources. This can generate problems in the supply of water fit for human consumption and for agricultural irrigation. De Oca et al. [53] find that the changes in the physical and chemical composition derived from human actions and the changes in the uses of the land have given rise to a reduction in the essential nutrients of the water, which can have an impact on the health of the consumers. In terms of agriculture, Saldaña-Robles et al. [54] conclude that irrigation with water contaminated with arsenic leads to an accumulation of this substance in the soil and its concentration increases in the crops, affecting their growth and yields.

The yellow cluster includes studies focused on the use of remote sensors and satellite images to estimate the yields and water consumption of the crops. For example, these technologies are used to calculate the evapotranspiration of forage maize crops, which enables a more efficient planning of the use of water resources, particularly in arid and semi-arid areas where the water is a limiting factor for agricultural production [55]. Reyes-González et al. [56] develop evapotranspiration maps based on remote sensing multi-spectral vegetation indexes to quantify the water consumption of crops, according to their growth phase. López-Hernández et al. [57] show that the determination of productivity through evapotranspiration can help increase the yields of the crops, as it enables the application of irrigation efficiently in accordance with their needs. Palacios-Vélez et al. [58] used satellite images to estimate the Normalized Difference Vegetation Index (NDVI) and evapotranspiration with the objective of conducting an anticipated estimate of the yield of the wheat crop.

The purple cluster examines the research on the effects that the changes in the land uses and pollution can generate on the biodiversity and conservation of natural spaces and water bodies. The loss of pastures due to the expansion of irrigated crop land is putting the survival of many species at risk as it has transformed their habitat [59]. Andrade-Herrera et al. [60] conclude that the intensification of the agricultural activity and the greater use of pesticides have led to a loss of biodiversity as a result of soil pollution. Vanderplank et al. [61] find that the intrusion of sea water in the aquifers as a result of unsustainable extraction, principally for agricultural irrigation, has had indirect effects on the adjacent ecosystems, leading to the loss of more than 20 native plants in the valley of San Quintín.

The light blue cluster studies erosion, which is one of the main causes of the degradation of the soil and depends on many factors, such as the type of land and soil, the land use, or the climate [62]. Silva-García et al. [63] carried out a study to determine the loss of soil as a consequence of water erosion in the Lake Chapala basin, concluding that it was produced mainly in the seasonal crops and that the organic material suffers the greatest losses. Meanwhile, López-Santos et al. [64] found that the implementation of actions to control soil erosion, such as correct rainwater management or the incorporation of organic material, is still limited among farmers.

The brown cluster shows a research line based on two crops that are fundamental in the Mexican diet: maize (zea mays) and wheat (Triticum aestivum) [15]. In this research field, certain agronomic practices are studied, which can improve the efficiency of the use of water and reduce water pollution. Paquini-Rodríguez et al. [65] conducted a study with varieties of wheat in different scenarios and found that using a lower amount of water could obtain the same yields. Honsdorf et al. [66] carried out a study with wheat in different agronomic environments, conventional tillage, and permanent raised beds in order to determine the importance of tillage in crops. Rangel-Fajardo et al. [67] analysed 25 varieties of maize with the objective of identifying their tolerance to water stress during germination. Grahmann et al. [68] found that it is necessary to promote practices that reduce nitrate pollution since the results of their study revealed that 19% of the nitrate applied in a wheat crop and 34% in a maize crop was lost by leaching.

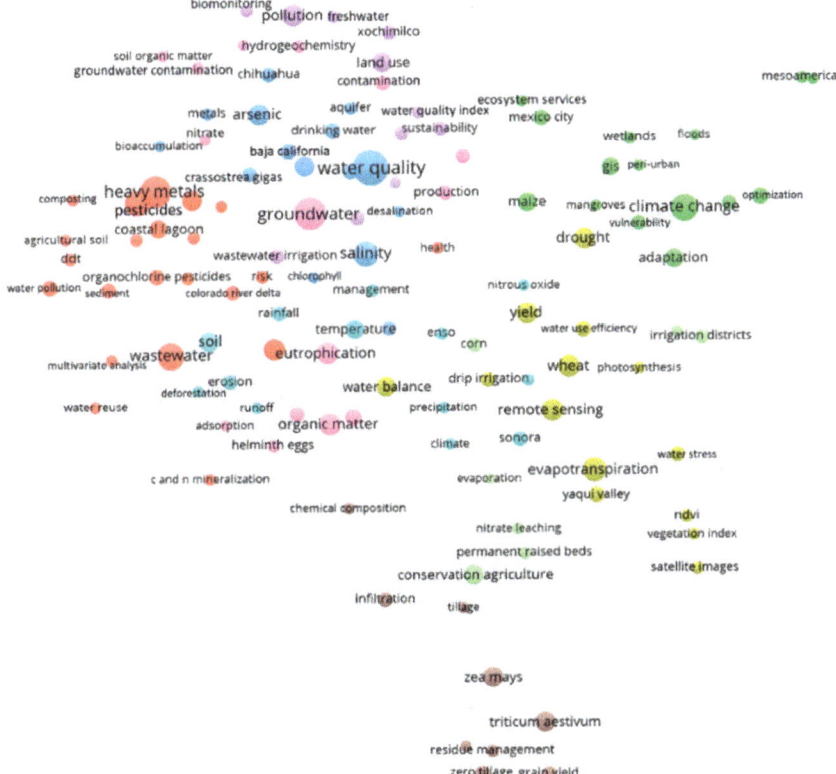

Figure 2. Trends in main keywords related to agricultural water management in Mexico (AWMM) research.

The pink cluster studies underground waters. A large part of the Mexican territory is arid or semi-arid, which means that many areas depend largely on underground water sources that are overexploited. Therefore, it is necessary to carry out actions that allow this situation to be controlled and reversed. For example, Saíz-Rodríguez et al. [69] conducted a study to identify possible locations of artificial recharging of the aquifers in the Valley of Guadalupe (Baja Califormia) while González-Trinidad et al. [70] did the same for the State of Zacatecas. On the other hand, with respect to agricultural activity, incorporating conservation practices and increasing the organic material of the soil can favour the infiltration of rainwater and increase the productivity of the soil, reducing the water needs of the crops [71]. Furthermore, the quality of the underground waters is also being affected by salinisation and pollution due to the use of waste water for agricultural irrigation and fertilizers [72]. To do this, it is necessary to design a plan for the use of the aquifers and create action plans that enable the reversal of the salinisation processes to which the aquifers are subjected and, therefore, avoid situations of collapse over the long term [73].

The light green cluster refers to conservation agriculture, which comprises a series of techniques such as minimum tillage, the permanent cover of the soil, and the diversification of the crops, which enable a more efficient use of the natural resources [74]. The application of conservation agriculture together with the efficient management of fertilizers can increase the yields and quality of the production of the crops [75]. Fuentes et al. [76] carried out a study on the maize crop and concluded that the application of conservation agriculture can increase the carbon content of the soil and reduce CO_2 emissions. Therefore, conservation agriculture can also favour a better control of plagues, as it improves the quality and reduces the erosion of the soil, creating an ideal habitat for organisms [77].

If we analyse the research on SAWMM, we find four differentiated clusters (Figure 3) with three focused on the fields of sustainability and a fourth based on a more technical perspective of the research.

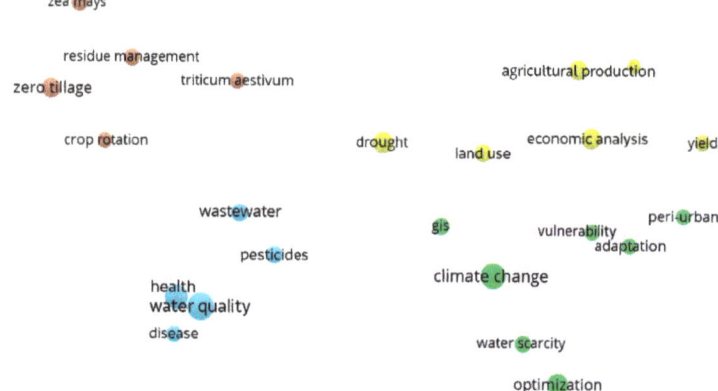

Figure 3. Trends in main keywords related to sustainable agricultural water management in Mexico (SAWMM) research.

The green cluster refers to the environmental dimension of sustainability. It is focused on the research of climate change effects on the availability and management of water resources. This confirms that the environmental perspective of sustainability receives more attention than the rest of the dimensions. In this respect, farmers must take into account the climate variations as part of their production system in order to guarantee the survival of their economic activity and food security [78]. Furthermore, it will also be necessary to identify and study the agricultural areas most prone to variations in order to be able to design specific adaptation plans that minimise their vulnerability to climate change [79].

The blue cluster studies the social dimension, particularly with respect to health. The changes in the uses of the land and the spread of certain anthropogenic practices have led to the contamination of natural resources, which can affect the quality of life and the health of people. The presence of emerging contaminants (faecal sterols, alcaphenols, and pesticides) has been detected in wells in agricultural and urban areas [80]. Contreras et al. [81] carried out a study that compared the incidence of diarrhoeal diseases in children under the age of five in areas that use untreated waste water for irrigation and in which well water is used, concluding that diarrhoea is more frequent in the cases where waste water is used. The accumulation of heavy metals in the soil can put public health at risk since these elements concentrate in the water sources and are absorbed by plants, affecting the quality and security of food [82].

The yellow cluster focuses on the economic dimension, as the increase in demand for water and the possible effects derived from climate change can endanger the survival of agriculture [83]. For example, Bautista-Capetillo et al. [84] found that droughts led to losses for the region of Zacatecas with a value of 478 million dollars in a period of 10 years. Granados et al. [85] conducted a study in Guanajuato in which they concluded that the variability of rainfall has given rise to a loss in the productivity of maize and bean crops, which has reduced the revenue and quality of life of the area.

The brown cluster focuses on the study of the most ideal agronomic practices for maize and wheat crops in order to improve production and efficiency in water use to guarantee the sustainability of these crops.

4. Conclusions

The objective of this article is to analyse the dynamics of the research on the use of water in agriculture in Mexico and especially its sustainable management. To achieve it, a bibliometric analysis has been carried out on a sample of 1490 articles in the research on AWMM and 436 articles in the case of the research on SAWMM. For each of the lines of research, a productivity analysis has been developed based on the number of articles, the journals, the subject categories, the authors, affiliation, and collaboration relations. The principal topics developed in each of them have also been analysed according to the keywords used.

The results reveal that both lines of research have gained importance in recent years. Although research focusing on the use of water in agriculture in Mexico with a focus on sustainability is still in its infancy, it has become a priority field. This result is consistent with the trend observed on a global level in research in this field, particularly related to the fulfilment of some of the sustainable development objectives of the United Nations. In both cases, the principal subject categories are Environmental Sciences, Agricultural and Biological Sciences, and Earth and Planetary Sciences. This enables us to affirm that, in both cases, there is a predominance of research from a technical and environmental perspective. In the case of the research on SAWMM, the social and economic dimensions of sustainability received greater attention than in the case of the research on AWMM. However, it is necessary to promote research from these two approaches and also all three dimensions of sustainability together.

The analysis of the collaboration networks established by Mexico has enabled us to determine that the number of studies carried out through an international collaboration is higher in the case of research on SAWMM than in the general research on AWMM. In this way, we can see that, similarly to other fields of study, sustainability is not only more multidisciplinary, but it is studied to a greater extent through international collaboration between institutions.

The analysis of keywords reveals nine clusters in the overall subject, focused on topics such as the pollution of water bodies, climate change, the quality of water, the application of technology in order to make a more efficient use of water, biodiversity, erosion, agronomic practices that reduce water consumption, underground water sources, and conservation agriculture. With regard to research on SAWMM, three clusters have been found focused on the three dimensions of sustainability and a fourth analysing more technical aspects of agriculture. The topics on climate change and the technical aspects to improve water efficiency are common in both lines of research.

Author Contributions: The four authors have equally contributed to this paper. All authors have read and agreed to the published version of the manuscript.

Funding: This research received no external funding.

Acknowledgments: This work was partially supported by the Spanish Ministry of Economy and Competitiveness and the European Regional Development Fund by means of the research project ECO2017-82347-P.

Conflicts of Interest: The authors declare no conflict of interest.

References

1. Fróna, D.; Szenderák, J.; Rákos, M.H. The Challenge of Feeding the World. *Sustainability* **2019**, *11*, 5816. [CrossRef]
2. Oberle, B.; Bringezu, S.; Hatfield-Dodds, S.; Hellweg, S.; Schandl, H.; Clement, J. Global Resources Outlook 2019: Natural Resources for the Future We Want. Available online: http://pure.iiasa.ac.at/id/eprint/15879/1/unep_252_global_resource_outlook_2019_web.pdf (accessed on 27 July 2020).
3. Ceratti, M. Dos Planetas Más Para Poder Vivir en Este. Available online: https://www.bancomundial.org/es/news/feature/2016/08/09/objetivo-desarrollo-sostenible-ods-12-consumo (accessed on 27 July 2020).
4. Aznar-Sánchez, J.A.; Piquer-Rodríguez, M.; Velasco-Muñoz, J.F.; Manzano-Agugliaro, F. Worldwide research trends on sustainable land use in agriculture. *Land Use Policy* **2019**, *87*, 104069. [CrossRef]
5. Food and Agriculture Organization of the United Nations. *High Level Expert Forum—How to Feed the World in 2050*; Office of the Director, Agricultural Development Economics Division: Rome, Italy, 2009.
6. Baguma, D.; Loiskandl, W. Rainwater harvesting technologies and practises in rural Uganda: A case study. *Mitig. Adapt. Strat. Glob. Chang.* **2010**, *15*, 355–369. [CrossRef]
7. Dias, C.S.L.; Rodrigues, R.G.; Ferreira, J.J. What's new in the research on agricultural entrepreneurship? *J. Rural. Stud.* **2019**, *65*, 99–115. [CrossRef]
8. Foley, J.A.; Ramankutty, N.; Brauman, K.A.; Cassidy, E.S.; Gerber, J.S.; Johnston, M.; Mueller, N.D.; O'Connell, C.; Ray, D.K.; West, P.C.; et al. Solutions for a cultivated planet. *Nature* **2011**, *478*, 337–342. [CrossRef] [PubMed]
9. Velasco-Muñoz, J.F.; Aznar-Sánchez, J.A.; Belmonte-Ureña, L.J.; López-Serrano, M.J. Advances in Water Use Efficiency in Agriculture: A Bibliometric Analysis. *Water* **2018**, *10*, 377. [CrossRef]
10. Cunningham, S.A.; Attwood, S.J.; Bawa, K.S.; Benton, T.G.; Broadhurst, L.M.; Didham, R.K.; McIntyre, S.; Perfecto, I.; Samways, M.J.; Tscharntke, T.; et al. To close the yield-gap while saving biodiversity will require multiple locally relevant strategies. *Agric. Ecosyst. Environ.* **2013**, *173*, 20–27. [CrossRef]
11. Mancosu, N.; Snyder, R.L.; Kyriakakis, G.; Spano, D. Water Scarcity and Future Challenges for Food Production. *Water* **2015**, *7*, 975–992. [CrossRef]
12. FAO (Food and Agricultural Organization). *El sistema alimentario en México—Oportunidades Para el Campo Mexicano en la Agenda 2030 de Desarrollo Sostenible*; FAO: Ciudad de México, Mexico, 2019. Available online: http://www.fao.org/3/CA2910ES/ca2910es.pdf (accessed on 27 July 2020).
13. SIAP (Food, Agricultural and Fisheries Information Service). *2019 Food & Agriculture Overview*; SIAP: Mexico City, Mexico, 2019. Available online: https://nube.siap.gob.mx/gobmx_publicaciones_siap/pag/2019/Agricultural-Atlas-2019 (accessed on 27 July 2020).
14. WTO (World Trade Organization). World Trade Statistical Review 2019. Available online: https://www.wto.org/english/res_e/statis_e/wts2019_e/wts19_toc_e.htm (accessed on 27 July 2020).
15. Sosa-Baldivia, A.; Ruíz-Ibarra, G. Food availability in Mexico: An analysis of agricultural production over the last 35 years and its projection for 2050. *Pap. Poblac.* **2017**, *23*, 207–230.
16. The World Bank. 2020. Available online: https://data.worldbank.org/indicator/SL.AGR.EMPL.ZS?end=2019&locations=MX&start=1991 (accessed on 29 July 2020).
17. Riojas, C. La naturaleza de las articulaciones regionales en México a través del tiempo. *Amerika* **2011**, *4*. [CrossRef]
18. Conagua (Comisión Nacional del Agua). *Estadísticas del Agua en México*; Edición 2018; Conagua: Ciudad de México, México, 2018. Available online: http://sina.conagua.gob.mx/publicaciones/EAM_2018.pdf (accessed on 29 July 2020).
19. Secretaria de Medio Ambiente y Recursos Naturales (Semarnat); Comisión Nacional del Agua (Conagua). Programa Nacional Hídrico 2019–2014. 2019. Available online: http://187.191.71.192/portales/resumen/48732 (accessed on 29 July 2020).
20. Gómez-Merino, F.C.; Hernández-Anguiano, A.M. El Contexto del Sector Agroalimentario en México. In *Líneas Prioritarias de Investigación. Informe de Gestión 2009–2011*; Hernández-Anguiano, A.M., Gómez-Merino, F.C., Pérez-Hernández, L.M., Villanueva-Jiménez, J.A., Eds.; Colegio de Postgraduados: Estado de México, México, 2013; ISBN 978-607-715-135-7.
21. Delgado-Carranza, C.; Bautista, F.; Ihl, T.J.; Palma-López, D. Duración del periodo de lluvias y aptitud de tierras para la agricultura de temporal. *Ecosistemas Recur. Agropecu.* **2017**, *4*, 485–497. [CrossRef]

22. ENA (Encuesta Nacional Agropecuaria). 2017. Available online: https://www.inegi.org.mx/programas/ena/2017/ (accessed on 29 July 2020).
23. Romero, A.A.; Rivas, A.I.M.; Díaz, J.D.G.; Mendoza, M.; Ángel, P.; Salas, E.N.N.; Blanco, J.L.; Álvarez, A.C.C. Crop yield simulations in Mexican agriculture for climate change adaptation. *Atmósfera* **2020**, *33*, 215–231. [CrossRef]
24. García-Hernández, J.; Leyva-Morales, J.B.; Martínez-Rodríguez, I.E.; Hernández-Ochoa, M.I.; Aldana-Madrid, M.L.; Rojas-García, A.E.; Betancourt-Lozano, M.; Pérez-Herrera, N.E.; Perera-Ríos, J.H. Estado actual de la investigación sobre plaguicidas en México. *Rev. Int. Contam. Ambient.* **2018**, *34*, 29–60. [CrossRef]
25. Bonilla-Moheno, M.; Redo, D.J.; Aide, T.M.; Clark, M.L.; Grau, H.R. Vegetation change and land tenure in Mexico: A country-wide analysis. *Land Use Policy* **2013**, *30*, 355–364. [CrossRef]
26. Komiyama, H.; Takeuchi, K. Sustainability science: Building a new discipline. *Sustain. Sci.* **2006**, *1*, 1–6. [CrossRef]
27. Yarime, M.; Takeda, Y.; Kajikawa, Y. Towards institutional analysis of sustainability science: A quantitative examination of the patterns of research collaboration. *Sustain. Sci.* **2009**, *5*, 115–125. [CrossRef]
28. Velasco-Muñoz, J.F.; Aznar-Sánchez, J.A.; Batlles-Delafuente, A.; Fidelibus, M.D. Sustainable Irrigation in Agriculture: An Analysis of Global Research. *Water* **2019**, *11*, 1758. [CrossRef]
29. Garfield, E.; Sher, I.H. New factors in the evaluation of scientific literature through citation indexing. *Am. Doc.* **1963**, *14*, 195–201. [CrossRef]
30. Huang, L.; Zhang, Y.; Guo, Y.; Zhu, D.; Porter, A.L. Four dimensional Science and Technology planning: A new approach based on bibliometrics and technology roadmapping. *Technol. Forecast. Soc. Chang.* **2014**, *81*, 39–48. [CrossRef]
31. Aznar-Sánchez, J.A.; García-Gómez, J.J.; Velasco-Muñoz, J.F.; Carretero-Gómez, A. Mining Waste and Its Sustainable Management: Advances in Worldwide Research. *Minerals* **2018**, *8*, 284. [CrossRef]
32. Aznar-Sánchez, J.A.; Belmonte-Ureña, L.J.; López-Serrano, M.J.; Velasco-Muñoz, J.F. Forest Ecosystem Services: An Analysis of Worldwide Research. *Forests* **2018**, *9*, 453. [CrossRef]
33. Albort-Morant, G.; Henseler, J.; Leal-Millán, A.; Carrión, G.A.C. Mapping the Field: A Bibliometric Analysis of Green Innovation. *Sustainability* **2017**, *9*, 1011. [CrossRef]
34. Opejin, A.K.; Aggarwal, R.M.; White, D.D.; Jones, J.L.; Maciejewski, R.; Mascaro, G.; Sarjoughian, H.S. A Bibliometric Analysis of Food-Energy-Water Nexus Literature. *Sustainability* **2020**, *12*, 1112. [CrossRef]
35. Durieux, V.; Gevenois, P.A. Bibliometric Indicators: Quality Measurements of Scientific Publication. *Radiology* **2010**, *255*, 342–351. [CrossRef] [PubMed]
36. Robinson, D.K.R.; Huang, L.; Guo, Y.; Porter, A.L. Forecasting Innovation Pathways (FIP) for new and emerging science and technologies. *Technol. Forecast. Soc. Chang.* **2013**, *80*, 267–285. [CrossRef]
37. Figueroa-Rodríguez, K.A.; Álvarez-Ávila, M.D.C.; Castillo, F.H.; Rindermann, R.S.; Sandoval, B.F. Farmers' Market Actors, Dynamics, and Attributes: A Bibliometric Study. *Sustainability* **2019**, *11*, 745. [CrossRef]
38. Kumar, A.; Mallick, S.; Swarnakar, P. Mapping Scientific Collaboration: A Bibliometric Study of Rice Crop Research in India. *J. Sci. Res.* **2020**, *9*, 29–39. [CrossRef]
39. Gavel, Y.; Iselid, L. Web of Science and Scopus: A journal title overlap study. *Online Inf. Rev.* **2008**, *32*, 8–21. [CrossRef]
40. Aznar-Sánchez, J.A.; Belmonte-Ureña, L.J.; Velasco-Muñoz, J.F.; Manzano-Agugliaro, F. Economic analysis of sustainable water use: A review of worldwide research. *J. Clean. Prod.* **2018**, *198*, 1120–1132. [CrossRef]
41. Ngwenya, S.; Boshoff, N. Participation of 'international national organisations' in Africa's research: A bibliometric study of agriculture and health in Zimbabwe. *Scientometrics* **2020**, *124*, 533–553. [CrossRef]
42. Velasco-Muñoz, J.F.; Aznar-Sánchez, J.A.; Belmonte-Ureña, L.J.; Román-Sánchez, I.M. Sustainable Water Use in Agriculture: A Review of Worldwide Research. *Sustainability* **2018**, *10*, 1084. [CrossRef]
43. Alonso, S.; Cabrerizo, F.J.; Herrera-Viedma, E.; Herrera, F. h-Index: A review focused in its variants, computation and standardization for different scientific fields. *J. Informetr.* **2009**, *3*, 273–289. [CrossRef]
44. Falagas, M.E.; Kouranos, V.D.; Arencibia-Jorge, R.; Karageorgopoulos, D.E. Comparison of SCImago journal rank indicator with journal impact factor. *FASEB J.* **2008**, *22*, 2623–2628. [CrossRef] [PubMed]
45. Aznar-Sánchez, J.A.; Velasco-Muñoz, J.F.; Belmonte-Ureña, L.J.; Manzano-Agugliaro, F. Innovation and technology for sustainable mining activity: A worldwide research assessment. *J. Clean. Prod.* **2019**, *221*, 38–54. [CrossRef]

46. Robledo-Zacarías, V.H.; Velázquez-Machuca, M.A.; Montañez-Soto, J.L.; Pimentel-Equihua, J.L.; Vallejo-Cardona, A.A.; López-Calvillo, M.D.; Venegas-González, J. Hydrochemistry and emerging contaminants in industrial urban wastewater in Morelia, Michoacán, Mexico. *Rev. Int. Contam. Ambie.* **2017**, *33*, 221–235. [CrossRef]
47. Gilabert-Alarcón, C.; Salgado-Méndez, S.O.; Daesslé, L.W.; Mendoza-Espinosa, L.G.; Villada-Canela, M. Regulatory Challenges for the Use of Reclaimed Water in Mexico: A Case Study in Baja California. *Water* **2018**, *10*, 1432. [CrossRef]
48. Castresana, G.P.; Flores, V.T.; Reyes, L.L.; Aldana, F.H.; Vega, R.C.; Perales, J.L.M.; Suastegui, W.A.G.; Diaz, A.; Silva, A.H. Atoyac River Pollution in the Metropolitan Area of Puebla, México. *Water* **2018**, *10*, 267. [CrossRef]
49. Rodríguez-Aguilar, B.A.; Martínez-Rivera, L.M.; Peregrina-Lucano, A.A.; Ortiz-Arrona, C.I.; Cárdenas-Hernández, O.G. Analysis of pesticide residues in the surface water of the Ayuquila-Armeria River watershed, Mexico. *Terra Latinoam.* **2019**, *37*, 151–161. [CrossRef]
50. Hernández-Bedolla, J.; Solera, A.; Paredes-Arquiola, J.; Pedro-Monzonís, M.; Andreu, J.; Sánchez-Quispe, S.T. The Assessment of Sustainability Indexes and Climate Change Impacts on Integrated Water Resource Management. *Water* **2017**, *9*, 213. [CrossRef]
51. Molina-Navarro, E.; Hallack-Alegría, M.; Martínez-Pérez, S.; Ramírez-Hernández, J.; Mungaray-Moctezuma, A.; Sastre-Merlín, A. Hydrological modeling and climate change impacts in an agricultural semiarid region. Case study: Guadalupe River basin, Mexico. *Agric. Water Manag.* **2016**, *175*, 29–42. [CrossRef]
52. Hernandez-Ochoa, I.M.; Pequeno, D.N.L.; Reynolds, M.; Babar, M.A.; Sonder, K.; Milan, A.M.; Hoogenboom, G.; Robertson, R.; Gerber, S.; Rowland, D.L.; et al. Adapting irrigated and rainfed wheat to climate change in semi-arid environments: Management, breeding options and land use change. *Eur. J. Agron.* **2019**, *109*, 125915. [CrossRef]
53. De Oca, R.M.G.F.; Ramos-Leal, J.A.; Solache-Ríos, M.J.; Martínez-Miranda, V.; Fuentes-Rivas, R.M. Modification of the Relative Abundance of Constituents Dissolved in Drinking Water Caused by Organic Pollution: A Case of the Toluca Valley, Mexico. *Water Air Soil Pollut.* **2019**, *230*, 171. [CrossRef]
54. Saldaña-Robles, A.; Abraham-Juárez, M.R.; Saldaña-Robles, A.L.; Saldaña-Robles, N.; Ozuna, C.; Gutiérrez-Chávez, A.J. The Negative Effect of Arsenic in Agriculture: Irrigation Water, Soil And Crops, State of the Art. *Appl. Ecol. Environ. Res.* **2018**, *16*, 1533–1551. [CrossRef]
55. Reyes-González, A.; Reta-Sánchez, D.G.; Sánchez–Duarte, J.I.; Ochoa-Martínez, E.; Rodríguez-Hernández, K.; Preciado-Rangel, P. Estimation of evapotranspiration of forage corn supported with remote sensing and in situ measurements. *Terra Latinoam.* **2019**, *37*, 279–290. [CrossRef]
56. Reyes-González, A.; Kjaersgaard, J.; Trooien, T.; Hay, C.; Ahiablame, L. Estimation of Crop Evapotranspiration Using Satellite Remote Sensing-Based Vegetation Index. *Adv. Meteorol.* **2018**, *2018*, 1–12. [CrossRef]
57. López-Hernández, M.; Arteaga-Ramírez, R.; Ruiz-García, A.; Vázquez-Peña, M.A.; López-Resano, J.I. Productividad del agua normalizada para el cultivo de maíz (Zea mays) en Chapingo, México. *Agrociencia* **2019**, *53*, 811–820.
58. Palacios-Vélez, E.; Palacios-Sánchez, L.; Espinosa-Espinosa, J.L. Early estimation of the wheat crop yield in irrigation district 038, Río Mayo, Sonora, México/Estimación temprana del rendimiento de la cosecha de trigo en el distrito de riego 038, Río Mayo, Sonora, México. *Tecnol. Cienc. Agua* **2019**, *10*, 225–240. [CrossRef]
59. Pool, D.B.; Panjabi, A.O.; Macías-Duarte, A.; Solhjem, D.M. Rapid expansion of croplands in Chihuahua, Mexico threatens declining North American grassland bird species. *Biol. Conserv.* **2014**, *170*, 274–281. [CrossRef]
60. Andrade-Herrera, M.; Escalona-Segura, G.; González-Jáuregui, M.; Reyna-Hurtado, R.A.; Vargas-Contreras, J.A.; Osten, J.R.-V. Presence of Pesticides and Toxicity Assessment of Agricultural Soils in the Quintana Roo Mayan Zone, Mexico Using Biomarkers in Earthworms (*Eisenia fetida*). *Water Air Soil Pollut.* **2019**, *230*, 59. [CrossRef]
61. Vanderplank, S.; Ezcurra, E.; Delgadillo, J.; Felger, R.; McDade, L.A. Conservation challenges in a threatened hotspot: Agriculture and plant biodiversity losses in Baja California, Mexico. *Biodivers. Conserv.* **2014**, *23*, 2173–2182. [CrossRef]
62. Estrada-Herrera, I.R.; Hidalgo-Moreno, C.; Guzmán-Plazola, R.; Almaraz Suárez, J.J.; Navarro-Garza, H.; Etchevers-Barra, J.D. Soil quality indicators to evaluate soil fertility. *Agrociencia* **2017**, *51*, 813–831.

63. Silva-García, J.T.; Cruz-Cárdenas, G.; Ochoa-Estrada, S.; Estrada-Godoy, F.; Nava-Velázquez, J.; Álvarez-Bernal, D. Loss of soil from water erosion in the basin Chapala Lake, Michoacan, Mexico. *Tecnol. Cienc. Agua* **2017**, *8*, 117–128. [CrossRef]
64. López-Santos, A.; Bueno-Hurtado, P.; Arreola-ávila, J.G.; Emmanuel Pérez-Salinas, J. Conservation activities of soils identified through indices kappa indices in northeast of Durango, Mexico. *Agrociencia* **2017**, *51*, 591–605.
65. Paquini-Rodríguez, S.L.; Benítez-Riquelme, I.; Villaseñor-Mir, H.E.; Muñoz-Orozco, A.; Vaquera-Huerta, H. Gains in yield and its components under normal and limited irrigation of Mexican wheat cultivars. *Rev. Fitotec. Mex.* **2016**, *39*, 367–378.
66. Honsdorf, N.; Mulvaney, M.J.; Singh, R.P.; Ammar, K.; Burgueño, J.; Govaerts, B.; Verhulst, N. Genotype by tillage interaction and performance progress for bread and durum wheat genotypes on irrigated raised beds. *Field Crop. Res.* **2018**, *216*, 42–52. [CrossRef]
67. Rangel-Fajardo, M.A.; Gómez-Montiel, N.; Tucuch-Haas, J.I.; De la Cruz Basto-Barbudo, D.; Villalobos-González, A.; Burgos-Díaz, J.A. Polyethylene glicol 8000 to identify corn tolerant to water stress during germination. *Agron. Mesoam.* **2019**, *30*, 255–266. [CrossRef]
68. Grahmann, K.; Verhulst, N.; Palomino, L.M.; Bischoff, W.-A.; Govaerts, B.; Buerkert, A. Ion exchange resin samplers to estimate nitrate leaching from a furrow irrigated wheat-maize cropping system under different tillage-straw systems. *Soil Tillage Res.* **2018**, *175*, 91–100. [CrossRef]
69. Saiz-Rodríguez, J.A.; Banda, M.A.L.; Salazar-Briones, C.; Ruiz-Gibert, J.M.; Mungaray-Moctezuma, A. Allocation of Groundwater Recharge Zones in a Rural and Semi-Arid Region for Sustainable Water Management: Case Study in Guadalupe Valley, Mexico. *Water* **2019**, *11*, 1586. [CrossRef]
70. González-Trinidad, J. Dynamics of Land Cover Changes and Delineation of Groundwater Recharge Potential Sites in the Aguanaval Aquifer, Zacatecas, Mexico. *Appl. Ecol. Environ. Res.* **2017**, *15*, 387–402. [CrossRef]
71. Aguilar-García, R.; Ortega-Guerrero, M.A. Analysis of the water dynamics in the unsaturated zone, in a soil subject to conservation practices: Implications for aquifer management and adaptation to climatic change. *Rev. Mex. Cienc. Geol.* **2017**, *34*, 91–104.
72. Celestino, A.E.M.; Ramos-Leal, J.; Cruz, D.A.M.; Tuxpan, J.; Bashulto, J.D.L.; Ramírez, J.M. Identification of the Hydrogeochemical Processes and Assessment of Groundwater Quality, Using Multivariate Statistical Approaches and Water Quality Index in a Wastewater Irrigated Region. *Water* **2019**, *11*, 1702. [CrossRef]
73. Mahlknecht, J.; Merchán, D.; Rosner, M.; Meixner, A.; Ledesma-Ruiz, R. Assessing seawater intrusion in an arid coastal aquifer under high anthropogenic influence using major constituents, Sr and B isotopes in groundwater. *Sci. Total. Environ.* **2017**, *587–588*, 282–295. [CrossRef] [PubMed]
74. Fonteyne, S.; Gamiño, M.-A.M.; Tejeda, A.S.; Verhulst, N. Conservation Agriculture Improves Long-term Yield and Soil Quality in Irrigated Maize-oats Rotation. *Agronomy* **2019**, *9*, 845. [CrossRef]
75. Santillano-Cázares, J.; Núñez-Ramírez, F.; Ruíz-Alvarado, C.; Cárdenas-Castañeda, M.E.; Ortiz-Monasterio, J.I. Assessment of Fertilizer Management Strategies Aiming to Increase Nitrogen Use Efficiency of Wheat Grown Under Conservation Agriculture. *Agronomy* **2018**, *8*, 304. [CrossRef]
76. Fuentes, M.; Hidalgo, C.; Etchevers, J.; De León, F.; Guerrero, A.; Dendooven, L.; Verhulst, N.; Govaerts, B. Conservation agriculture, increased organic carbon in the top-soil macro-aggregates and reduced soil CO_2 emissions. *Plant Soil* **2012**, *355*, 183–197. [CrossRef]
77. Rivers, A.; Barbercheck, M.; Govaerts, B.; Verhulst, N. Conservation agriculture affects arthropod community composition in a rainfed maize—Wheat system in central Mexico. *Appl. Soil Ecol.* **2016**, *100*, 81–90. [CrossRef]
78. Paredes-Tavares, J.; Gómez-Albores, M.A.; Mastachi-Loza, C.A.; Delgado, C.D.; Becerril-Piña, R.; Martínez-Valdés, H.; Bâ, K.M. Impacts of Climate Change on the Irrigation Districts of the Rio Bravo Basin. *Water* **2018**, *10*, 258. [CrossRef]
79. Ahumada-Cervantes, R.; Angulo, G.V.; Rodríguez-Gallegos, H.B.; Flores-Tavizón, E.; Felix-Gastelum, R.; Romero-Gonzalez, J.; Granados-Olivas, A. An indicator tool for assessing local vulnerability to climate change in the Mexican agricultural sector. *Mitig. Adapt. Strat. Glob. Chang.* **2017**, *22*, 137–152. [CrossRef]
80. González-Acevedo, Z.I.; Zarate, M.A.G.; Flores-Lugo, I.P. Emerging contaminants and nutrients in a saline aquifer of a complex environment. *Environ. Pollut.* **2019**, *244*, 885–897. [CrossRef]

81. Contreras, J.D.; Meza, R.; Siebe, C.; Rodríguez-Dozal, S.; López-Vidal, Y.; Castillo-Rojas, G.; Amieva, R.I.; Solano-Gálvez, S.G.; Mazari-Hiriart, M.; Silva-Magaña, M.A.; et al. Health risks from exposure to untreated wastewater used for irrigation in the Mezquital Valley, Mexico: A 25-year update. *Water Res.* **2017**, *123*, 834–850. [CrossRef]
82. Castro-González, N.P.; Calderón-Sánchez, F.; Moreno-Rojas, R.; Tamariz-Flores, J.V.; Reyes-Cervantes, E. Heavy metals pollution level in wastewater and soils in the alto balsas sub-basin in Tlaxcala and Puebla, Mexico. *Rev. Int. Contam. Ambient.* **2019**, *35*, 335–348. [CrossRef]
83. Duchin, F.; López-Morales, C. Do Water-Rich Regions Have a Comparative Advantage in Food Production? Improving the Representation of Water for Agriculture in Economic Models. *Econ. Syst. Res.* **2012**, *24*, 371–389. [CrossRef]
84. Bautista-Capetillo, C.; Carrillo, B.; Picazo, G.; Júnez-Ferreira, H. Drought Assessment in Zacatecas, Mexico. *Water* **2016**, *8*, 416. [CrossRef]
85. Granados, R.; Soria, J.; Cortina, M. Rainfall variability, rainfed agriculture and degree of human marginality in North Guanajuato, Mexico. *Singap. J. Trop. Geogr.* **2017**, *38*, 153–166. [CrossRef]

Publisher's Note: MDPI stays neutral with regard to jurisdictional claims in published maps and institutional affiliations.

© 2020 by the authors. Licensee MDPI, Basel, Switzerland. This article is an open access article distributed under the terms and conditions of the Creative Commons Attribution (CC BY) license (http://creativecommons.org/licenses/by/4.0/).

Review

Experimental Economics in Agriculture: A Review of Worldwide Research

Ernesto Mesa-Vázquez, Juan F. Velasco-Muñoz, José A. Aznar-Sánchez *and Belén López-Felices

Research Centre on Mediterranean Intensive Agrosystems and Agrifood Biotechnology (CIAIMBITAL), Department of Economy and Business, University of Almería, 04120 Almería, Spain; ermeva@ual.es (E.M.-V.); jfvelasco@ual.es (J.F.V.-M.); blopezfelices@ual.es (B.L.-F.)
* Correspondence: jaznar@ual.es

Abstract: Over the last two decades, experimental economics has been gaining relevance in the research of a wide range of issues related to agriculture. In turn, the agricultural activity provides an excellent field of study within which to validate the use of instruments employed by experimental economics. The aim of this study is to analyze the dynamics of the research on the application of experimental economics in agriculture on a global level. Thus, a literature review has been carried out for the period between the years 2000 and 2020 based on a bibliometric study. The main results show that there has been a growing use of experimental economics methods in the research on agriculture, particularly over the last five years. This evolution is evident in the different indicators analyzed and is reflected in the greater scientific production and number of actors involved. The most relevant topics within the research on experimental economics in agriculture focus on the farmer, the markets, the consumer, environmental policy, and public goods. These results can be useful for policy makers and researchers interested in this line of research.

Keywords: sustainable development; agricultural economics; environmental economics; bibliometric analysis

1. Introduction

Experimental economics is a branch of Economics that enables the controlled study of experimental subjects, markets, economic institutions, and ground rules [1]. According to Vernon Smith, the Nobel prize-winner for Economics in 2002, "Experimental economics applies laboratory methods to study the interactions of human beings in social contexts governed by explicit or implicit rules" [2]. Experimental economics has become consolidated as a body of knowledge in its own right, similarly to other areas which were initially questioned as to their usefulness, such as game theory, mathematical economics, or econometrics [3].

Experimental economics enables policy makers to test whether certain public policies or actions could have significant effects before implementing them. In this way, the experiments can guide economic policy measures before they are applied [4]. The experiments enable researchers to observe groups of people participating in a specific problem, clearly specifying the decisions to make, avoiding uncontrolled effects or noise that can distort their decision-making process, simulating a context of the real economy [5]. During the experiment, the experimental subjects are offered appropriate incentives so that they act according to their own criterion, obtaining a compensation at the end of the experiment based on the result of their actions [6,7]. In this way, the researchers know how and why both markets and agents react to changes in the rules throughout the different stages of the experiment. Experimental economics provides important indications with respect to economic behavior in a wide variety of sub-disciplines of Economics, such as Game Theory, Consumer Behavior, Industrial Organization, Public Finance, Labour Economics, and Agricultural Economics [8,9].

Over the last twenty years, experimental economics has been gaining prominence in the research of a wide range of issues related to agriculture. This study field is relevant for policy makers when designing measures that enhance social well-being by improving the efficiency of the markets and creating a regulatory framework better adapted to reality, based on an improved understanding of the behavior of the agents involved (suppliers, demand, markets, and institutions) [10]. In addition, agriculture is considered to be an ideal scenario within which to implement the tools of experimental economics, with a mutually beneficial bidirectional relationship prevailing between the two [11]. On the one hand, experimental economics can be used to gain a deeper understanding of issues of interest to the agricultural sector. In this way, the experimental methods have been proven to be efficient in contributing greater knowledge about the behavior of the agents involved, such as farmers, producers, consumers, markets, and economic institutions [10,12]. On the other hand, agriculture constitutes an ideal area of research to test the validity of the experimental instruments, contributing to the development of this study field, and the debate existing in the literature on certain areas [13].

Although we can confirm a growing interest in the use of experimental economics methods in agriculture, to date, there are no known studies that analyze the dynamics that this line of research has followed on a global scale. Therefore, in order to contribute to covering this gap, this study conducts a review of the literature produced between the years 2000 and 2020 through a bibliometric analysis. The results obtained will allow us to identify the principal actors that constitute the driving agents of the knowledge and the most relevant topics within this line of research. Therefore, this article could serve as a reference for both policy makers and researchers interested in this line of research.

2. Methodology

In order to fulfil the proposed objective, a bibliometric analysis is considered to be the most appropriate methodology. Garfield developed this methodology in the 1950s with the aim of identifying, organizing, and evaluating the principal components of an area of specific knowledge [14,15]. Bibliometric analysis has been gaining ground in disciplines as diverse as economics, agronomics, biology, engineering, medicine, or psychology [16,17]. The possibilities that it offers include different mapping techniques to represent the bibliographic information available in different databases and statistical and mathematical methods to determine the trends in an area of research [18,19]. In order to conduct a bibliometric analysis, we can use different approaches, considered as being traditional [20]: co-occurrence, co-citation and bibliographic coupling analysis. The co-occurrence approach is understood as the study of the joint occurrences of two terms in a given text, with the purpose of identifying the conceptual and thematic structure of a thematic field. In the process of co-occurrence analysis, once the terms to be analyzed have been selected, co-occurrence matrices are constructed, with which similarity measures are calculated [21]. The similarity measures serve as input to different kinds of multivariate analysis, among which we can find clustering analysis and multidimensional scaling analysis. Therefore, this approach has been considered appropriate for the development of this work, given that the proposed objective is to analyze the structure of the body of scientific literature on Experimental Economics applied to Agriculture (EEA).

Furthermore, this methodology provides various types of indicators that measure different aspects of the bibliographic information [22]: quantity indicators measure productivity; relevance indicators reveal the impact of the publications; and structural indicators analyze the connections between the different elements of the same field of research. In order to conduct this study, three types of indicators have been used together with the traditional co-occurrence approach. Article counts were used to measure the output of the different actors. To assess the relative importance of research in this area, the quality indicators of citation counts, the h-index and the SCImago Journal Rank (SJR) impact factor were used. The second of these, the h-index, is defined as the total (h) of N papers with at least h citations each [23]. The SJR, on the other hand, measures the number of weighted citations, where the weighting of citations depends on the subject and the prestige of the cited journal. Finally, the analysis of the co-occurrence structure of keywords has allowed us to identify the main themes in EEA research.

Once the methodological tool to be used has been determined, the following stage was the sample selection of studies to be analyzed. Regarding the database selection for the extraction of the paper sample, studies have been carried out to measure the overlap between databases and the use impact of different data sources for specific research fields on bibliometric indicators. A higher number of journals indexed by Scopus compared to WoS has been demonstrated [24]. In terms of overlap, 84% of WoS titles are also indexed in Scopus, while only 54% of Scopus titles are indexed in WoS [25]. This was the main reason for selecting Scopus for this work. It is therefore considered that the use of this database ensures that a representative sample of papers on EEA research is extracted [26,27]. The selection of the sample of articles to analyze in this study was made in April 2021 based on the following parameters: TÍTULO-ABS-CLAVE ("experimental economic*") Y TÍTULO-ABS-CLAVE (agricultur* OR crop* OR cultivation OR agrosystem OR agroecosystem OR farm*). The search covered the period 2000–2020. Only articles published until 2020 have been included to enable the comparison of complete annual periods [28,29]. It is important to remember that different search queries may generate different results. The final sample included a total of 105 documents.

With respect to the preparation, processing and analysis of the information, after being downloaded, the data were refined so as to eliminate duplications, omissions and errors and to detect any incomplete information [30]. Furthermore, a search for articles on agriculture was undertaken according to the same criteria in order to determine the relative importance of the use of experimental economics within the general field. The variables analyzed were the number of articles, their year of publication, subject area, the name of the journals and the institutions and countries of affiliation of the researchers. The tools used for processing the information were Excel (version 2016, Microsoft, Redmond, DC, USA), and SciMaT (v1.1.04, research group of Soft Computing y Sistemas Inteligentes de Información, University of Granada, Granada, Spain). Figure 1 summarizes the methodological development of this study.

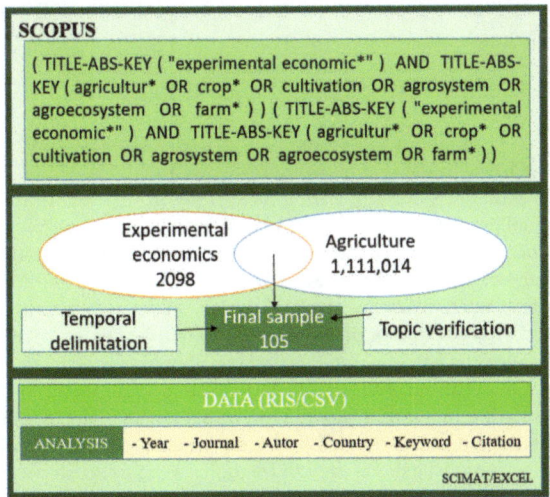

Figure 1. Summary of the methodology.

3. Results and Discussion

3.1. Evolution of the General Characteristics of Research on Experimental Economics in Agriculture

Table 1 shows the evolution of the main variables related to the research on experimental economics in agriculture (EEA) in the period 2000–2020. The number of articles has increased irregularly throughout the period, with a minimum value of 0 in 2002 and a maximum of 10 in 2013. It is important to note that this line of research has been experiencing strong growth in recent years, as more than 75% of the articles in the sample were published in the last decade and almost 40% in the last five years. In order to verify whether the increase in the number of publications is due to the overall trend in the research as a whole, the annual variation in the number of articles published with respect to the overall research has been calculated, taking the first year of the period analyzed as a base (Figure 2). The average annual growth in the number of studies on agriculture was 9.1% while that of articles on EEA was 11.6%. Although this line of research can be considered to still be in its infancy and exhibits an irregular evolution, these data suggest that EEA will become an increasingly relevant line within the research on agriculture.

Table 1. Main characteristics of the EEA research.

Year	Documents	Authors	Journals	Countries	Citation	Average Citation
2000	1	1	1	1	0	0.0
2001	1	5	1	1	1	0.5
2002	0	0	0	0	1	1.0
2003	1	6	1	2	4	2.0
2004	4	11	4	3	9	2.1
2005	3	7	1	1	12	2.7
2006	1	1	1	1	27	4.9
2007	2	6	1	2	28	6.3
2008	4	11	3	5	26	6.4
2009	9	27	5	4	26	5.2
2010	6	16	2	6	47	5.7
2011	9	27	9	9	69	6.1
2012	1	2	1	1	76	7.8
2013	10	26	7	9	127	8.7
2014	6	15	6	5	133	10.1
2015	7	29	7	7	132	11.0
2016	7	16	7	6	166	12.3
2017	8	27	8	10	208	13.7
2018	9	31	9	12	199	14.5
2019	8	31	8	9	237	15.8
2020	8	18	8	8	272	17.1

Throughout the whole period under study, a total of 242 authors participated in the 105 documents that make up the sample. This variable has grown from one author in 2000 to 18 in 2020, with a maximum of 31 in the years 2018 and 2019. The average number of authors per article varied considerably, with the minimum being one author per study in 2000 and the maximum being six per study in 2003. It should be noted that the number of studies undertaken by each author is very low. More than 80% of the researchers participated in just one study. Only 3.7% of the authors participated in four or more studies and only one author participated in more than ten. The average number of documents per publication has remained almost constant at one, with an average number of documents per publication for the whole period being 1.9. In total, the 105 documents were published in 55 different publications. With respect to the countries that participated in the studies, for the whole period analyzed there were a total of 30. The number of countries has also experienced a general growth trend, but with irregular oscillations throughout the period. Thus we can observe 1 in 2000, 8 in 2020, and a maximum value of 12 in 2018. Overall, the documents in the sample accumulated a total of 1800 citations for the whole period. This figure increased from one in 2001 (first year with citations) to 272 in 2020. The average number of citations obtained per document increased from 0.5 in 2001 to 17.1 in 2020.

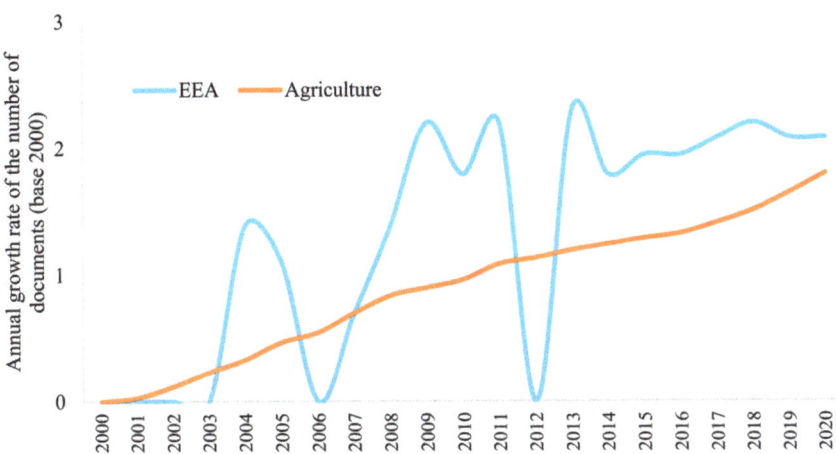

Figure 2. Comparative trends between EEA and agriculture research.

3.2. Distribution of the Research on EEA by Subject Area, Type of Document, and Language

Figure 3 shows the distribution of the documents published based on the classification by subject area established by Scopus. It is necessary to point out that the same study may be classified in more than one category concurrently. As expected, the categories that include a higher number of studies are Economics, Econometrics, and Finance with 74.3% of the total of the sample and Agriculture and Biological Sciences with 66.7%. These two categories coincide with the established search parameters (economics and agriculture). However, these disciplines are not the only ones that intervene in the studies making up the sample. There are also studies from the perspectives of Environmental Sciences accounting for 21%; Social Sciences with 18.1%; and Business, Management, and Accounting with 4.8%. Table 2 reveals the type of document and language in which the studies on EEA were published. It can be observed that 80% of the studies were published in the format of a scientific article. These are followed by conference papers, accounting for 10.5% and literature review studies, representing 6.7%. The rest are books, data papers and notes, each representing just 1%. With respect to the language, 95.2% of the studies in the sample were published in English, which is the dominant language in this line or research, as could be expected. The other languages found were German and French, representing 2.9% and 1.9% of the studies respectively.

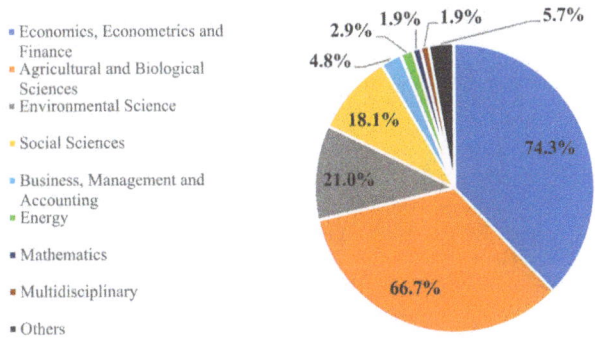

Figure 3. Distribution of EEA research by subject area.

Table 2. Document type and language related to EEA research.

Document Type	%	Language	%
Article	80.0	English	95.2
Conference Paper	10.5	German	2.9
Review	6.7	French	1.9
Book	1.0		
Data Paper	1.0		
Note	1.0		

3.3. Most Relevant Journals in Research on EEA

Table 3 shows the most prolific journals in EEA in the period 2000–2020 and the principal characteristics of their publications. This group includes all of the journals that published two or more studies on this field of research during the period studied. Of the 55 journals that published studies on EEA, this group accounts for 27.3% of the total. The remaining 72.7% only published one study on this subject matter. Overall, the journals in the table published 65 documents included in the sample, which represent 61.9% of the total. Furthermore, this group accumulates 84.6% of the total citations of the documents of the sample. Therefore, these journals can be considered to be the basic core of the publications that promote research on EEA.

Table 3. Major characteristics of the most active journals related to EEA research.

Journal	Documents	SJR [1]	H Index	Country	Citation	Average Citation	1st Article	Last Article
American Journal of Agricultural Economics	16	1.949 (Q1)	13	UK	847	52.9	2001	2020
European Review of Agricultural Economics	10	1.400 (Q1)	8	UK	284	28.4	2009	2019
Agricultural and Resource Economics Review	7	0.475 (Q2)	5	USA	95	13.6	2004	2011
Journal of Agricultural and Resource Economics	5	0.548 (Q2)	3	USA	68	13.6	2008	2018
Agricultural Economics	3	1.200 (Q1)	2	UK	12	4.0	2016	2019
Applied Economics	3	0.569 (Q2)	3	UK	36	12.0	2013	2015
Environmental and Resource Economics	3	1.270 (Q1)	3	Netherlands	32	10.7	2010	2019
Food Policy	3	2.092 (Q1)	3	UK	41	13.7	2000	2015
Journal of Agricultural Economics	3	1.157 (Q1)	2	USA	27	9.0	2013	2020
Applied Economic Perspectives and Policy	2	1.400 (Q1)	2	UK	49	24.5	2011	2015
Cahiers Agricultures	2	0.381 (Q2)	1	France	3	1.5	2011	2018
Canadian Journal of Agricultural Economics	2	0.505 (Q2)	2	USA	7	3.5	2008	2016
Ecological Economics	2	1.917 (Q1)	2	Netherlands	12	6.0	2015	2018
German Journal of Agricultural Economics	2	0.146 (Q4)	1	Germany	2	1.0	2014	2015
World Development	2	2.386 (Q1)	2	UK	7	3.5	2017	2019

[1] Scimago Journal Rank 2020.

The journal that published the most articles is the *American Journal of Agricultural Economics* with a total of 16. This journal has 847 citations, an average number of citations per article of 52.9 and an H index of 13. Its impact factor in the Scimago Journal Rank (SJR) in 2020 was 1.949 and it has been publishing on EEA since 2001, when it published its first text on this subject field. *European Review of Agricultural Economics* holds the second place with 10 studies. Its H index is 8, it has 284 citations and an average number of citations per article of 28.4. Its SJR impact factor is 1.400. In third place is *Agricultural and Resource Economics Review* with 7 documents. This journal last published on EEA in 2011. It has an H index of 5, it has 95 citations in total and an average number of citations per study of 13.6.

The most veteran journal in the table is *Food Policy*, given that it published its first article on EEA in 2000. This journal shares fifth place, with three texts published on EEA.

Its H index is 3, it has 41 citations and 13.7 citations per article, on overage. The SJR factor in this case is 2.092, but the last study published by this journal on this subject area was in 2015. *World Development* is the journal with the highest SJR in the table with 2.386. This journal shares tenth place, with two texts published on EEA. It accumulates a total of 7 citations and an average of 3.5 citations per article. It first published on this subject area in 2017 and is therefore the most recent incorporation. This explains the low number of citations obtained by its publications on EEA.

3.4. Most Relevant Countries in Terms of Research on EEA

Table 4 shows the most prolific countries in terms of research on EEA for the period 2000–2020 and the main characteristics of their studies. This group of countries is highly heterogeneous as they are located in every continent except for Africa. We should take into account that there is wide disparity with respect to the incorporation of the different countries into the research on this field of study. Furthermore, not all of them published studies in 2020, taken as a reference given that it is the last year of the period analyzed. The USA is the country that has published most documents on this subject area, with 48. Germany is next with 21, followed by France with 16. The rest of the countries have published less than 10 studies on this subject area. The pioneer countries in this line of research were France, the USA, the Netherlands, and Norway, in this order. The latest country to publish on EEA is China, given that its first study in the sample was published in 2017. With regard to the relevance of the research, measured through the number of citations of the studies, the USA leads the table with a total of 1236. France follows with 315 and then Germany with 198. However, in terms of the average number of citations per article, the UK is the most prominent country with 27.3. This is followed by the USA with 25.8 and France with 19.7.

Table 4. Main characteristics of the most active countries related to EEA research.

Country	Documents	Citation	Average Citation	H Index	1st Article	Last Article
USA	48	1236	25.8	18	2001	2020
Germany	21	198	9.4	8	2011	2019
France	16	315	19.7	9	2000	2019
Netherlands	6	46	7.7	3	2003	2020
Canada	5	38	7.6	3	2008	2016
Sweden	5	31	6.2	3	2013	2019
Australia	4	24	6.0	3	2007	2020
China	4	11	2.8	2	2017	2020
Norway	4	12	3.0	3	2004	2020
UK	3	82	27.3	2	2010	2019

Table 5 shows the results of analyzing of the collaboration networks established between the most active countries with respect to research on EEA. An average of 53.2% of studies were conducted through international collaboration by the group of the 10 countries. Four countries, on average, made up the collaboration networks. The UK is the country with the highest percentage of studies carried out through international collaboration, that is, 100%. This is followed by China and Norway with 75% and Sweden with 60%. The USA has the largest collaboration network with 17 different collaborators. Next is France with eight and Germany with seven. Australia is at the other end of the scale as it has not published any studies in collaboration with institutions from other countries. The table also includes the main collaborators of each country. Finally, the table shows the average number of citations of the studies, differentiating whether they were conducted through international collaboration or not. On average, in the former case the studies have a total of 10.8 citations, while in the latter case they have 8.8 citations. However, as the table shows, the result varies depending on the country.

Table 5. Major characteristics in the collaboration of the most active countries related to EEA research.

Country	Percentage of Collaboration	Number of Collaborators	Main Collaborators	Average Citation	
				Collaboration	Non Collaboration
USA	37.5	17	Norway, Canada, China, Netherlands, Spain, UK	19.4	29.6
Germany	38.1	7	France, Belgium, Jordan	15.0	6.0
France	56.3	8	Germany, Austria, Belgium	20.4	18.7
Netherlands	50.0	3	USA, Ethiopia, UK	13.7	1.7
Canada	40.0	1	USA	3.5	10.3
Sweden	60.0	3	Germany, Indonesia, UK	2.3	12.0
Australia	0.0	0	-	0.0	6.0
China	75.0	2	USA, France	3.7	0.0
Norway	75.0	1	USA	2.7	4.0
UK	100.0	5	USA, Colombia, India, Netherlands, Sweden	27.3	0.0

3.5. Most Relevant Institutions in the Research on EEA

The most active institutions in terms of research on EEA in the period 2000–2020 and the principal characteristics of the studies can be seen in Table 6. These institutions belong to Germany, the USA, France, and the Netherlands. Worth noting is the small number of documents per institution, given that only two of them have published more than 10 studies on this subject area. The Georg-August-Universität Göttingen in Germany is in first position with 13 documents. This institution has 74 citations, an average of 5.7 citations per study and an H index of 6. The institution with the second highest number of studies is Cornell University in the USA with 11 publications. This university has 223 citations, an average of 20.3 citations per document and an H index of 8. In third place is the University of Delaware, also in the United States, with seven documents in total, 150 citations, an average of 21.4 citations per study and an H index of 6. The Oklahoma State University-Stillwater has the highest total number of citations of those included in the table, with 362 and the highest average number of citations per study, with 72.4. This institution shares sixth position with five studies published on EEA.

Table 6. Major characteristics of the most active institutions related to EEA research.

Institution	Country	Documents	Citation	Average Citation	H Index	Percentage of Collaboration	Average Citation	
							Collaboration	Non Collaboration
Georg-August-Universität Göttingen	Germany	13	74	5.7	6	15.4	10.5	4.8
Cornell University	USA	11	223	20.3	8	18.2	4.5	23.8
University of Delaware	USA	7	150	21.4	6	28.6	4.5	28.2
Centre de recherche Île-de-France-Versailles-Grignon	France	6	220	36.7	6	50.0	37.3	36.0
Economie Publique	France	6	133	22.2	6	66.7	30.3	6.0
Oklahoma State University-Stillwater	USA	5	362	72.4	5	0.0	0.0	72.4
French National Institute for Agricultural Research	France	5	125	25.0	4	80.0	29.8	6.0
Arizona State University	USA	4	56	14.0	3	50.0	18.0	10.0
Wageningen University and Research	Netherlands	4	23	5.8	3	50.0	10.0	1.5
University of Wyoming	USA	4	7	1.8	2	25.0	3.0	1.3
Technical University of Munich	Germany	4	115	28.8	4	100.0	28.8	0.0

With regard to the international collaboration of the institutions, the average percentage of studies conducted in collaboration is 44%. Most noteworthy are the Technical University of Munich in Germany with 100% of its studies carried out through international collaboration and Oklahoma State University-Stillwater with 0%. On average, the studies jointly conducted among different institutions obtained a total of 16.1 citations. Meanwhile, the studies carried out autonomously had 17.3 citations on average. Again, similarly to the countries, there is wide disparity between the institutions in this field of study.

3.6. Most Relevant Authors in Research on EEA

Table 7 shows those authors who conducted the most research on EEA together with the most noteworthy characteristics of their studies. This group comprises 12 authors belonging to 10 institutions in four different countries. Six of the institutions are also on the list of the most active institutions (see Table 6). New institutions appear in this section due to the establishment of collaboration networks between authors, which helps to place some of them in prominent positions. The most published author, with 15 documents is Oliver Musshoff from Georg-August-Universität Göttingen. This author has 100 citations in total, an average of 6.7 citations per document and an H index of 7. Musshoff shares authorship with other prominent authors from the same institution, such as Daniel Hermann and Syster C. Maart-Noelck, in fifth and seventh position, with six and four studies respectively. Another notable author in the table with whom he conducts research is Jens Rommel, from the Swedish Sveriges lantbruksuniversitet, who shares the 10th position with three documents.

Table 7. Major characteristics of the most active authors related to EEA research.

Author	Documents	Citation	Average Citation	H Index	Country	Affiliation	First Article	Last Article
Musshoff, Oliver	15	100	6.7	7	Germany	Georg-August-Universität Göttingen	2013	2019
Messer, Kent D.	8	169	21.1	6	USA	University of Delaware	2005	2019
Kaiser, Harry. M.	7	153	21.9	5	USA	Cornell SC Johnson College of Business	2005	2015
Lusk, Jayson L.	7	520	74.3	7	USA	Purdue University	2001	2019
Hermann, Daniel	6	24	4.0	4	Germany	Georg-August-Universität Göttingen	2015	2019
Marette, Stephan	6	127	21.2	6	France	Economie Publique	2008	2017
Maart-Noelck, Syster C.	4	57	14.3	2	Germany	Georg-August-Universität Göttingen	2013	2014
Roosen, Jutta	4	115	28.8	4	Germany	TUM School of Management	2008	2013
Bastian, Christopher T.	3	4	1.3	2	USA	University of Wyoming	2009	2019
Blanchemanche, Sandrine	3	94	31.3	3	France	French National Institute for Agricultural Research	2008	2013
Rommel, Jens	3	9	3.0	3	Sweden	Sveriges lantbruksuniversitet	2018	2019
Schulze, William D.	3	68	22.7	3	USA	Cornell University	2004	2010

In second place in terms of the number of documents published is Kent D. Messer from the University of Delaware, with eight. This author has 169 citations, an average of 21.1 citations per article and an H index of 6. In this field of research he collaborates mainly with Harry M. Kaiser and William D. Schulze, from Cornell University and Johnson College of Business, with whom he shares five and two studies respectively. Jayson L. Lusk, shares the third position in terms of the number of documents with seven in total. This researcher from Purdue University accumulates the most citations with 520, the highest average number of citations per study with 74.3 and an H index of 7. He is the most veteran author of the table, publishing his first study on EEA in 2001.

3.7. Relevant Topics in Research on EEA

An analysis of the keywords enables the most relevant topics within the research on EEA to be identified. These topics focus on the consumer, the farmer, the markets, environmental policy, and public goods.

With regard to consumer behavior, willingness-to-pay (WTP) is widely used as an analytical tool. In this respect, Stenger [31] elicits the WTP in a laboratory setting of the subjects for products that offer greater food safety as they have not been grown on land irrigated with wastewater. On the other hand, Lusk et al. [32] conduct a field experiment that reveals that consumers prefer to pay a higher price for tender steaks in a blind tasting of different meats. Toler et al. [33] find that concern for equity can explain why consumers prefer to shop at farm markets rather than traditional grocery stores, with a greater WTP premium for local food products. Economic experiments also measure the acceptance of technology in the consumption of food products. In this respect, Bieberstein et al. [34] conclude, after a laboratory experiment, that consumers are reluctant to accept both food and packaging produced with nanotechnology.

Another aspect of consumer behavior is its stance towards risk. Taking food safety as a base, Lusk and Coble [35] carried out a laboratory experiment in which they elicited the risk perception and risk preferences of the subjects in relation to the consumption of genetically modified food. Another aspect of consumer behavior studied is how the information presented affects decision making, which is interesting for advertising and marketing campaigns. This is the case of Marette et al. [8] in which, through a field experiment, the weight that certain messages related to health in the choice of the consumer was tested.

Second, the analysis of farmer behavior within the studies of the sample is noteworthy. In relation to decision making that affects management and investment, De Koeijer et al. [36] investigate the relationship between the complexity of farm management and technical farm performance, applied to the management of nitrogen in arable farms. The results enable the identification of the weak aspects in the management of individual farms, laying the foundations on which to work to improve their management. The behavior of producers is also studied with respect to the management of financing needs. This is the case of Messer et al. [37], who investigate, in a laboratory, the effectiveness of alternative voluntary financing mechanisms of agricultural commodities as opposed to generic advertising programmes. In a study of the behavior of agricultural entrepreneurs, Musshoff et al. [38] conduct a within-subject experiment in a laboratory setting to determine how de-investment decisions are affected not only by economic reasons, but also by non-monetary factors (emotions, attachment to farming, and different facets of psychological inertia). Given that classical investment theory and the real options approach do not correctly explain the behavior of investors, Maart-Noelck and Musshoff [39] perform a laboratory experiment with farmers which reveals how they learn from their previous investments as well as considering that waiting is of great value in decision-making.

Other studies explain the risk attitude of farmers. In this sense, Warnick et al. [40] conduct a field experiment that analyses risk and ambiguity aversion in rural Peru, showing how the latter has a negative effect on the probability of farmers planting more than one variety of the main crop. In a field experiment setting, Bocquého et al. [41] find that farmers are averse to risk and are doubly sensitive to losses than gains. Brunette et al. [42] carry out a lab experiment focused on forest parks and the influence of the risk attitude of the forest owners on the harvesting decision, due to the interest of policy makers to promote a public-private initiative. Gars and Ward [43] show how we use our own personal experience and that of others in order to learn a new technology. This is the case of the adoption of a hybrid rice in India, where farmers' risk and uncertainty preferences are elicited using lottery based experiments. Pollard et al. [44] conduct a field experiment with farmers in Scotland to test the results obtained through laboratory experiments that find that cooperation is low in a context of uncertainty and work with different sources of uncertainty. Senapati [45] continues with the study of farmer's risk attitudes in terms of irrigated and rain-fed farming in India. This lab experiment shows that factors such as age, the level of education, the farm size and the HL lottery have a positive and significant effect on the risk behavior of the farmers in the sample.

The third relevant topic within the research on EEA is the functioning of the markets. In this respect, Wu and Roe [46] justify why it is appropriate for growers and processors to use fixed performance contracts instead of tournament contracts in the regulation of agricultural production contracts. Also, in relation to failure markets, Yesuf and Bluffstone [47] carry out a field experiment in the rural areas of Ethiopia to study the determinants of risk aversion in communities that largely depend on rain-fed agriculture/livestock production, which involves a high level of risk if mechanisms to transfer this risk to third parties are not available. In relation to the marketing of the products, Kanter et al. [48] undertake an experiment involving milk and show how labeling new products ("rBST-free", "organic food") stigmatizes the conventional products that are already on the market. Dillaway et al. [49] study the impacts of media information on the purchasing decision of products, using an experimental case study on food safety. Based on a laboratory experiment with a within-subject design, Wu et al. [50] provide insights into how domestic agricultural producers seek to differentiate themselves through labeling with the place of origin and local messages in response to growing international competition.

A fourth topic in the research on EEA is environmental policy. The study by Palm-Forster et al. [51], provides a reference of policy-making for the design of programmes that mitigate environmental damages and enhance the environmental benefits produced by agricultural landscapes. Murphy and Stevens [52] explain how experimental economics helps to improve the effectiveness of calibrating and estimating the aggregates used in environmental valuation. Along the same lines, Poe et al. [53] refer to experimental economics results to justify the design of policies to improve mechanisms for controlling environmental pollution. Lybbert [54] experimentally analyses the willingness of poor farmers in India to use "pro-poor seeds" that stabilize crop yields and limit yield losses and better withstand climate fluctuations and biotic stresses. Bougherara and Combris [55] use a mixed within-subject and between-subject laboratory experiment to study the WTP for products that are labeled 'eco-friendly'. Important issues arise related to fair practices, farmers and local production, the purchase of 'organic food' being associated with interest, not only of an individual's well-being but also of that of the group, which allows distinction to be made between altruist and selfish behavior. Cecchini et al. [56] focus on the interest of consumers in agricultural and ecologically sustainable products (which translates in a willingness to pay a higher price), highlighting the use of 'certificates' to guide consumers.

A fifth relevant topic within the research on EEA is the study of public goods. In this respect, not only does experimental economics contribute its wealth of benefits to agriculture, but it also helps to confirm the experimental methods. Along these lines, Chang et al. [57] conduct a field experiment to study consumer behavior, in which different scenarios are contemplated (hypothetical vs. non-hypothetical) in the purchase of certain products (ground beef, wheat flour). The results confirm previous evidence of experimental economics, indicating that non-hypothetical scenarios have a higher predictive power to elicit consumer behavior. With a laboratory experiment, Lusk and Norwood [58] analyze the altruism of consumers expressed towards animal well-being (a positive externality), measuring the public-good value of farm animal welfare. An important point in this topic is the distribution of water, a problem of cooperation that is studied in Abbink et al. [59], derived from the collapse of the USSR and the conflicting interests of Kyrgyzstan, Uzbekistan, and Kazakhstan. With a multi-round laboratory experiment, based on a three-player Trust Game with non-binding contracts, they show the difficulty of establishing cooperation between the actors involved. Examining the guidelines of some of the journals of the Agricultural and Applied Economics Association, they find that researchers are able to use some forms of deception. In this respect, they evaluate 10 potentially deceptive experimental techniques, discussing arguments both in favor of and against the practices used.

Finally, it has been possible to establish the evolution of the main research topics identified throughout the analysed period. From 2000 to 2004 the main research topics focused on food safety [29], consumer demand [30,50], as well as arable farming and environmental practices [34,51]. From 2005 to 2009, the central addressed issues were how consumer information affects consumers [35], especially related to biotechnology [52], and product labeling [46]. From 2010 to 2014, the research focus shifted to water management [57], investment management [36], and farmer training [39]. During this period, animal welfare [56] also appeared as a concern, as well as the continuing study of ambiguity and risk [38,39]. Finally, from 2015 to 2020, studies related to agroenvironmental policy [42,49,54] and the attention drawn to developing countries [40,41,43] will definitely become a greater importance [40,41,43].

4. Conclusions

The objective of this study was to analyze the dynamics of the research on the application of experimental economics to the field of agriculture over the last two decades. In this respect, the principal drivers of the subject area have been analyzed in depth together with the most relevant research topics. The results of the study reveal that there has been a growing use of experimental economics methods in the research on agriculture, particularly over the last five years. This progress is evident in the different indicators analyzed and is reflected in the greater scientific production and number of actors involved. It has been found that the number of articles published on the use of Experimental Economics as a tool for analysis is increasing to a greater extent than articles on agriculture in general terms. Thus, this field of study is becoming a relevant research line within agriculture.

The main categories that included articles on the use of experimental economics in agriculture were Economics, Econometrics and Finance (74.3%); and Agricultural and Biological Sciences (66.7%). The preferred format for the publication of research papers is the scientific article, with 80% of the total. The dominant language in this field of study is English, with 95.2% of the papers published. The countries that published most articles were the USA, Germany, and France, although the countries that published the articles with the highest impact were the UK, USA, and France. Countries publishing the most articles with international collaborations were the UK, China, and Norway. Australia, USA, and Germany are the countries that less use this formula. The small size of the international collaborative networks is noteworthy. This is due to the incipient nature of research in this field, as well as the still small number of published papers.

The most relevant institutions within this field of study belong to the following four countries: Germany, USA, France, and the Netherlands. The emergence of this field is evidenced by the small number of works carried out per institution. The same observation applies to the main authors in the application of experimental economics to the field of agriculture. It is precisely this fact that makes it possible to identify incipient networks of collaboration between authors, who, in these initial stages, tend to belong to the same institution or to a small group of national institutions.

Among the principal contributions of experimental economics applied to agriculture, particularly noteworthy is the interest in better characterizing the behavior of the agents and institutions that interact in the agricultural environment. In terms of the consumer, the research has identified different factors that can increase or reduce the willingness of the subjects to consume certain products (depending on where the products are produced, how consumers perceive the information received from the media or depending on their labeling). With respect to agricultural entrepreneurs, the intrinsic motivations of their decision are identified as are the determinants of the degree of their aversion to risk. Premises are also obtained referring to the functioning of the markets in accordance with the regulations that prevail. With respect to environmental policy, experimental economics offers interesting results regarding measures to prevent environmental pollution and how to promote certain crops in specific geographical areas that generate more stable yields. Finally, the research offers results concerning public goods, which are of interest to policy makers when establishing cooperation strategies between countries for managing common resources.

Author Contributions: Conceptualization, E.M.-V., J.F.V.-M., J.A.A.-S. and B.L.-F.; methodology, E.M.-V., J.F.V.-M., J.A.A.-S. and B.L.-F.; software, E.M.-V., J.F.V.-M., J.A.A.-S.; validation, E.M.-V., J.F.V.-M., J.A.A.-S.; formal analysis, E.M.-V., J.F.V.-M., J.A.A.-S. and B.L.-F.; investigation, E.M.-V., J.F.V.-M., J.A.A.-S. and B.L.-F.; resources, J.F.V.-M., J.A.A.-S. and B.L.-F.; data curation, E.M.-V., J.F.V.-M., J.A.A.-S.; writing—original draft preparation, E.M.-V., J.F.V.-M., J.A.A.-S. and B.L.-F.; writing—review and editing, E.M.-V., J.F.V.-M., J.A.A.-S. and B.L.-F.; supervision, E.M.-V., J.F.V.-M., J.A.A.-S. and B.L.-F.; All authors have read and agreed to the published version of the manuscript.

Funding: This research received no external funding.

Acknowledgments: This work was partially supported by the Spanish Ministry of Economy and Competitiveness and the European Regional Development Fund by means of the research project ECO2017-82347-P; from Junta de Andalucía and FEDER aid (project P18-RT-2327, Consejería de Transformación Económica, Industria, Conocimiento y Universidades); and by the FPU19/04549 Predoctoral Contract to Belén López-Felices.

Conflicts of Interest: The authors declare no conflict of interest.

References

1. Doyon, M.; Rondeau, D.; Mbala, R. Keep It Down: An Experimental Test of the Truncatedk-Double Auction. *Agric. Resour. Econ. Rev.* **2010**, *39*, 193–212. [CrossRef]
2. Smith, V. What is Experimental Economics? Interdisciplinary Center for Economic Science (ICES) at George Mason University: Arlington, TX, USA, 2003.
3. Brañas-Garza, P.; Paz-Espinosa, M. Economía Experimental y del Comportamiento. *Pap. Psicólogo* **2011**, *32*, 185–193.
4. Birol, E.; Meenakshi, J.V.; Oparinde, A.; Perez, S.; Tomlins, K. Developing country consumers' acceptance of biofortified foods: A synthesis. *Food Secur.* **2015**, *7*, 555–568. [CrossRef]
5. Canavari, M.; Drichoutis, A.C.; Lusk, J.L.; Nayga, R.M., Jr. How to run an experimental auction: A review of recent advances. *Eur. Rev. Agric. Econ.* **2019**, *46*, 862–922. [CrossRef]
6. Whitaker, J.B. Whispering in the Ears of Princes—Using Experimental Economics to Evaluate Agricultural and Natural Resource Policies: Discussion. *Am. J. Agric. Econ.* **2008**, *90*, 1216–1217. [CrossRef]
7. Costanigro, M.; Onozaka, Y. A Belief-Preference Model of Choice for Experience and Credence Goods. *J. Agric. Econ.* **2019**, *71*, 70–95. [CrossRef]
8. Marette, S.; Roosen, J.; Blanchemanche, S.; Verger, P. The choice of fish species: An experiment measuring the impact of risk and benefit information. *J. Agric. Resour. Econ.* **2008**, *33*, 1–18.
9. Galdeano-Gómez, E.; Pérez-Mesa, J.C.; Aznar-Sánchez, J.A. Internationalisation of SMEs and simultaneous strategies of cooperation and competition: An exploratory analysis. *J. Bus. Econ. Manag.* **2016**, *17*, 1114–1132. [CrossRef]
10. Nguyen, Q.; Leung, P. Do fishermen have different attitudes toward risk? An application of prospect theory to the study of vietnamese fishermen. *J. Agric. Resour. Econ.* **2009**, *34*, 518–538.
11. Roosen, J.; Marette, S. Making the 'right' choice based on experiments: Regulatory decisions for food and health. *Eur. Rev. Agric. Econ.* **2011**, *38*, 361–381. [CrossRef]
12. Disdier, A.-C.; Marette, S. Globalisation issues and consumers' purchase decisions for food products: Evidence from a laboratory experiment. *Eur. Rev. Agric. Econ.* **2012**, *40*, 23–44. [CrossRef]
13. Hermann, D.; Musshoff, O. Measuring time preferences: Comparing methods and evaluating the magnitude effect. *J. Behav. Exp. Econ.* **2016**, *65*, 16–26. [CrossRef]
14. Garfield, E.; Sher, I.H. New factors in the evaluation of scientific literature through citation indexing. *Am. Doc.* **1963**, *14*, 195–201. [CrossRef]
15. Huang, L.; Zhang, Y.; Guo, Y.; Zhu, D.; Porter, A. Four dimensional Science and Technology planning: A new approach based on bibliometrics and technology roadmapping. *Technol. Forecast. Soc. Chang.* **2014**, *81*, 39–48. [CrossRef]
16. Aznar-Sánchez, J.A.; Velasco-Muñoz, J.F.; López-Felices, B.; Román-Sánchez, I.M. An Analysis of Global Research Trends on Greenhouse Technology: Towards a Sustainable Agriculture. *Int. J. Environ. Res. Public Health* **2020**, *17*, 664. [CrossRef] [PubMed]
17. Velasco-Muñoz, J.F.; Aznar-Sánchez, J.A.; Batlles-Delafuente, A.; Fidelibus, M.D. Rainwater Harvesting for Agricultural Irrigation: An Analysis of Global Research. *Water* **2019**, *11*, 1320. [CrossRef]
18. Santos, J.; Maldonado, M.; Santos, R. Inovação e Conhecimento Organizacional: Um Mapeamento Bibliométrico das Publicações Científicas até 2009. *Rev. Organ. Contexto* **2011**, *7*, 31–58. [CrossRef]
19. Albort-Morant, G.; Henseler, J.; Leal-Millán, A.; Cepeda-Carrión, G. Mapping the Field: A Bibliometric Analysis of Green Innovation. *Sustainability* **2017**, *9*, 1011. [CrossRef]
20. Robinson, D.K.; Huang, L.; Guo, Y.; Porter, A. Forecasting Innovation Pathways (FIP) for new and emerging science and technologies. *Technol. Forecast. Soc. Chang.* **2013**, *80*, 267–285. [CrossRef]
21. Galvez, C. Análisis de co-palabras aplicado a los artículos muy citados en Biblioteconomía y Ciencias de la Información (2007–2017). *Transinformação* **2018**, *30*, 277–286. [CrossRef]
22. Durieux, V.; Gevenois, P.A. Bibliometric Indicators: Quality Measurements of Scientific Publication. *Radiology* **2010**, *255*, 342–351. [CrossRef] [PubMed]
23. Li, W.; Zhao, Y. Bibliometric analysis of global environmental assessment research in a 20-year period. *Environ. Impact Assess. Rev.* **2015**, *50*, 158–166. [CrossRef]
24. Mongeon, P.; Paul-Hus, A. The journal coverage of Web of Science and Scopus: A comparative analysis. *Science* **2015**, *106*, 213–228. [CrossRef]
25. Gavel, Y.; Iselid, L. Web of Science and Scopus: A journal title overlap study. *Online Inf. Rev.* **2008**, *32*, 8–21. [CrossRef]

26. Aznar-Sánchez, J.A.; Belmonte-Ureña, L.J.; Velasco-Muñoz, J.F.; Manzano-Agugliaro, F. Economic analysis of sustainable water use: A review of worldwide research. *J. Clean. Prod.* **2018**, *198*, 1120–1132. [CrossRef]
27. Aznar-Sánchez, J.A.; Piquer-Rodríguez, M.; Velasco-Muñoz, J.F.; Manzano-Agugliaro, F. Worldwide research trends on sustainable land use in agriculture. *Land Use Policy* **2019**, *87*, 104069. [CrossRef]
28. Aznar-Sánchez, J.A.; Velasco-Muñoz, J.F.; Belmonte-Ureña, L.J.; Manzano-Agugliaro, F. Innovation and technology for sustainable mining activity: A worldwide research assessment. *J. Clean. Prod.* **2019**, *221*, 38–54. [CrossRef]
29. Cascajares, M.; Alcayde, A.; Salmerón-Manzano, E.; Manzano-Agugliaro, F. Transfer of Agricultural and Biological Sciences Research to Patents: The Case of EU-27. *Agronomy* **2021**, *11*, 252. [CrossRef]
30. Bonilla, C.A.; Merigó, J.M.; Torres-Abad, C. Economics in Latin America: A bibliometric analysis. *Science* **2015**, *105*, 1239–1252. [CrossRef]
31. Stenger, A. Experimental valuation of food safety: Application to sewage sludge. *Food Pol.* **2000**, *25*, 211–218. [CrossRef]
32. Lusk, J.L.; Fox, J.A.; Schroeder, T.C.; Mintert, J.; Koohmaraie, M. In-store valuation of steak tenderness. *Am. J. Agric. Econ.* **2001**, *83*, 539–550. [CrossRef]
33. Toler, S.; Briggeman, B.C.; Lusk, J.L.; Adams, D.C. Fairness, farmers markets, and local production. *Am. J. Agric. Econ.* **2009**, *91*, 1272–1278. [CrossRef]
34. Bieberstein, A.; Roosen, J.; Marette, S.; Blanchemanche, S.; Vandermoere, F. Consumer choices for nano-food and nano-packaging in France and Germany. *Eur. Rev. Agric. Econ.* **2012**, *40*, 73–94. [CrossRef]
35. Lusk, J.L.; Coble, K.H. Risk perceptions, risk preference, and acceptance of risky food. *Am. J. Agric. Econ.* **2005**, *87*, 393–405. [CrossRef]
36. de Koeijer, T.; Wossink, G.; Smit, A.; Janssens, S.; Renkema, J.; Struik, P. Assessment of the quality of farmers' environmental management and its effects on resource use efficiency: A Dutch case study. *Agric. Syst.* **2003**, *78*, 85–103. [CrossRef]
37. Messer, K.D.; Schmit, T.M.; Kaiser, H.M. Optimal Institutional Mechanisms for Funding Generic Advertising: An Experimental Analysis. *Am. J. Agric. Econ.* **2005**, *87*, 1046–1060. [CrossRef]
38. Musshoff, O.; Odening, M.; Schade, C.; Maart-Noelck, S.C.; Sandri, S. Inertia in disinvestment decisions: Experimental evidence. *Eur. Rev. Agric. Econ.* **2012**, *40*, 463–485. [CrossRef]
39. Maart-Noelck, S.C.; Musshoff, O. Investing Today or Tomorrow? An Experimental Approach to Farmers' Decision Behaviour. *J. Agric. Econ.* **2012**, *64*, 295–318. [CrossRef]
40. Warnick, J.C.E.; Escobal, J.; Laszlo, S.C. Ambiguity Aversion and Portfolio Choice in Small-Scale Peruvian Farming. *BE J. Econ. Anal. Policy* **2011**, *11*, 1–56. [CrossRef]
41. Bocquého, G.; Jacquet, F.; Reynaud, A. Expected utility or prospect theory maximisers? Assessing farmers' risk behaviour from field-experiment data. *Eur. Rev. Agric. Econ.* **2013**, *41*, 135–172. [CrossRef]
42. Brunette, M.; Foncel, J.; Kéré, E.N. Attitude towards Risk and Production Decision: An Empirical Analysis on French Private Forest Owners. *Environ. Model. Assess.* **2017**, *22*, 563–576. [CrossRef]
43. Gars, J.; Ward, P.S. Can differences in individual learning explain patterns of technology adoption? Evidence on heterogeneous learning patterns and hybrid rice adoption in Bihar, India. *World Dev.* **2019**, *115*, 178–189. [CrossRef]
44. Pollard, C.R.J.; Redpath, S.; Bussière, L.F.; Keane, A.; Thompson, D.B.A.; Young, J.C.; Bunnefeld, N. The impact of uncertainty on cooperation intent in a conservation conflict. *J. Appl. Ecol.* **2019**, *56*, 1278–1288. [CrossRef]
45. Senapati, A.K. Evaluation of risk preferences and coping strategies to manage with various agricultural risks: Evidence from India. *Heliyon* **2020**, *6*, e03503. [CrossRef]
46. Wu, S.; Roe, B. Behavioral and Welfare Effects of Tournaments and Fixed Performance Contracts: Some Experimental Evidence. *Am. J. Agric. Econ.* **2005**, *87*, 130–146. [CrossRef]
47. Yesuf, M.; Bluffstone, R.A. Poverty, risk aversion, and path dependence in low-income countries: Experimental evidence from Ethiopia. *Am. J. Agric. Econ.* **2009**, *91*, 1022–1037. [CrossRef]
48. Kanter, C.; Messer, K.D.; Kaiser, H.M. Does Production Labeling Stigmatize Conventional Milk? *Am. J. Agric. Econ.* **2009**, *91*, 1097–1109. [CrossRef]
49. Dillaway, R.; Messer, K.D.; Bernard, J.C.; Kaiser, H.M. Do consumer responses to media food safety information last? *Appl. Econ. Perspect. Policy* **2011**, *33*, 363–383. [CrossRef]
50. Wu, S.; Fooks, J.R.; Messer, K.D.; Delaney, D. Consumer demand for local honey. *Appl. Econ.* **2015**, *47*, 4377–4394. [CrossRef]
51. Palm-Forster, L.H.; Ferraro, P.J.; Janusch, N.; Vossler, C.A.; Messer, K.D. Behavioral and Experimental Agri-Environmental Research: Methodological Challenges, Literature Gaps, and Recommendations. *Environ. Resour. Econ.* **2019**, *73*, 719–742. [CrossRef]
52. Murphy, J.J.; Stevens, T.H. Contingent valuation, hypothetical bias, and experimental economics. *Agric. Res. Econ. Rev.* **2004**, *33*, 182–192. [CrossRef]
53. Poe, G.L.; Schulze, W.D.; Segerson, K.; Suter, J.F.; Vossler, C.A. Exploring the Performance of Ambient-Based Policy Instruments When Nonpoint Source Polluters Can Cooperate. *Am. J. Agric. Econ.* **2004**, *86*, 1203–1210. [CrossRef]
54. Lybbert, T.J. Indian farmers' valuation of yield distributions: Will poor farmers value 'pro-poor' seeds? *Food Pol.* **2006**, *31*, 415–441. [CrossRef]
55. Bougherara, D.; Combris, P. Eco-labelled food products: What are consumers paying for? *Eur. Rev. Agric. Econ.* **2009**, *36*, 321–341. [CrossRef]

56. Cecchini, L.; Torquati, B.; Chiorri, M. Sustainable agri-food products: A review of consumer preference studies through experimental economics. *Agric. Econ.* **2018**, *64*, 554–565. [CrossRef]
57. Chang, J.B.; Lusk, J.L.; Norwood, F.B. How Closely Do Hypothetical Surveys and Laboratory Experiments Predict Field Behavior? *Am. J. Agric. Econ.* **2009**, *91*, 518–534. [CrossRef]
58. Lusk, J.L.; Norwood, F.B. Speciesism, altruism and the economics of animal welfare. *Eur. Rev. Agric. Econ.* **2011**, *39*, 189–212. [CrossRef]
59. Abbink, K.; Moller, L.C.; O'Hara, S. Sources of Mistrust: An Experimental Case Study of a Central Asian Water Conflict. *Environ. Resour. Econ.* **2009**, *45*, 283–318. [CrossRef]

Article

A Look at the Past, Present and Future Research Trends of Artificial Intelligence in Agriculture

José Luis Ruiz-Real [1,*], Juan Uribe-Toril [1], José Antonio Torres Arriaza [2] and Jaime de Pablo Valenciano [1]

[1] Faculty of Economics and Business, University of Almeria, Ctra. De Sacramento, s/n, 04120 Almería, Spain; juribe@ual.es (J.U.-T.); jdepablo@ual.es (J.d.P.V.)
[2] Department of Computer Science, University of Almeria, Ctra. De Sacramento, s/n, 04120 Almería, Spain; jtorres@ual.es
* Correspondence: jlruizreal@ual.es; Tel.: +34-950-015742

Received: 21 October 2020; Accepted: 20 November 2020; Published: 22 November 2020

Abstract: Technification in agriculture has resulted in the inclusion of more efficient companies that have evolved into a more complex sector focused on production and quality. Artificial intelligence, one of the relevant areas of technology, is transforming the agriculture sector by reducing the consumption and use of resources. This research uses a bibliometric methodology and a fractional counting method of clustering to analyze the scientific literature on the topic, reviewing 2629 related documents recorded on the Web of Science and Scopus databases. The study found significant results regarding the most relevant and prolific authors (Hoogenboom), supporting research organizations (National Natural Science Foundation of China) and countries (U.S., China, India, or Iran). The identification of leaders in this field gives researchers new possibilities for new lines of research based on previous studies. An in-depth examination of authors' keywords identified different clusters and trends linking Artificial Intelligence and green economy, sustainable development, climate change, and the environment.

Keywords: artificial intelligence; agriculture; bibliometric analysis; cross world research

1. Introduction

Artificial Intelligence, hereinafter AI, is one of the disruptive technologies that has changed processes and developments in the field of science, technology, and business in recent years. The development of AI has resulted in streams of research: the analysis of events and their correlation over time, and the search for relationships between phenomena that may cause general deductive or inferential rules. In both cases, this research aims to explain past episodes and predict future events.

AI began in the 1950s, demonstrating that a new form of computing was possible, with an approach derived from known cognitive processes and from neurobiology. The initial purpose of the AI was to automate, through computers, non-analytical human knowledge, from symbolic computation processes, connectionist ones, or a combination of both. Although initially considered a branch of computer sciences with limited application and restricted by the capabilities of the hardware of the time, AI has since evolved into a vital element for the development of many services and industrial sectors in the 21st century.

AI is a discipline of computer science that studies algorithms to develop computer solutions that copy the cognitive, physiological, or evolutionary phenomena of nature and human beings. Unlike the traditional model, it does not require knowledge of specific paths to the resolution of problems. Rather, it is the data, examples of solutions, or relationships between these that facilitate the resolution of diverse problems. AI exhibits, in certain aspects, "an intelligent behavior" that can be confused with that of a human expert in the development of certain tasks [1].

At present, AI has been redirected, following the definition of double analysis previously mentioned, towards the construction of solutions to problems with large volumes of data which change over time. This type of data can present inaccuracies and, in some cases, contradictions. Currently, the systems for approaching functions using iterative techniques, and the neural network architectures interconnected with each other, make up most of the techniques, which are grouped under the terms "Machine Learning" and "Deep Learning". The field of application for AI has been extended, having as a common denominator the analysis of large volumes of data or complex data structures which are dependent on time and unknown factors.

Agriculture is a sector that includes studies in science, engineering, and its economic derivatives. AI has not neglected this sector and there are numerous studies that have focused on it. McKinion and Lemmon and Murase [2,3] make comprehensive reviews of the use of deductive techniques based on expert systems in the field of agriculture. Other works highlight the applications of expert systems and decision support systems for the simulation of processes and the management of supply operations [4–6].

In other studies, AI has been used in quality control processes, whether or not they are supported by artificial vision, as in the case of Nair and Mohandas [7], or in processes of justification of food policy decisions, such as the case study by Bryceson and Slaughter [8], where the use of AI is analyzed as a collaborative tool between the different actors that supply the agri-food chain, using distributed computing processes.

Some studies investigate the price behavior of agri-food products [9–14]. In these cases, artificial neural networks and machine learning techniques are applied to limit the price variations of these commodities. In the field of science, aspects of climate are studied by Hewitson and Crane [15] and Mellit [16], who try to model and predict solar radiation using neural networks.

Interest in the application of AI to the world of agriculture and its multiple facets has been growing in recent years as it has proven to be a powerful tool for data analysis.

The expansion and intensification of industrial and technological agriculture have increased production, lowering the number of people suffering from poor nutrition and meeting the increased demand for richer and more resource-intense diets. Industrial agricultural activities also generate employment, improve economic growth, and boost the service sector in industrial regions [17].

Agriculture 3.0 brought robotics and automation to the agricultural world, as evidenced by agricultural machinery that performs complete cycles of agricultural work such as planting, spraying, and harvesting [18–20]. Now, it is the turn of agriculture 4.0, which, along with intelligent farms and the interconnection of machines and systems, seeks to adapt production ecosystems by optimizing the use of water, fertilizers, and phytosanitary products, giving rise to what is known as precision agriculture [21–23]. Combined with genetic engineering and the use of data, it can solve an important part of agriculture by maximizing efficiency in the use of resources and adapting to climate change and other challenges. To this end, the use of big data in decision-making is essential [24,25]. The technification of agriculture and the inclusion of concepts of Industry 4.0 by agri-food companies has also generated greater interest in AI.

At the same time, bibliometric studies that connect the different disciplines are of growing interest in the analysis of the impact of these synergies and their future within the research community. In general, bibliometric studies are of great interest to academia, as it is a clear indicator of interest in a particular field. An example of this is the paper by Gu [26], which shows the structure and model of the scientific production of researchers worldwide and the relationships between quality, references, and synergies among authors.

More specifically, in the field of AI, the work of Cobo et al. [27] analyzes its evolution through various bibliometric indicators based on citations of scientific production related to knowledge-based systems. Similar studies related to the agri-food industry or agriculture are not as common, nor are they as documented. However, the growth of interest in both fields is palpable.

Thus, the number of publications related to these two fields continues to grow. In Google Scholar, for example, there has been a sustained increase in the number of publications in the last five years, which leads us to think that this synergy will be maintained, and interest in studying the implications of one of these disciplines in another will be relevant in the short and medium term.

This work is a formal study of the scientific production of these lines of research, revealing the importance that this synergy currently has for the scientific community. A content analysis was carried out using two databases—WoS and Scopus—to determine the volume of publications, the scientific journals in which they are published, the most relevant researchers from the point of view of the quality of their publications and the volume of these, as well as a study of the geographical origin of these works to determine the interest in these issues at a global level.

2. Materials and Methods

To analyze the evolution of AI in the agricultural industry in scientific publications, a bibliometric analysis was carried out. This study is based on a systematic bibliographical analysis of the literature related to a central topic, following a sequence of steps [28]: (a) definition of the search criteria, keywords, and time; (b) selection of databases; (c) adjustment of research criteria; (d) full export of results; (e) analysis and discussion of results (Figure 1).

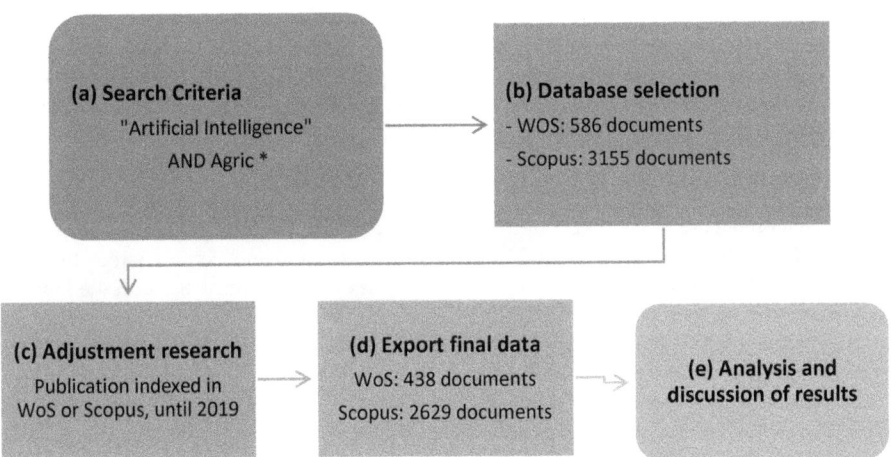

Figure 1. Stages of bibliometric analysis, * search extends.

Two terms, "Artificial Intelligence" and "agric*", were selected for this research, and were focused on papers published until 2019. The quotation marks were used to retrieve correct and exact expressions, while the asterisk was used in "agric*" to retrieve all potential derivatives of the words. Therefore, "agric" is used as the root term of many expressions, such as "agriculture", "agricultural", "agriculturist", and "agriculturalist". Following this, publications from robust and reliable databases were identified. Garfield [29] first described a citation index for science. Two online databases were selected for this work: the publication of indexes in the Web of Science (WoS) Core Collection and Scopus. These are the most frequently used databases and both are multidisciplinary, recording scientific articles, reviews, and books, but also other documents such as meetings, proceedings, editorials, and letters. In addition, these databases provide access to the full texts of the documents.

The preliminary results of this search, without the application of filters, retrieved 3155 documents in Scopus and 586 documents in WoS. These results were adjusted and subsequently filtered, redefining the date until 2019. After collecting the documents including the selected terms for this research (in the title, keywords, and/or abstract), the results were checked one by one in order to verify their relevance to the objectives of this study. After debugging the databases, the initial query of these terms in the titles, abstracts, and keywords resulted in 2629 documents in Scopus and 438 in WoS, including articles, proceedings papers, reviews, editorial materials, book chapters, notes, software reviews, and letters.

After obtaining the final results, the data were exported into ".txt" format. For the analysis and discussion of the results, this research considered: number of annual publications and citations, languages, countries, journals, organizations publishing and entities funding research on this topic, and trends.

Bibliometric analyses are based on two criteria: the scientific publication, as an indicator of research output [30], and citations, as a proxy of their scientific impact [31]. Therefore, different bibliometric indicators were used in this analysis: impact of papers, based on the number of citations; and frequency, through the Hirsch index (h-index and averages), proposed by Hirsch [32] and defined as the number of papers with citation number ≥ h.

3. Discussion of Results

The use of AI in agriculture is an increasingly widespread phenomenon, encompassing different areas of the sector and linked, therefore, to multiple topics. Thus, to ensure that this research, focused on scientific publications, is as complete as possible and can cover all these related fields, the following keywords have been used in the search: "artificial intelligence" and "agric*".

In this research, the following elements have been analyzed: the annual evolution of the volume of publications and citations, the most influential countries in publications related to this field, the most outstanding journals, the most relevant authors, the most prolific universities related to these topics, the main entities supporting these publications, the main areas of knowledge involved, as well as the trends and terms that indicate future lines of research.

3.1. Evolution in the Number of Publications per Year

Research on the use of AI in agriculture is relatively recent. The first publication is from 1976 (Scopus), with the Proceedings of the IEEE Conference on Decision and Control, which took place in Florida (U.S.) [33]. In WoS, there are no publications on this topic until 1989, with the following two articles: "Some lessons for Artificial-Intelligence and agricultural systems simulation" [34]; and "Agassistant—An Artificial-Intelligence system for discovering patterns in agricultural knowledge and creating diagnostic advisory systems" [35].

In addition, during the first few years, there was hardly any scientific production in this field. Figure 2 shows the historical evolution of publications including the terms "artificial intelligence" and "agric*" in the title, abstract, or keywords in the WoS and Scopus databases. In 2003 and 2004, there was a certain takeoff in the volume of publications, although it is not until 2005 when the first turning point is appreciated, with a significant increase in the number of scientific works. This notwithstanding, it has only been in the last decade that the interest in this topic has acquired greater relevance and the number of investigations has grown significantly, with figures that surpass 200 annual documents in Scopus. The significant growth in 2019 is quite notable, reaching 489 publications (Scopus), which clearly shows the interest that this field currently has among the scientific community.

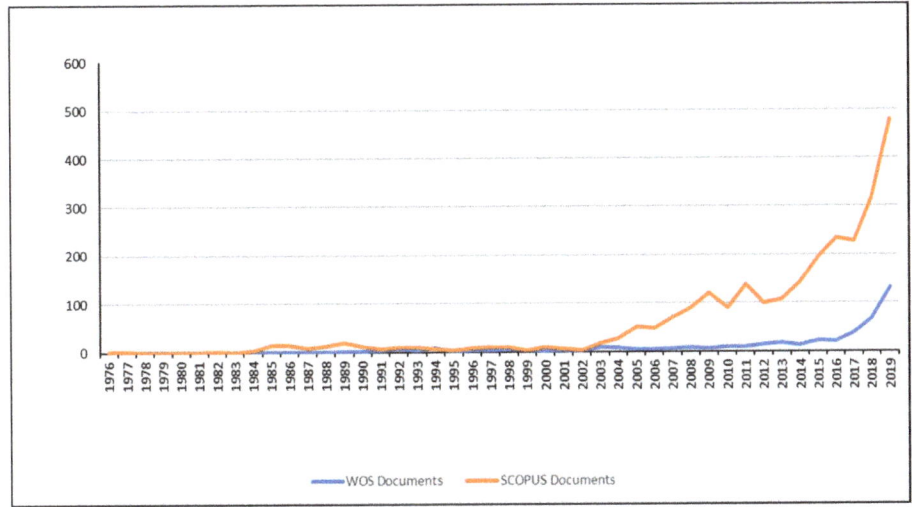

Figure 2. Annual evolution of publications.

A similar evolution to that of the volume of publications is observed in the number of citations per year, although with slight differences. Thus, the highest figures are in recent years, particularly in the 21st century, with several years showing more than 1000 annual citations. The highest number of citations (2743) is in 2017 (Figure 3).

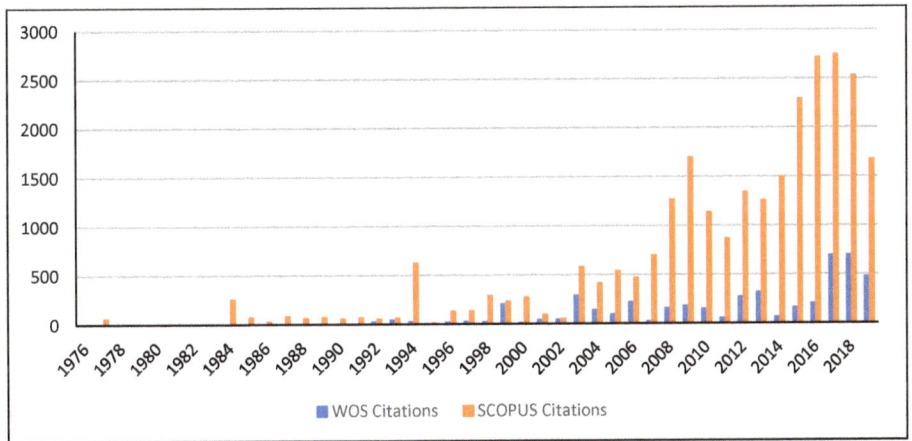

Figure 3. Number of citations per year.

3.2. Most Influential Countries

China is the most influential country in AI in agriculture (Scopus), with 489 publications, closely followed by the U.S., with 449 publications. India is in third place, with 291 publications. These three countries account for 47% of the total publications in Scopus in this field (Table 1). The importance that this topic has in these countries could be one of the main reasons why they occupy this privileged position in the ranking, showing a clear commitment to improving the profitability and efficiency of the agricultural sector. A second group of influential countries consists of Spain (127 publications), Germany (106), Australia (105), and the U.K. (95). Finally, the role played by countries such as Iran, Malaysia, and Egypt, where agriculture represents a significant percentage of its economic activity, is also noteworthy.

Table 1. Ranking of countries attending the number of publications and citations.

	WOS				SCOPUS			
Country	P	C	C/P	h-i	P	C	TC/P	h-i
U.S.	67	1169	17.45	18	449	6661	14.83	41
China	64	507	7.92	13	489	2882	5.89	26
Iran	36	702	19.50	16	86	1342	23.96	21
India	32	311	9.72	7	291	2064	7.09	23
Brazil	26	155	5.96	5	87	693	7.97	14
U.K.	23	462	20.09	10	95	1482	15.60	19
Spain	21	270	12.86	7	127	2370	18.66	25
Germany	16	151	9.44	6	106	1471	13.88	21
Japan	16	87	5.44	4	62	531	8.56	12
Malaysia	14	225	16.07	9	51	590	11.57	16
Turkey	15	329	21.93	7	39	612	15.69	13
Australia	13	143	11.00	9	105	1459	13.90	21
Egypt	13	82	6.31	6	38	220	5.79	8
Italy	12	339	28.25	7	82	2003	24.43	24

P—total number of publications; C—total number of citations; C/P—average citations per publication; h-i—Hirsch index.

When analyzing indicators such as h-index and total citations, the ranking undergoes some significant variations. In that scenario, the U.S. leads the ranking (h: 41; 6661 citations), followed by China (26; 2882), Spain (25; 2370), and Italy (24; 2003). Italy also leads the indicator of citations per article, with an average of 24.43, which also shows its prominent role in this area of knowledge, as well as the importance of the sector in its economy and the level of development of AI in this field.

Although China dominates in terms of the number of publications on the topics analyzed in this research, the most frequent language of publication is English, representing 96.96%. Thus, this is the language most used on these issues by the most relevant journals.

3.3. Most Influential Journals

When analyzing the most influential journals publishing on issues related to the use of AI in agriculture, it is observed that there are numerous journals in different areas of knowledge. On the one hand, those mainly specialized in Information and Communications Technology (ICT) (such as computers, engineering, etc.) and on the other hand, those journals more focused on aspects such as agriculture, environment, resources management, hydrology, etc. This shows the importance of this topic, which arouses interest in different fields, as it is a cross-cutting issue, resulting from the combination of several areas of knowledge.

As shown in Tables 2 and 3, according to the h-index, Computers and Electronics in Agriculture is the most influential journal for AI in Agriculture (33 in Scopus; 8 in WoS). This ranking is followed by Agricultural Water Management (14) and Lecture Notes in Computer Science (9).

Table 2. Journals and impact (WoS).

R	Journal	IF	P	C	C/P	h-i
1	Computers and Electronics in Agriculture	3.858	25	613	24.52	11
2	Science of the Total Environment	6.551	7	241	34.43	7
3	Sensors	3.275	6	184	30.67	4
4	Water Resources Management	2.924	6	123	20.50	5
5	Environmental Science and Pollution Research	3.056	5	17	3.40	2
6	Agricultural Water Management	4.021	4	87	21.75	4
7	Environmental Earth Science	2.180	4	82	20.5	3
8	Remote Sensing	4.509	4	44	11.00	3
9	Biosystem Engineering	3.215	4	37	9.25	4
10	IEEE Access	3.745	4	21	5.25	2

R—ranking; IF—JCR index 2019; P—total number of publications; C—total number of citations; C/P—average citations per publication; h-i—Hirsch index.

Table 3. Journals and impact (Scopus).

R	Journal	IF	P	C	C/P	h-i
1	Computers and Electronics in Agriculture	1.058	137	3,643	26.56	33
2	Lecture Notes in Computer Science	0.427	91	370	4.07	11
3	Nongye Gongcheng Xuebao Transact. of the Chinese Society of Agric. Eng.	0.438	39	230	5.90	10
4	IFIP Advances in Information and Communication Technology	0.209	38	74	1.95	4
5	Advances in Intelligent Systems and Computing	0.184	37	99	2.68	6
6	Agricultural Water Management	1.369	35	713	20.37	18
7	Intern. Archiv. of the Photogrammetry Remote Sensing and Spatial Inf. Science	0.367	28	73	2.61	4
8	Procedia Computer Science	0.342	21	230	10.95	7
9	Communications in Computer and Information Science	0.188	21	31	1.48	3
10	Paper American Society of Agricultural Engineers	0.177	18	20	1.11	2

R—ranking; IF—SJR index 2019; P—total number of publications; C—total number of citations; C/P—average citations per publication; h-i—Hirsch index.

Likewise, Computers and Electronics in Agriculture leads the ranking of the volume of publications in both databases (137, Scopus; 25, WoS). The following journals with the largest number of publications are Lecture Notes in Computer Science (91) and Transactions of the Chinese Society of Agricultural Engineering (39). With regards to the number of citations, once again, Computers and Electronics in Agriculture occupies first place in the ranking, with 3643 citations. This journal is followed by Agricultural Water Management (713) and Lecture Notes in Computer Science (370).

Among the most relevant journals in this field, those with the greatest impact factor (taking into account the 2019 JCR index on WoS) are: Science of the Total Environment (6.551), Remote Sensing (4.509), and Agricultural Water Management (4.021).

3.4. Most Relevant Authors and Cited References

When studying the relevance of authors in a specific topic, bibliometric analysis can take into account several indicators. This work focused on two aspects: the volume of publications, which shows the involvement of the researcher in the field; and the impact of publications, with reference to the number of citations, that is, by counting the number of other papers referencing it.

With regards to the number of publications and h-index, according to the Scopus database (Table 4), Professor Gerrit Hoogenboom from the Department of Agricultural and Biological Engineering at the University of Florida (U.S.) is the most relevant author, since his 15 publications about this topic have

483 citations, having an h-index of nine. Some other relevant authors in this topic are: James W. Jones (13 publications, h-index of 58), Director of the Florida Climate Institute and Professor Emeritus in the Agricultural and Biological Engineering Department at the University of Florida (U.S.); Fu, Z. at Beijing Laboratory of Food Quality and Safety (China) (12 publications, h-index of 22); John R. Barret (9 publications, h-index of 6) and J.C. Ascough (8 publications, h-index of 29), both at the United States Department of Agriculture, Agricultural Research Service (U.S.).

Table 4. Most relevant authors (Scopus).

R	Author	P	C	C/P	h-i
1	Hoogenboom, G.	15	483	32.20	52
2	Jones, J.W.	13	209	16.07	58
3	Fu, Z.	12	41	3.41	22
4	Barrett, J.R.	9	14	1.55	6
5	Ascough, J.C.	8	71	8.87	29
6	He, Y.	7	106	15.14	55
7	Corrales, J.C.	7	18	2.57	10
8	Fraisse, C.W.	7	62	8.86	19
9	Thorp, K.R.	6	196	32.66	29
10	McClendon, R.W.	6	139	23.17	20
10	McMaster, G.S.	6	45	7.50	25

R—ranking; P—total number of publications; C—total number of citations; C/P—average citations per publication; h-i—author Hirsch index on Scopus.

According to the number of citations, "WEKA: A machine learning workbench" [36] leads the ranking with 578 citations. This research focuses on WEKA, a workbench that is intended to aid in the application of machine learning techniques to real-world problems, mainly those arising from agriculture. The second research with the highest number of citations is "Big Data in Smart Farming—A review" [37] published in Agricultural System. The third study (338) is "The regularized iteratively reweighted MAD method for change detection in multi- and hyperspectral data" [38]. This paper describes new extensions to the multivariate alteration detection (MAD) method for change detection in bi-temporal, multi-, and hypervariate data, and provides examples using SPOT High Resolution Visible data from an agricultural region in Kenya and a small rural area in Germany.

The next work with the highest number of citations (254) is "Colour and shape analysis techniques for weed detection in cereal fields" [39] with 224 citations and which deals with the development of near-ground image capture and processing techniques in order to detect broad leaf weeds in cereal crops, showing the potential of using image processing techniques to generate weed maps. Following this is the paper "Making computers think like people" [40]. In this research, fuzzy sets and fuzzy logic are qualitatively described, and the application of fuzzy concepts to expert systems and computer vision is also discussed. The sixth work with the highest number of citations (227) is "Application of ANN for reservoir inflow prediction and operation" [41], which analyzes the influence of the land use and the plant species in the waterbed on the water quality of a high-altitude wetland in India.

Finally, in order to understand the total strength of the co-authorship links with other authors, the minimum number of documents of an author was set at two articles. In total, 836 authors out of 5870 meet this threshold. For each of them, the total strength of the co-authorship links with other authors was calculated. Thus, the largest set of connected authors consists of 19 links. Some clusters may be highlighted, such as the one led by Hoogenboom, G. (pink color); Wang, L. and Zhang, X. (mustard color); Chen, X. and Li, Y. (green); Fu, Z. (brown); Wang, M. and Li, X. (red); Liu, X. (orange); Huang, Q. and Yu, X. (purple); and Wang, Y. and Liu, Y. (grey) (Figure 4).

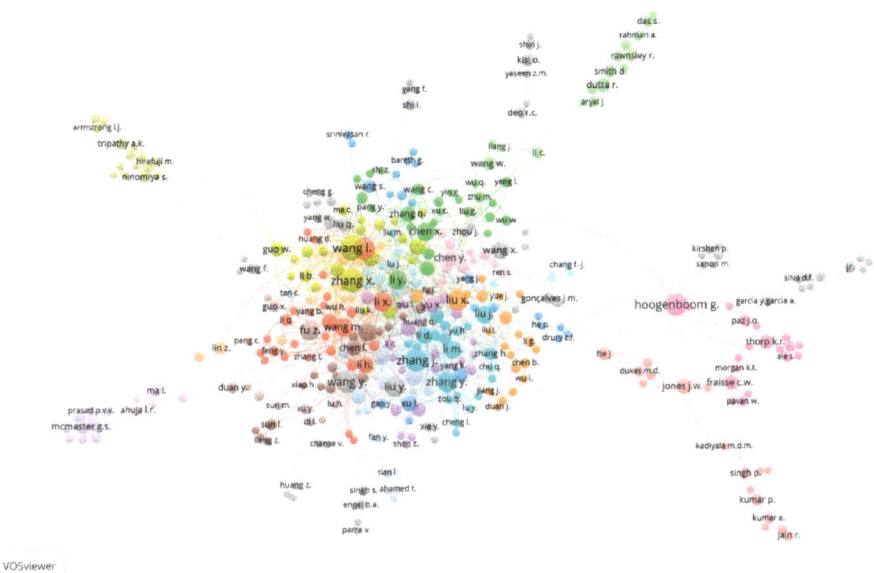

Figure 4. Co-citation map based on bibliographic data.

3.5. Institutions and Funding Sponsors

There are several public and private institutions disseminating knowledge through research and scientific publications on AI in agriculture. The twelve most relevant institutions (Table 5) have produced 14.97% of the total publications on this topic. This ranking is clearly led by entities from China and the U.S. Six of the twelve main organizations are from China (China Agricultural University, Chinese Academy of Sciences, Ministry of Education China, Chinese Academy of Agricultural Sciences, Ministry of Agriculture of the People's Republic of China, and Zhejiang University), four from the U.S. (University of Florida, United States Department of Agriculture, Texas A&M University, and USDA Agricultural Research Service), one from Iran (University of Tehran), and there is also one from Europe, specifically from Spain (Consejo Superior de Investigaciones Científicas).

Table 5. Institutions promoting research (Scopus).

R	Institution	T	P	C	C/P	h-i
1	China Agricultural University	China	61	213	3.49	8
2	University of Florida	U.S.	50	1,025	20.50	17
3	Chinese Academy of Sciences	China	36	416	11.55	12
4	United States Department of Agriculture	U.S.	34	459	13.5	10
5	University of Tehran	Iran	33	546	16.54	16
6	USDA Agricultural Research Service	U.S.	30	345	11.50	10
7	Ministry of Education China	China	29	209	7.20	8
8	Ministry of Agriculture of the People's Republic of China	China	24	206	8.58	7
9	Texas A&M University	U.S.	23	235	10.21	8
10	Chinese Academy of Agricultural Sciences	China	20	259	12.95	7
11	Consejo Superior de Investigaciones Científicas	Spain	18	437	24.27	9
12	Zhejiang University	China	17	204	12.00	6

R—ranking; P—total number of publications; T—Territory. C—total number of citations; C/P—average citations per publication; h-i—Hirsch index.

The institution with the largest volume of publications is China Agricultural University (CAU) (61 publications). CAU, from Beijing, is a Double First-Class University according to the Chinese Ministry of Education, specialized in agriculture, engineering, economics, management, and social sciences. However, taking into account the h-index and the number of citations, the institutions that lead the ranking are the University of Florida (UF) (h-index of 17; 1025 citations), University of Tehran (h-index of 16; 546 citations), United States Department of Agriculture (h-index of 9; 399 citations), and Chinese Academy of Sciences (h-index of 10; 459 citations). All these institutions are highly committed to AI research in the agricultural industry.

With regard to the U.S. entities, the University of Florida's Institute of Food and Agricultural Sciences (UF/IFAS) is a federal state–county partnership dedicated to "developing knowledge in agricultural, human and natural resources and making that knowledge accessible to sustain and enhance the quality of human life". It also has the UF-IFAS Space Agricultural and Biotechnology Research and Education (SABRE) Center and the Precision Agriculture Laboratory, focused on the development of agriculture through the use of ICT. The United States Department of Agriculture (USDA) provides support on food, agriculture, natural resources, rural development, nutrition, and related issues based on public policy. The National Institute of Food and Agriculture (NIFA) is part of the USDA, supporting AI activities through a variety of programs and areas: Agricultural Systems and Engineering; Natural Resources and Environment; and Economics and Rural Communities. In addition, some subsections of the Agriculture and Food Research Initiative (AFRI) Foundational and Applied Science program provide funding in AI: Agriculture Systems and Technology; Bioenergy, Natural Resources, and Environment; and Agricultural Economics and Rural Community program areas.

The University of Tehran (Iran) has two main centers related to AI and agriculture: Faculty of Agricultural Science and Engineering; and Research in Artificial Intelligence, Robotics, and Information Science.

Another important indicator to understand the impact of the publications of an organization is the average number of citations, a ranking led by the Consejo Superior de Investigaciones Científicas (24.27) and the University of Florida (20.50), followed by the University of Tehran (16.54).

With reference to funding agencies sponsoring research on AI in agriculture, the top ten is led by National Natural Science Foundation of China (65 publications) (Table 6). China occupies two of the top ten positions in this ranking, with the following entities also publishing research on this topic: National Basic Research Program of China (973 Program) (18 publications), promoted by the People's Republic of China to achieve technology in several scientific fields; National Science Foundation of China (NSFC) (65), an organization directly affiliated to China's State Council.

Table 6. Funding agencies sponsoring research.

R	Funding Agencies	T	P
1	National Natural Science Foundation of China	China	65
2	European Commission	Europe	36
3	National Science Foundation	U.S.	24
3	Conselho Nacional de Desenvolvimento Científico e Tecnológico (CNPq)	Brasil	20
4	U.S. Department of Agriculture	U.S.	20
5	National Basic Research Program of China (973 Program)	China	18
6	National Institute of Food and Agriculture	U.S.	15
8	European Regional Development	Europe	14
9	Coordenação de Aperfeiçoamento de Pessoal de Nível Superior	Brazil	13
10	Ministry of Science and Technology	Taiwan	13

R—ranking; T—territory; P—total number of publications.

There is, therefore, an obvious commitment by the Chinese government to encourage and support Chinese universities and research centers to advance in studying the potential of AI in the agricultural industry. Being aware of the strong demand for AI, the Chinese government has planned a robust support system in education, research, and AI applications, and this becomes visible in the development of numerous plans that encourage research centers to apply AI to improve efficiency in agricultural production.

The U.S. is also relevant in this ranking, with three agencies in the top ten, one being the National Science Foundation (NSF) (24 publications), a government agency that supports fundamental research and education in all the non-medical fields of science and engineering. In some fields, such as mathematics, computer science, economics, and the social sciences, the NSF is the major source of federal backing; in addition, there is the U.S. Department of Agriculture (USDA) (20) and the National Institute of Food and Agriculture (NIFA) (15), which is also part of USDA. In the U.S., AI spending in agriculture industry increased at 66.0% during 2018, reaching USD 122.9 million.

Another relevant country is Brazil, with two agencies in this ranking: Conselho Nacional de Desenvolvimento Científico e Tecnológico (CNPq) (20 publications) and Coordenação de Aperfeiçoamento de Pessoal de Nível Superior (13), a public foundation for providing funds and programs to support research, education, and innovation of private and public institutions and companies. Finally, the European Commission (36) is the most important European entity supporting research on AI in agriculture, having AI and robotics among its priorities, in order to put it at the service of European citizens and economy. The Commission has increased its annual investments in AI by 70% with the research and innovation program, Horizon 2020. It will reach EUR 1.5 billion by 2018–2020 [42].

3.6. Trends

Due to the transversal nature of the topics analyzed, there are a wide variety of fields dealing with them. Figure 5 shows a map based on bibliographic data on the co-occurrence of the authors' keywords by using a fractional counting method, which is useful to comprehend trends in the research.

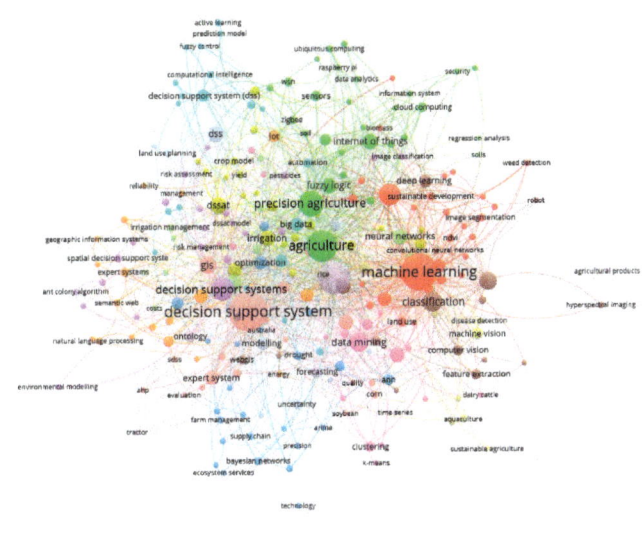

Figure 5. Map based on the co-occurrence on the authors' keywords.

This map also identifies the main interactions between the most frequent terms in this research, as well as the existing clusters. This cluster analysis showed 14 different groups, with 255 items. Thus, with reference to the map, academia is currently focused on terms such as: decision support system, machine learning, artificial intelligence, agriculture, precision agriculture, decision support systems, remote sensing, data mining, or image processing. These terms are grouped into clusters, such as: Decision Support System (expert system; Geographic Information System—GIS; evaluation); Machine Learning (sustainable development; image segmentation; land use; the normalized difference vegetation index—NDVI; weed detection); Agriculture (precision agriculture; fuzzy logic; Internet of Things; neural networks; sensors; cloud computing); Classification (computer vision; feature extraction; robot); Data mining (clustering) and Irrigation (Decision Support System for Agrotechnology Transfer—DSSAT; optimization). Due to the large amount of dates and potential technical difficulties to share the whole information, this information is available for readers upon request.

A trend map was also created using a fractional counting method based on bibliographic data on the co-occurrence of the authors' keywords. This map uses different colors to highlight the most commonly used authors' keywords in each of the last few years since 2011. In order to identify the most relevant trends, the most recent keywords are marked in yellow. Of the 1254 keywords, the minimum occurrences of a keyword was set at four. In total, 255 authors out of 4976 authors met this threshold. Figure 6 shows that current research on AI in agriculture is focused on topics such as: machine learning; Internet of things; deep learning; big data; sensors; cloud computing; drought; and robot.

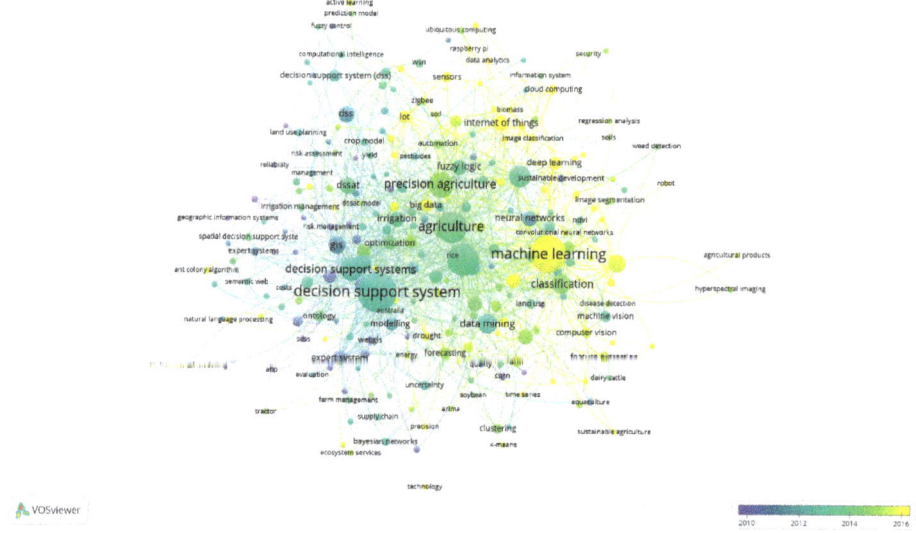

Figure 6. Map based on the co-occurrence on the authors´ keywords, and evolution since 2010.

4. Conclusions

After many years of continuous growth and technification in agriculture, with more efficient companies, the two main concerns of the sector are focused on production and quality. This is due to the appearance of factors and concerns such as population growth, climate change, and food security. Therefore, the future of the sector involves advances in the application of technology developed in various fields such as computer vision, Big Data, and AI, which engender the development of multiple companies dedicated to precision agriculture.

The use of AI in agriculture is a huge step forward for the sector, allowing it to enter a new stage of development since in addition, it can drastically reduce the consumption and use of resources. AI results in even more professional crop management, resulting in a more profitable and more sustainable agricultural sector.

In analyzing the historical evolution of publications about AI in agriculture, it can be observed that the first works are relatively recent. The oldest in Scopus is from 1976, with the publication of the proceedings of the IEEE Conference on Decision and Control. Given that the IEEE is one of the world's largest technical professional organizations dedicated to advancing technology for the benefit of humanity, the first publications began in engineering, computing, and technology information research and addressed agriculture as a relevant sector in which to develop numerous applications, thus covering the specific needs of this industry. This late start in publications on these topics is even more evident in WoS, a database in which the first works do not appear until 1989, with articles focused on using AI in agricultural systems simulation, and creating diagnostic advisory systems. During the initial years, there was little scientific production on these topics, and it was not until 2008 when the volume of publications reached higher values, as proof of the current scientific interest on this topic. Regarding the annual number of citations, the evolution is similar, with some slight differences.

China is the most influential country in AI in agriculture, in terms of the volume of publications, closely followed by the U.S. The three most prolific countries in this field (China, U.S., and India) represent 47% of all publications, an indication of the importance that these countries grant to improving the profitability and efficiency of the agricultural sector. There is a second group of influential countries, made up of Spain, Germany, Australia, and the U.K. Likewise, Italy leads in terms of the number of citations per article, which also shows the importance of this industry and the level of development of this field. Finally, there is another group of countries, such as Iran, Malaysia, and Egypt, in which agriculture plays an important role in their economies.

Since AI in agriculture is a very broad field, in which many areas of knowledge are involved, there are several journals publishing on this topic. On the one hand, there are journals directly related to technology, computers, engineering, etc. whilst, on the other hand, there are another group of journals related to agriculture, environment, resources management, hydrology, etc. This is a consequence of the cross-cutting character of this topic. Thus, according to their h-index, the most relevant journals are: Computers and Electronics in Agriculture (which also leads the volume of publications and the number of citations), Agricultural Water Management, and Lecture Notes in Computer Science.

In order to study the most relevant authors, this bibliometric research analyzes both quantitative (volume of publications) and qualitative indicators (number of citations). Thus, with reference to the number of publications and h-index, Professor Gerrit Hoogenboom is the most influential author, followed by J.W. Jones, Z. Fu, J.R. Barret, and J.C. Ascough. In order to understand the total strength of the co-authorship links with other authors, different clusters were also identified in this research. The publication with the highest number of citations is "WEKA: A machine learning workbench" [36], as the result of an international conference. Other papers with a high number of citations are "The regularized iteratively reweighted MAD method for change detection in multi- and hyperspectral data" [38] and "Colour and shape analysis techniques for weed detection in cereal fields" [39], published in the journal Computers and Electronics in Agriculture.

There are numerous public and private organizations with a high commitment to research on AI in the agricultural industry. This ranking is led by institutions from China (China Agricultural University, Chinese Academy of Sciences, Ministry of Education China, Chinese Academy of Agricultural Sciences, Ministry of Agriculture of the People's Republic of China, Zhejiang University) and the U.S. (United States Department of Agriculture, University of Florida, Texas A&M University). Other institutions such as the University of Tehran (Iran) and the Consejo Superior de Investigaciones Científicas (Spain) also have relevant positions.

In addition, many funding agencies support research in this field. Once again, China and the U.S. lead this ranking, in which Brazil and the European Union also play an important role. China's research on AI began later than the U.S. and Europe, but its development is rapid and intense, since the inclusion, in 1986, of AI Research and Development in basic research funding, mainly through two agencies: The National Natural Science Foundation of China (NSFC) and the National Basic Research Program of China (973 programs) for applied research. Thus, the Chinese government is promoting research related to AI, not only focused on the agriculture industry, but on various topics, such as: intelligent application systems, neural networks, human–computer interaction, computer vision, or genetic algorithms. Since 2000, the Chinese public administrations committed definitively to research AI, increasing funds destined to this research. Therefore, the Chinese institutions encourage and support Chinese research centers in order to identify the potential of AI in the agricultural industry and to apply it to production. Of note also is China's Ministry of Education's AI Innovation Action Plan for College and Universities, wherein more than 70 Chinese universities and colleges have introduced AI-related majors.

The U.S. is also relevant in this ranking, having been one of the leading pioneers in this field. Its current commitment to research on this topic is very high, with the funds allocated to it increasing significantly every year. Some of the most important American entities are the National Science Foundation, Department of Agriculture (USDA), and National Institute of Food and Agriculture (NIFA). Brazil also has two relevant entities supporting research on this field, one of them at the national level (Conselho Nacional de Desenvolvimento Científico e Tecnológico), and the other at a regional level (Fundação de Amparo à Pesquisa do Estado de São Paulo). Finally, the European Commission also has AI and robotics among its priorities under the research and innovation of the program Horizon 2020.

The use of AI in agriculture is quite transversal, with several areas of knowledge around this topic. By using a fractional counting method, based on bibliographic data on the co-occurrence of authors' keywords, this research identifies different clusters and trends: machine learning; Internet of things; deep learning; big data; sensors; cloud computing; drought; and robot. Other emergent lines of research are life cycle assessment (LCA), green economy, sustainable development, climate change, and the environment.

This research has some limitations, which may be the basis for future research. Some of them are related to the bibliometric analysis, a research method which is essentially quantitative in nature. However, completing it with a qualitative analysis is important to attain a better view of the research field analyzed. Some authors may be very influential in a specific field, even with only a few articles. The opposite can also happen, wherein a certain author with only one work published in the field analyzed may have a high number of citations. This is the reason why we not only consider the volume of publications, but also qualitative features and standardized metrics, such as the number of citations or the h-index. In any event, this methodology could be completed with other quantitative or qualitative tools (e.g., knowledge maps or visuals). It could also be of interest to implement a systematic literature review using other tools such as a meta-analysis.

Author Contributions: Conceptualization, J.L.R.-R. and J.U.-T.; methodology, J.U.-T.; validation, J.L.R.-R. and J.U.-T.; formal analysis, J.L.R.-R., J.U.-T., J.A.T.A. and J.d.P.V.; writing—original draft preparation, J.L.R.-R., J.U.-T., J.A.T.A. and J.d.P.V.; writing—review and editing, J.L.R.-R., J.U.-T., J.A.T.A. and J.d.P.V.; supervision, J.L.R.-R. All authors have read and agreed to the published version of the manuscript.

Funding: This research received no external funding.

Conflicts of Interest: The authors declare no conflict of interest.

References

1. King, B.A.; Hammond, T.; Harrington, J. Disruptive Technology: Economic Consequences of Artificial Intelligence and the Robotics Revolution. *J. Strateg. Innov. Sustain.* **2017**, *12*, 53–67.
2. McKinion, J.M.; Lemmon, H.E. Expert Systems for Agriculture. *Comput. Electron. Agric.* **1985**, *1*, 31–40. [CrossRef]
3. Murase, H. Artificial Intelligence in Agriculture. *Comput. Electron. Agric.* **2000**, *29*, 178. [CrossRef]
4. Attonaty, J.M.; Chatelin, M.H.; Garcia, F. Interactive Simulation Modeling in Farm Decision-Making. *Comput. Electron. Agric.* **1999**, *22*, 157–170. [CrossRef]
5. El Yasmine, A.S.L.; Ghani, B.A.; Trentesaux, D.; Bouziane, B. Supply Chain Management Using Multi-Agent Systems in the Agri-Food Industry. In *Service Orientation in Holonic and Multi-Agent Manufacturing and Robotics*; Springer: Cham, Switzerland, 2014; pp. 145–155.

6. Thomopoulos, R.; Croitoru, M.; Tamani, N. Decision Support for Agri-Food Chains: A Reverse Engineering Argumentation-Based Approach. *Ecol. Inf.* **2015**, *26*, 182–191. [CrossRef]
7. Nair, B.B.; Mohandas, V.P. Artificial Intelligence Applications in Financial Forecasting-a Survey and Some Empirical Results. *Front. Artif. Intell. Appl.* **2015**, *9*, 99–140. [CrossRef]
8. Bryceson, K.; Slaughter, G. Integrated Autonomy A Modeling-Based Investigation of Agrifood Supply Chain Performance. In Proceedings of the 2009 11th International Conference on Computer Modelling and Simulation, Cambridge, UK, 25–27 March 2009.
9. Patrício, D.I.; Rieder, R. Computer Vision and Artificial Intelligence in Precision Agriculture for Grain Crops: A Systematic Review. *Comput. Electron. Agric.* **2018**, *153*, 69–81. [CrossRef]
10. Kaur, M.; Gulati, H.; Kundra, H. Data Mining in Agriculture on Crop Price Prediction: Techniques and Applications. *Int. J. Comput. Appl.* **2014**, *99*, 1–2. [CrossRef]
11. Kohzadi, N.; Boyd, M.S.; Kermanshahi, B.; Kaastra, I. A Comparison of Artificial Neural Network and Time Series Models for Forecasting Commodity Prices. *Neurocomputing* **1996**, *10*, 169–181. [CrossRef]
12. Limsombunchai, V. House Price Prediction: Hedonic Price Model vs. Artificial Neural Network. In Proceedings of the New Zealand Agricultural and Resource Economics Society Conference, Blenheim, New Zealand, 25–26 June 2004; pp. 25–26.
13. Li, G.Q.; Xu, S.W.; Li, Z.M. Short-Term Price Forecasting for Agro-Products Using Artificial Neural Networks. *Agric. Agric. Seletivo Proc.* **2010**, *1*, 278–287. [CrossRef]
14. Dahikar, S.S.; Rode, S.V. Agricultural Crop Yield Prediction Using Artificial Neural Network Approach. *Int. J. Innov. Res. Electr. Electron. Instrum. Control Eng.* **2014**, *2*, 683–686.
15. Hewitson, B.C.; Crane, R.G. Self-Organizing Maps: Applications to Synoptic Climatology. *Clim. Res.* **2002**, *22*, 13–26. [CrossRef]
16. Mellit, A. Artificial Intelligence Technique for Modelling and Forecasting of Solar Radiation Data: A Review. *Int. J. Artif. Intell. Soft Comput.* **2008**, *1*, 52–76. [CrossRef]
17. Aznar-Sánchez, J.A.; Piquer-Rodríguez, M.; Velasco-Muñoz, J.F.; Manzano-Agugliaro, F. Worldwide research trends on sustainable land use in agriculture. *Land Use Policy* **2019**, *87*, 104069. [CrossRef]
18. Ren, G.; Lin, T.; Ying, Y.; Chowdhary, G.; Ting, K.C. Agricultural robotics research applicable to poultry production: A review. *Comput. Electron. Agric.* **2020**, *169*, 105216. [CrossRef]
19. Fountas, S.; Mylonas, N.; Malounas, I.; Rodias, E.; Hellmann Santos, C.; Pekkeriet, E. Agricultural Robotics for Field Operations. *Sensors* **2020**, *20*, 2672. [CrossRef]
20. Lowenberg-DeBoer, J.; Huang, I.Y.; Grigoriadis, V.; Blackmoret, S. Economics of robots and automation in field crop production. *Precis. Agric.* **2020**, *21*, 278–299. [CrossRef]
21. Rose, D.C.; Wheeler, R.; Winter, M.; Lobley, M.; Chivers, C.A. Agriculture 4.0: Making it work for people, production, and the planet. *Land Use Policy* **2020**, *100*, 104933. [CrossRef]
22. Zhai, F.Z.; FernánMartínez, J.; Beltran, V.; Martínez, N.L. Decision support systems for agriculture 4.0: Survey and challenges. *Comput. Electron. Agric.* **2020**, *170*, 105256. [CrossRef]
23. Roy, S.K.; De, D. Genetic Algorithm based Internet of Precision Agricultural Things (IopaT) for Agriculture 4.0. *Internet Things* **2020**. [CrossRef]
24. Ryan, M. Agricultural Big Data Analytics and the Ethics of Power. *J. Agric. Environ. Ethics* **2020**, *33*, 49–69. [CrossRef]
25. Mokarram, I.M.; Khosravi, M.R. 2020. A cloud computing framework for analysis of agricultural big data based on Dempster–Shafer theory. *J. Supercomput.* **2020**. [CrossRef]
26. Gu, Y. Global Knowledge Management Research: A Bibliometric Analysis. *Scientometrics* **2004**, *61*, 171–190. [CrossRef]
27. Cobo, M.J.; Martinez, M.A.; Gutierrez-Salcedo, M.; Fujita, H.; Herrera-Viedma, E. 25 Years at Knowledge-Based Systems: A Bibliometric Analysis. *Knowl. Based Syst.* **2015**, *80*, 3–13. [CrossRef]
28. Brereton, P.; Kitchenham, B.A.; Budgen, D.; Turner, M.; Khalil, M. Lessons from applying the systematic literature review process within the software engineering domain. *J. Syst. Softw.* **2007**, *80*, 571–583. [CrossRef]
29. Garfield, E. Citation Index for Science. A New Dimension in Documentation through Association of Ideas. *Science* **1955**, *122*, 108–111. [CrossRef]
30. Moed, H.F. *Citation Analysis in Research Evaluation*; Springer: Dordrecht, The Netherlands, 2005.
31. Merton, R.K. The sociology of science: An episodic memoir. In *The Sociology of Science in Europe*; Merton, R.K., Gaston, J., Eds.; Southern Illinois University Press: Carbondale, IL, USA, 1977; pp. 3–141.

32. Hirsch, J.E. An index to quantify an individual's scientific research output. *Proc. Natl. Acad. Sci. USA* **2005**, *102*, 16569–16572. [CrossRef]
33. Kailath, T.; Morf, M.; Athans, M.; Ljung, L.; Chu, K.; Willsky, A.S.; Asher, R.B.; Caines, P.E.; Mitter, S.K.; Sandell, N.R.; et al. Research on the use of AI in agriculture relatively recent. In Proceedings of the IEEE Conference on Decision and Control 1976, Clearwater, FL, USA, 1 December 1976.
34. Maran, L.R.; Beck, H.W. Some lessons for Artificial-Intelligence and agricultural systems simulation. In *Advances in AI and Simulation: Proceedings of the SCS Multiconference on AI and Simulation*; Society for Computer Simulation: Tampa, FL, USA, 1989; pp. 98–102.
35. Fermanian, T.W.; Michalski, R.S.; Katz, B.; Kelly, J. Agassistant—An Artificial-Intelligence system for discovering patterns in agricultural knowledge and creating diagnostic advisory systems. *Agron. J.* **1989**, *81*, 306–312. [CrossRef]
36. Holmes, G.; Donkin, A.; Witten, I.H. WEKA: A machine learning workbench. In Proceedings of the Australian and New Zealand Conference on Intelligent Information Systems—Proceedings of ANZIIS '94, Brisbane, Australia, 29 November–2 December 1994; pp. 357–361.
37. Wolfert, S.; Ge, L.; Verdouw, C.; Bogaardt, M.J. Big data in smart farming—A review. *Agric. Syst.* **2017**, *153*, 69–80. [CrossRef]
38. Nielsen, A.A. The regularized iteratively reweighted MAD method for change detection in multi- and hyperspectral data. *IEEE Trans. Image Proc.* **2007**, *16*, 463–478. [CrossRef]
39. Pérez, A.J.; López, F.; Benlloch, J.V.; Christensen, S. Colour and shape analysis techniques for weed detection in cereal fields. *Comput. Electron. Agric.* **2000**, *25*, 197–212. [CrossRef]
40. Zadeh, L.A. Making computers think like people. *IEEE Spectr.* **1984**, *21*, 26–32. [CrossRef]
41. Jain, S.K.; Das, A.; Srivastava, D.K. Application of ANN for reservoir inflow prediction and operation. *J. Water Res. Plan. Manag.* **1999**, *125*, 263–271. [CrossRef]
42. European Commission. Available online: https://ec.europa.eu/digital-single-market/en/artificial-intelligence (accessed on 7 July 2019).

Publisher's Note: MDPI stays neutral with regard to jurisdictional claims in published maps and institutional affiliations.

© 2020 by the authors. Licensee MDPI, Basel, Switzerland. This article is an open access article distributed under the terms and conditions of the Creative Commons Attribution (CC BY) license (http://creativecommons.org/licenses/by/4.0/).

Article

Transfer of Agricultural and Biological Sciences Research to Patents: The Case of EU-27

Mila Cascajares [1], Alfredo Alcayde [1], Esther Salmerón-Manzano [2] and Francisco Manzano-Agugliaro [1,*]

1 Department of Engineering, University of Almeria, ceiA3, 04120 Almeria, Spain; milacas@ual.es (M.C.); aalcayde@ual.es (A.A.)
2 Faculty of Law, Universidad Internacional de La Rioja (UNIR), Av. de la Paz, 137, 26006 Logroño, Spain; esther.salmeron@unir.net
* Correspondence: fmanzano@ual.es; Tel.: +34-950-015-346; Fax: +34-950-015-491

Citation: Cascajares, M.; Alcayde, A.; Salmerón-Manzano, E.; Manzano-Agugliaro, F. Transfer of Agricultural and Biological Sciences Research to Patents: The Case of EU-27. *Agronomy* **2021**, *11*, 252. https://doi.org/10.3390/agronomy11020252

Academic Editor: Massimo Blandino
Received: 31 December 2020
Accepted: 27 January 2021
Published: 29 January 2021

Publisher's Note: MDPI stays neutral with regard to jurisdictional claims in published maps and institutional affiliations.

Copyright: © 2021 by the authors. Licensee MDPI, Basel, Switzerland. This article is an open access article distributed under the terms and conditions of the Creative Commons Attribution (CC BY) license (https://creativecommons.org/licenses/by/4.0/).

Abstract: Agriculture as an economic activity and agronomy as a science must provide food for a constantly growing population. Research in this field is therefore becoming increasingly essential. Much of the research is carried out in academic institutions and then developed in the private sector. Patents do not have to be issued through scientific institutions. Patents from scientific institutions are intended to have a certain economic return on the investment made in research when the patent is transferred to industry. A bibliometric analysis was carried out using the Scopus and SciVal databases. This study analyses all the research carried out in the field of agronomy and related sciences (Agricultural and Biological Sciences category of Scopus database) by EU-27 countries, which has been cited in at least one international patent. The data show that out of about 1 million published works only about 28,000 have been used as a source of patents. This study highlights the main countries and institutions in terms of this transfer. Among these, Germany, France and Spain stand out in absolute terms, but considering the degree of specialization. Regarding their specialization the institution ranking is led by Swedish University of Agricultural Sciences (58%), AgroParisTech (52%), Wageningen University & Research (48%), and INRAE (38%). It also analyses which journals used for this transfer are most important. For these publications more than 90% of the articles have had a higher-than-expected citation level for the year of publication, the type of publication and the discipline in which they are categorized. The most-obtained research fields can be distinguished as those related to genetics or mo-lecular biology, those related to specific foods, such as cheeses, milk, breads or oils, and, thirdly, the group covering food-related constituents such as caseins, probiotics, glutens, or starch.

Keywords: agronomy; SciVal; patents; Europe; bibliometrics; R&D; Scopus; patentometrics; Triple Helix

1. Introduction

Agronomy is based on scientific and technological principles, and must study the physical, chemical, biological, economic, and social factors that, in one way or another, influence crop production [1]. Its fundamental basis is focused on studying human intervention in nature from an agro-productive point of view, or in other words, studying the agro-ecosystem as a specific model of human intervention in nature, with the aim of producing food and raw materials [2]. In short agronomy may be defined as the science of soil management and crop production [3].

The essential issue in agronomy is the study of the relationship between soil, plant, and environment, with the aim of maximizing yields, and reducing production costs, but doing so with responsibility and not at any price [4]. To do this, it is necessary to plan the processes, as well as to implement different measures to obtain the maximum use of natural resources, in order to produce more and improve production standards [5]. All this must be done paying special attention to non-renewable natural resources, which are in danger due to their negligent and uncontrolled use by man [6]. At this point, it is agronomy, which must be in charge of developing sustainable plans, for the efficient use of these resources, in order not to aggravate this situation, such as the case of water re-use in agriculture [7].

Agronomy also deals with the selection of suitable crop varieties, i.e., those best suited to the particular conditions of the environment [8,9], as well as the adoption of the most effective production system [10], the choice of the most suitable growing techniques [11], the selection of appropriate plant protection measures [12,13], the adoption of the most efficient harvesting methods both in terms of quantity and quality [14,15], and the choice of the most appropriate post-harvest technologies [16,17]. This is done by considering the management of inputs, such as labor, seeds, fertilizers, facilities, and machinery [18].

Agronomy is certainly the fundamental basis of human nutrition [19]. The demographic pressure is increasing but the cultivation area remains static, therefore in order to feed the growing population it is necessary to exploit and maximize the yields of the production systems, and it is here that agronomy plays a fundamental role. Agronomy is a dynamic discipline, in continuous advance, which increases the knowledge of plants and their environment each day [20]. This leads to the development and implementation of new agricultural practices focused on fully exploiting the potential of the different production systems [21], as well as improving the production and processing processes of food from both a quantitative and qualitative point of view. In addition, agronomy must develop plans that enable integrated agricultural systems to be implemented, to achieve sustainable agricultural growth, that is to say without compromising the environment [22].

All these challenges are not possible without high-quality R&D that is broad and multidisciplinary, and above all geographically distributed [23]. It is well known that public research usually allocates its large resources to basic research, while companies focus on applied research, which they can market either directly or by selling the knowledge they have developed [24]. Regarding this last point, the key is the protection of these rights, generally via patents [25].

It is a consensus in all industrialized countries that patent law has a decisive influence on the organization of the economy, as it is a key element in promoting technological innovation [26]. This last aspect is of the utmost importance, as it largely regulates business investment in R&D. It should suffice to mention that one of the points to be reformed in the legislation of the applicant countries is the law governing patents when a country becomes a member of the European Union. For example, Spain's admission to the EU in 1986 led to the revocation of the 1929 patent law. European patent legislation is based on the Munich Convention of 5 October 1973 on the European Patent [27] and the Luxembourg Convention on the Community Patent of 15 December 1975 [28]. This European patent directive has been incorporated into almost all European patent legislation [29].

Without going into detail regarding European patent law, it should be noted that there are two categories of industrial property rights: patents for invention and utility models [30]. Patents give their holders a territorial right to prevent the commercial exploitation of the patented object without their consent for 20 years from the priority date, while for utility models this is limited to 10 years [31].

In short, patent laws must aim to promote the technological development of countries, starting from their industrial situation [32]. Particular attention has therefore been paid to the protection of national interests [33], especially by strengthening the obligations of patent holders so that the exploitation of patents takes place within their territory and a real transfer of technology takes place, but always in accordance with the Paris Union Convention of 20 May 1883, the text of which was revised in Stockholm on 14 July 1967 [34].

The issue of plant variety protection is particularly interesting. However, it is specified that a patent cannot be awarded for a particular variety of a plant or for essentially biological processes for obtaining plants such as crossing and selection. Some authors suggest that the right to patent agricultural innovations is increasingly placed in a political context [35].

Plant varieties can be protected by obtaining Plant Variety Protection (PVP) or Plant Variety Rights (PVR), provided that these varieties are new, distinct, uniform, and stable and have a name which is not liable to be confused with the names of other plants or with trademarks for Class 31 according to the Nice Classification [36].

In Spain, for example, the right obtained by entering the plant variety in the national register of commercial varieties does not correspond to this plant variety right but is distinct and complementary. To establish novelty there is a useful period of grace during which commercial acceptance can be verified. Plant variety titles grant their holder a territorial right to prevent the commercial exploitation of the variety without his consent for 30 years for vine, and potatoes varieties and tree species and 25 years for all other plant varieties, from the date the title is awarded [37].

In the plant breeding sector, patent protection of innovations is the prevalent strategy in the United States and China [38]. In Europe, however, plant breeders are choosing to protect new plant varieties [39]. According to the latest data provided by the International Union for the Protection of New Varieties of Plants (UPOV), the registration of plant varieties at the Community Office is the most widely used method worldwide, because it makes it possible to obtain protection in all EU Member States at a proportionately more attractive cost compared with the domestic route. The mission of UPOV is to provide and promote an effective system of plant variety protection, to encourage the development of new varieties of plants, for the benefit of society (https://www.upov.int/portal/index.html.en).

This article is organized as follows: first, a background section related to patentometrics and Triple Helix concept is introduced, then the data used and the methodology are described in the Materials and Methods section. The results are then analyzed and then discussed alongside other papers. This last section is organized as: global temporal trend; countries, affiliations, and collaborations; top journals used for the publications cited in patents; the quality of the articles; the open access and European funding agencies; topics of the publications cited in patents. Finally, the main conclusions of this research are drawn.

2. Background: Patentometrics and Triple Helix

Since the 2000s university patenting in the most advanced economies has been on the decline both as a percentage and in absolute terms [39]. We suggest that the institutional incentives for university patenting have disappeared with the new regime of university ranking, since patents or spin-offs are not counted in university rankings.

Patent statistics have long been of interest to innovation-conscious economists. The central question is whether or not patent statistics represent the real state of innovation [40]. The statistical analysis of patents can be named Patentometrics [41]. The first articles on this issue are quite recent, dating back to 2001 [40]. On the one hand, there are the statistics of the patents themselves, such as defining rank-ings for them based on citations [42], or as a patent h-index indicator to assess patenting quality [43]. A patent h-index has been introduced to evaluate the patenting activities of research organizations [44]. However, the h-index has been questioned for being insen-sitive to some exceptionally widely cited items, as can be seen from the large number of so-called h-indexes proposing to address this issue and to replace the original h-index; a review of these h-type indexes can be found in several studies, such as [45]. Patentometric indicators make it possible to quantify and qualify the performance of technological out-put on the basis of granted patents, e.g., in Brazil [46].

There is increasing interest in technology-based enterprises, for their capacity to contribute to economic and social development. To this end, patent-based indices have been developed with the aim of monitoring the impact of specific patents, or the state of technology in a given field, or comparing technology between countries. The comparative study between countries of patent production in a given field shows, according to some researchers, how advanced a technology is in the countries that are leaders in this field, and is called the specialisation index [47]. Therefore, the information contained in patent documentation has become one of the principal techniques for modeling technology scenarios for government, business and industry, research institutes or projects, [48]. Most of this work is based on patent databases such as United States Patent and Trademark Office (USPTO) [17] or European Patent Office (EPO), but one alternative that has proved to be valid and open access is Google Patents (www.google.com/patents), which includes over 8 million full-text patents [49,50].

Patenting is not only a significant method of university knowledge transfer, but also an important indicator for measuring academic R&D strength and knowledge utilization [43]. Because patents are a direct output of innovative activities, cross-border patents are used to analyze the trend of global collaborative creativity [51]. Usually two sets of documents, impact articles and patents have been used as approximation measures to analyze the research of the institutions, and in this way both the trajectories of the scientific and technical front are analyzed, and then the research into these can be categorized as basic science or applied technology [52]. e.g., Brazil, scientometric and patentometric indicators have been studied to assess the non-financial criteria associated with technology for the purposes of financial funding, as there is a growing interest in technology-based companies due to their ability to contribute to economic and social development [53]. Another issue of great relevance is the assessment of scientific publications and patent analysis production. This enables the definition of the growth rate of scientific and technological output in terms of the top countries, institutions and journals producing knowledge within the field as well as the identification of main areas of research and development [41].

A modern and competitive economic model needs science, as well as a strong public R&D system, funded in a stable way, and aligned with economic development. Science is gradually advancing towards a technological orientation rather than a theoretical orientation [54]. Triple Helix, is an academic theory that argues that the potential for development of the knowledge economy in regions or countries lies in the close collaboration of companies, universities and governments based on new institutional formulas designed for the production, transfer and application of knowledge. The theory of the triple helix introduced and developed by Etzkowitz and Leydesdorff [55] follows the same line, highlighting the role of government along with the other two helixes: universities and industry [56]. This is because innovation processes, as well as research and innovation policy decision-making processes, tend to increasingly involve the variety of components of the innovation system, i.e., academia, industry and stakeholders who are the end-users.

A triple helix model to study university-industry-government relationships is based on indicators such as: webometric, scientometric and technometric [55]. Patent-based metrics could be utilized in a Triple Helix context, and hybrid indicators could be developed by combining a patent with other data [55]. Most of the patented academic inventions are related to scientific research and are financed by public funds. These tend to be used in large companies rather than in start-ups founded by academic entrepreneurs [56]. Moreover, some studies show that scientific excellence and technology transfer activities are mutually reinforcing [57], so it is important to understand their relationship.

The first step in this context is to define the indicators, and then to establish a benchmarking framework. The European Commission has elaborated an evaluation re-port in this regard to benchmark the five aspects: human resources in RTD; public and private investment in RTD; scientific and technological productivity; impact of RTD on economic competitiveness and employment; promotion of RTD culture and public under-standing of science. These indicators are based on % of GDP or per million population.

In relation to agriculture, the Triple Helix model is not well studied, but it is worth noting the work done in this field in Korea and China, where they used bibliometric indi-cators. The raw inputs were the numbers (or %) of manuscripts with only academic au-thors, only industry authors, only government authors, only authors who are from aca-demia or industry, etc. [58].

Previous studies have focused only on the evolution of new technologies through the study of patents and have rarely explored the context of prior knowledge, i.e., the research on which these patents are based. The aim of this paper is therefore to analyze the potential contribution of research in the EU-27 countries as a driving force for technological innovation in the field of agricultural and biological sciences. To this end, bibliometric indicators will be used to analyze all the works published in this scientific field by the EU-27 countries that are cited in at least one patent. The Europe of 27 (EU27) is made up of the following countries: Austria, Belgium, Bulgaria, Croatia, Cyprus, Czech Republic, Denmark, Estonia, Finland, France, Germany, Greece, Hungary, Ireland, Italy, Latvia, Lithuania, Luxembourg, Malta, Netherlands, Poland, Portugal, Romania, Slovakia, Slovenia, Spain, and Sweden. Finally, the aim is to launch a visualized model that can be applied, as a tool for analyzing any scientific field in any country or group of countries, where the degree of transfer of the research carried out can be measured by means of patent citation. The Europe of 27 (EU27) is made up of the following countries: Austria, Belgium, Bulgaria, Croatia, Cyprus, Czech Republic, Denmark, Estonia, Finland, France, Germany, Greece, Hungary, Ireland, Italy, Latvia, Lithuania, Luxembourg, Malta, Netherlands, Poland, Portugal, Romania, Slovakia, Slovenia, Spain, and Sweden.

3. Materials and Methods

Science can be considered as what is published in scientific journals [59]. Scientific databases therefore play a key role in the progress of science since what has been published previously is the basis for new research. Within the existing scientific databases, Web Of Science (WOS) and Scopus can be considered to have leading positions in most branches of knowledge. There are many research studies that indicate that Scopus covers at least 80% of the content of the WOS database. Scopus has been used in considerable bibliometric studies in many branches of knowledge, such as those of Engineering [60], Environmental Science [61] or Agricultural and Biological Sciences [62,63].

To carry out this study, the publications in the scientific field of Agricultural and Biological Sciences indexed in Scopus in the period 1999–2019 in the geographical area of the European Union (the current 27 EU countries) were analyzed. Of the data obtained, the study focuses on those publications that have been cited at least once in patents. This limitation was made with SciVal; a tool closely linked to Scopus.

As one of the most important reference databases in the field of research, Scopus indexes around 25,000 journal titles from more than 5000 publishers. Although its contents date back to 1788, it was not until 1996 that these contents became the basis of SciVal, Elsevier's tool for metric analysis. SciVal provides access to the scientific output of more than 230 countries and 14,000 institutions. SciVal therefore makes it possible to visualize research performance, make comparisons, analyze trends, and evaluate collaborations [64]. As an analysis tool, SciVal has been employed in several publications, applying the metrics provided by this tool. e.g., studies on the progress of thermal spraying research were carried out between 1985 and 2015 [65,66] and supplemented by SciVal. Additionally, in 2016 Yu et al. [67] used SciVal in a comparison metric analysis with ResearchGate. In the domain of research in medical radiation science, Ekpo, Hogg and McEntee [68] analyzed international collaboration and institutional activity with metrics obtained from SciVal. Or as last example, the analysis of research results from Russian universities was also based on SciVal conducted in 2018 [69], and recently in 2019, a bibliometric analysis of big data was carried out using SciVal [70].

To achieve the direct download of data from Scopus and SciVal, the Scopus API Key was used, by means of this API it is possible to obtain more data than from a direct download (https://dev.elsevier.com/sc_apis.html). To visualize the results, Microsoft Excel was used as an analysis tool by means of dynamic tables and ArcGIS for the representation of the map.

Using these two tools, the data were obtained by carrying out two searches. See Figure 1 for an outline of the methodology. The first was in Scopus, of publications between 1999 and 2019, in the scientific field of Agricultural and Biological Sciences, in the EU-27; the second in SciVal, of publications between 1999 and 2019, in the scientific field of

Agricul-tural and Biological Sciences, in the EU-27, and which have been cited in patents. To ob-tain data on publications cited in patents, the bibliometric indicator "Patent-Cited Schol-arly Output" was selected for all publication types and for all patent offices. SciVal offers coverage of five of the largest patent offices: European patent office (EPO), US Patent Office (USPTO), UK Intellectual Property Office (UK IPO), Japan Patent Office (JPO) and World Intellectual Property Organization (WIPO) [71].

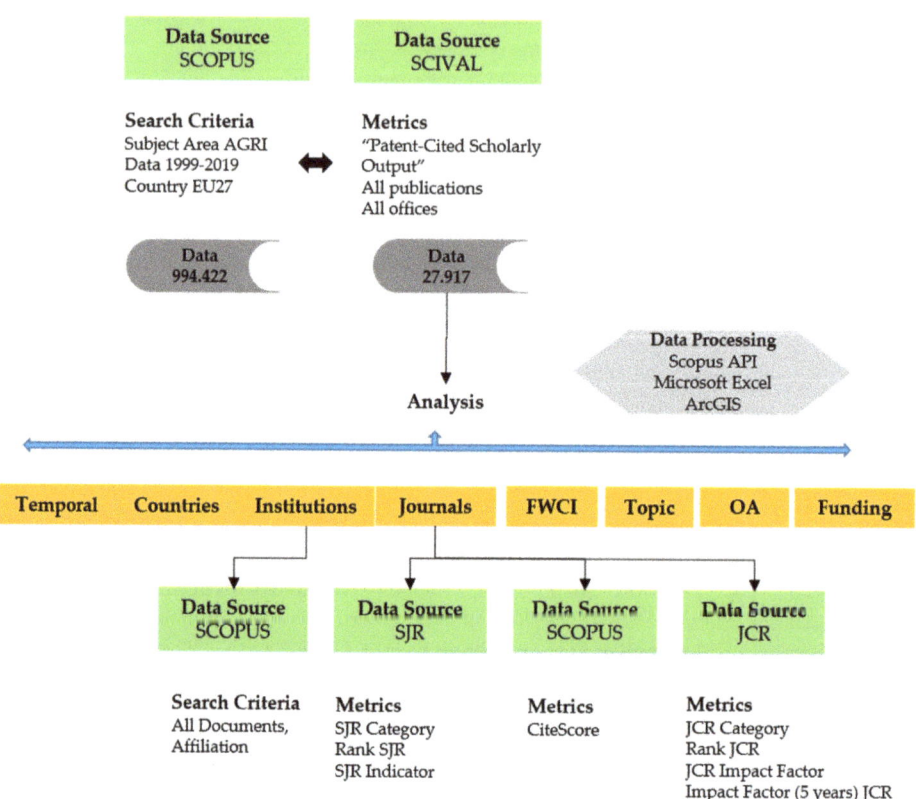

Figure 1. Methodology flowchart.

In order to establish the degree of specialization of institutions, an indicator called degree of specialization (ESP-AGRI) has been developed. The ESP-AGRI indicator shows the degree of specialization of the institution with respect to this scientific category. This indicator calculates the percentage of publications of the analyzed subject with respect to the total number of publications (N-AGRI) of a given institution.

To complete the analysis of the data, bibliometric indicators were obtained referring to the impact of the Top 20 journals in which the greatest number of papers have been published according to the search carried out. Thus, on the one hand, the indicators related to Scopus were extracted: SJR Category, Rank SJR, SJR Indicator and CiteScore, and, on the other hand, they were completed with the impact indicators of the other database referring to research, WOS–Journal Citation Reports (JCR). From JCR, JCR Category, Rank JCR, JCR Impact Factor and Impact Factor (5 years) JCR were extracted. These values were obtained by searching in JCR, SJR and Scopus.

SJR and JCR classify journals based on different categories within a certain scientific field, assessing the position within the category based on the total number of journals classified in that category, resulting in the quartile in which they are positioned within the category.

The SJR Indicator and JCR Impact Factor measure the quality of scientific publications based on the citations obtained in each publication. Both indicators are calculated by dividing the citations in the year being evaluated (in our case 2019) to articles published in previous years by the total number of articles and reviews published in that period. The difference between both indicators is that the SJR Indicator considers the three previous years, making the citation range is three years, while the JCR Impact Factor considers two years of citation. Based on the obtained result, it is possible to establish a ranking of journals that allows for determination of their quality.

At the end of 2016 [72], Scopus established a new indicator to measure the impact of a publication, CiteScore. Like the previous indicators, it measures the ratio of citations per article published in a given journal but extends the citation range to four years and includes citations of a larger typology of documents (articles, reviews, conference proceedings, book chapters and data documents) published on Scopus in that 4-year period.

Finally, the Impact Factor (5 years) JCR, shows the average number of times articles from the journal have been cited in the JCR year, from published in the last five years. The calculation is like the previous indicators; it is obtained by dividing the number of citations in the JCR year by the total number of articles published in the previous five years.

Citation as a basis for assessing the impact of publications has its roots in Eugene Garfield who developed the concept of the available citation index [73]. Both the JCR Impact Factor and the SJR Indicator provide a numerical value that needs to be interpreted in terms of several factors. The main consideration is the number of citations, which is directly linked to the area of research, the year of publication and the type of publication. Despite being the most widely used index in many bibliometric studies, the JCR Impact Factor is also the most discussed index because of its limitations such as asymmetry between numerator and denominator, differences between disciplines, insufficient citation range and the asymmetry of underlying citation distributions [74]. On the other hand, the SJR index tries to rectify these deviations by weighting the links based on the closeness of the citation, extending the number of years considered in the citation and setting thresholds for self-citation within the journal itself [75]. The CiteScore index also extends the range of years in the citation, but, although by including all types of documents the differences between the different types of documents are eliminated, some critics say that this index favors Elsevier's pub-lications, which tend to publish a higher proportion of types document other than articles than other publishers [76].

Regarding affiliations, Scopus has been the database used most often to calculate the percentage of publications indexed between 1999 and 2019 in the scientific field of Agricultural and Biological Sciences with respect to the total publications of the top 20 institutions that have published in the field. For this purpose, the total number of publications in the affiliation (documents, affiliation only) was considered.

On the other hand, it has been considered important to make an analysis of the research topics reflected in the publications that have been cited in patents. The Agricultural and Biological Sciences field covers many different subjects and SciVal uses the Topics to identify the predominant topics of interest. A Topic includes a set of documents with a common interest. They are clustered within SciVal based on direct citation analysis. Document reference lists are used for this purpose, so that a document can belong to only one Topic. However, as newly published documents are indexed, they are added to the Topics using their reference lists. This makes the Topics dynamic and most of them increase in size over time.

Topics with similar research interests are grouped into Topic Clusters forming broader research areas and, in both concepts, Topic and Topic Cluster, prominence can be measured by two parameters: the Topic Prominence Percentile and the Topic Cluster Prominence

Percentile. In both measures, prominence is calculated by SciVal by considering the number of citations received in the year with respect to citations received in the same and previous year, the number of views in Scopus in the year of publications in that and previous year, and the average number of citations in CiteScore in the year [77]. Prominence is therefore an indicator of the visibility and momentum of a given Topic, which is why it is important to analyze the percentage of publications in the Top20 journals that are in the first percentile (Top 10%). Note that these are indicators provided by the SciVal database.

While the Topics help us to see how visible the publications have been, it is the Field-Weighted Citation Impact (FWCI) that allows us to determine whether the publication has reached the level of citation that was expected of it. The FWCI considers the year of publication, the type of publication and the discipline in which it is categorized, so that if the FWCI value does not reach the benchmark we can say that it has not exceeded the prospects set for that publication. The benchmark is 1: a score equal to this or above it means that expectations were met in terms of citation; a score below it means they were not.

Since this study is based on the Europe of 27, it was considered interesting to analyze the sources of funding for research that are cited in patents. In this sense, together with other funding agencies, we wanted to see the role of European Commission through the different Research Framework Programs that were developed in this period (1999–2019): Fifth Framework Program 1998–2002 (FP5), Sixth Framework Program 2002–2006 (FP6), Seventh Framework Program (FP7) 2007–2013 and Horizon 2020 (H2020) 2014–2020.

Since the Budapest Declaration in 2002, there have been many public statements promulgating open access to scientific production without copyright restrictions. The European Commission itself requires open access publication of the results of research funded under its Framework Programs. Therefore, another element considered in this study is the impact of Open Access (OA).

4. Results and Discussion

For the search criteria in the Scopus database, and for the whole of the EU-27 in the Agricultural and Biological Sciences category, 994,422 records were obtained, while, for the same category in the SciVal database, and with the criterion of having been cited in at least one patent, there were 27,917 records.

4.1. Global Temporal Trend

Figure 2 shows the evolution of articles published by the EU-27 countries in the category of Agricultural and Biological Sciences (N-AGRI) from 1999 to 2019. It can be seen that in the last 8 years they have stabilized at just over 65,000 publications.

Furthermore, the evolution of the studies cited in patents (N-AGRI-CP) is shown, and until 2012, the articles cited stabilized at around 1500 studies. Research conducted in oth-er disciplines shows that the last 10 years of publications are not very significant in terms of citations by patents [78].

The series of data shown in Figure 2, up to 2009, shows great stability in the publications cited in patents. However, the relevant fact is that, at the beginning of the series, in 1999, publications cited were 6% of the total, but this figure slowly decreased to 3% of the total in 2010. This means that the research effort in relation to technological transfer and patents, has fallen by half in 10 years, from 1999 to 2009. The average overall transfer for the EU-27 countries for this period (1999–2009) was 5%.

Regarding EU funding, the different framework programs have had a positive impact on the increase in publications in the field under study, except for H2020, which seems to remain at the level reached in the previous scenario.

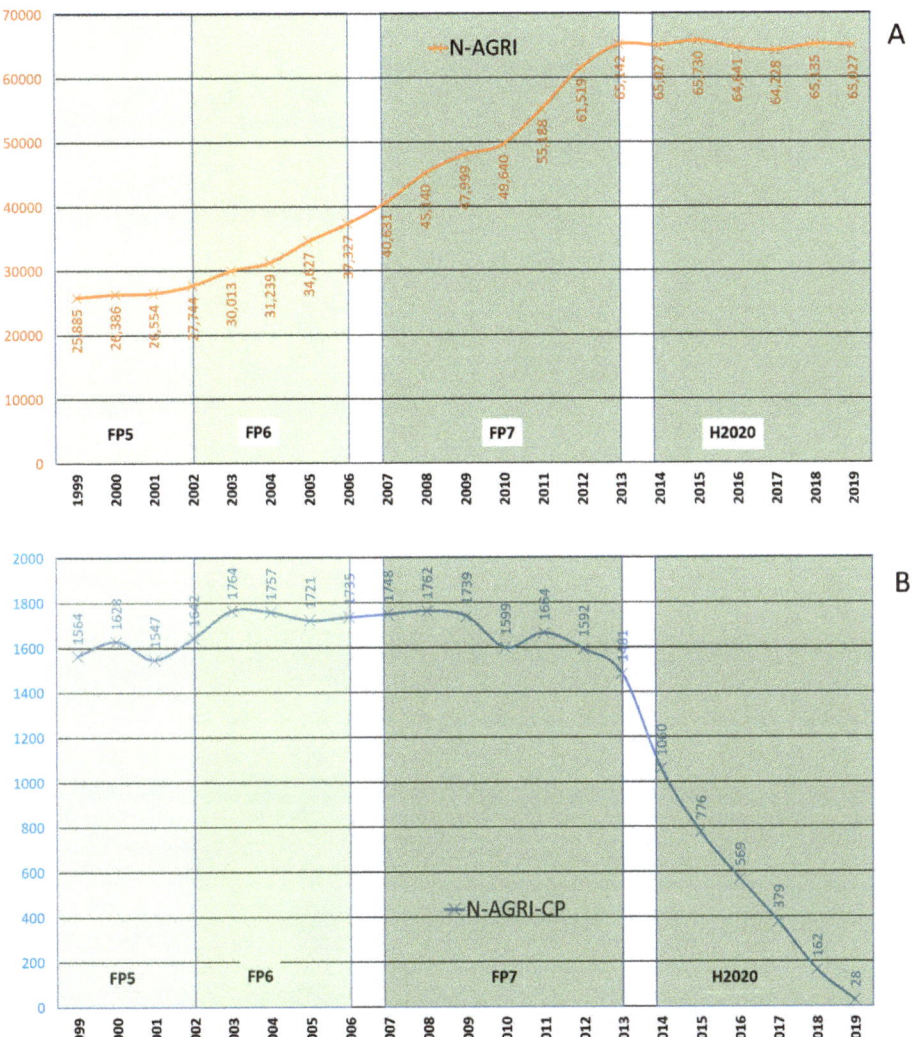

Figure 2. Agricultural and Biological Sciences publications: (**A**) total publication in Europe 27 (EU-27) (N-AGRI), (**B**) cited by international patents (N-AGRI-CP).

4.2. Countries, Affiliations, and Collaborations

In this section, publication data are counted for each of the authors of a publication when establishing countries, affiliations, and collaborations. This is the system used by the Scopus and SciVal databases. Figure 3 shows both the scientific production of the EU-27 countries in green, and the scientific collaboration with the other countries of the world in red. The higher color intensity indicates higher scientific production or collaboration with the EU-27. Of all these works, 40% are international collaborations with another 130 countries. These collaborations are mainly with the United States (4123), the United Kingdom (2373), Switzerland (878), Canada (707), Australia (586), Japan (520), China (465), Brazil (263), Israel (256), and Norway (255). This list of countries is not surprising as they are generally countries with a high research capacity, especially in the field of agricultural sciences. Others, such as Switzerland and Norway, have a geographical proximity to the EU-27, which makes them natural partners.

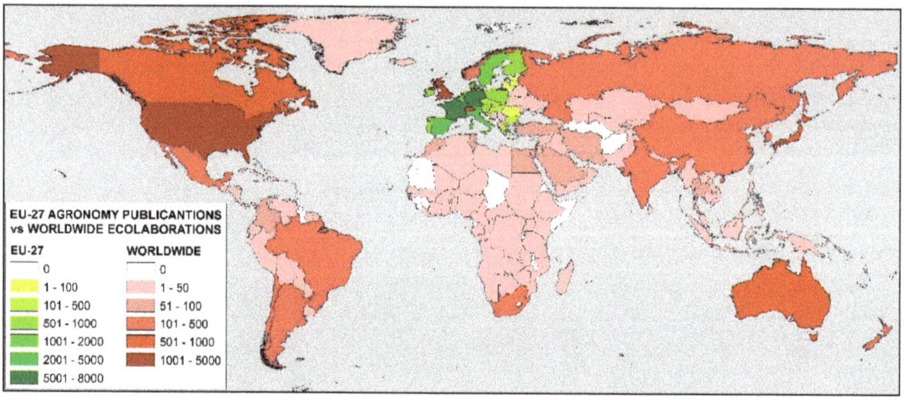

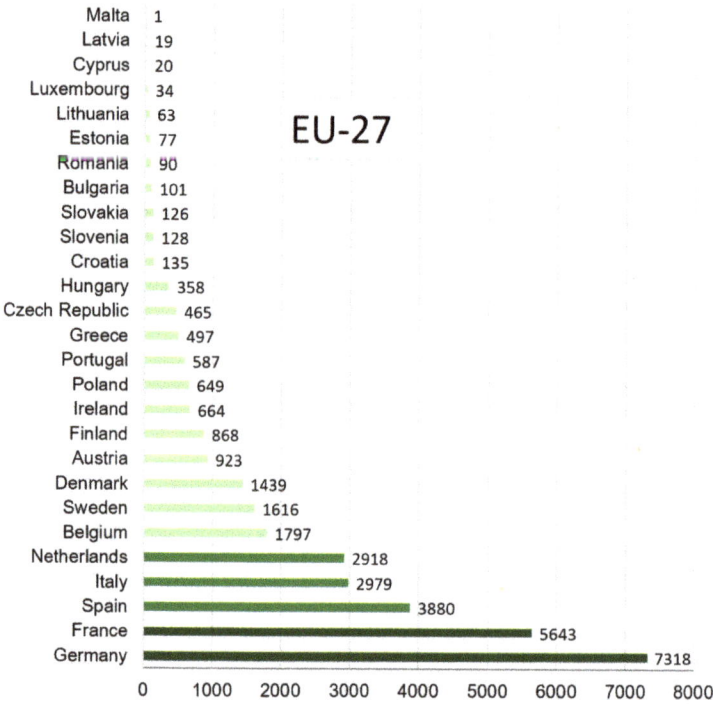

Figure 3. Worldwide production and collaboration of EU-27 publications cited in patents.

In Figure 4 the distribution by country of the scientific production in Agricultural and Biological Sciences that is cited in patents is shown. It is led by Germany with more than 7000 studies, followed by France with more than 5000, and in third place Spain with more than 3000. This list of outstanding countries continues with the Netherlands and Belgium with more than 2000 publications.

Figure 4. EU-27 publications cited in patents.

The research carried out by the countries is carried out in specific institutions, which are the real leaders in this research. Table 1 shows the top 20 institutions. This table reflects both the total works published by each institution in this period (N) and those in the category studied (N-AGRI), and of these those that were cited in patents (N-AGRI-CP). The ESP-AGRI indicator shows the degree of specialization of the institution with respect to this scientific category. The TIP-AGRI indicator measures the level of transfer of an institution, the relationship between publications indexed in the Agricultural and Biological Sciences category and publications that have been cited in patents.

Table 1. Agricultural and Biological Sciences Transference Index in Patents (TIP-AGRI).

Institutions	N-AGRI-CP [1]	N-AGRI [2]	N [3]	ESP-AGRI [4] (%)	TIP-AGRI [5] (%)
CNRS	2804	39,395	411,402	9.58	7.12
INRAE	2092	6356	16,563	38.37	32.91
CSIC	1458	22,974	110,344	20.82	6.35
Wageningen University & Research	1189	26,883	56,370	47.69	4.42
Institut National de la Santé et de la Recherche Médicale	884	8304	177,215	4.69	10.65
Université Paris-Saclay	734	6086	95,202	6.39	12.06
Ghent University	683	14,620	94,557	15.46	4.67
University of Copenhagen	662	15,892	99,175	16.02	4.17
National Research Council of Italy	442	14,586	139,335	10.47	3.03
Swedish University of Agricultural Sciences	437	15,912	27,592	57.67	2.75
KU Leuven	434	9573	120,699	7.93	4.53
Technical University of Munich	415	8138	104,312	7.80	5.10
Université de Montpellier	368	8414	49,926	16.85	4.37
University of Helsinki	366	13,856	84,064	16.48	2.64
Sorbonne Université	365	11,111	122,422	9.08	3.29
Utrecht University	362	8114	73,306	11.07	4.46
AgroParisTech	346	5693	11,001	51.75	6.08
Technical University of Denmark	340	6801	65,011	10.46	5.00
Universidad Autónoma de Madrid	340	5991	72,892	8.22	5.68
Institut Pasteur Paris	296	2459	22,126	11.11	12.04

[1] N-AGRI-CP Total number of publications classifies as Subject area Agricultural and Biological Sciences (ASJC) cited in patents. [2] N-AGRI Total number of publications published by the institution in period 1999–2019 classifies as Subject area Agricultural and Biological Sciences (ASJC). [3] N Total number of publications published by the institution in period 1999–2019. [4] ESP-AGRI = N-AGRI × 100/N. [5] TIP-AGRI = N-AGRI-CP × 100/N-AGRI.

From the data in Table 1, there are only three institutions specializing in this scientific category, considering those that have more than 30% of their scientific production in it. This specialization is led by Swedish University of Agricultural Sciences (58%), AgroParisTech (52%), Wageningen University & Research (48%), and INRAE (38%). The other institutions have a degree of specialization that is quite far away, between 4 and 20%.

The high level of transfer can be verified as oscillating from 2 to 33% of the total of works published in this category by each one of these institutions. In this regard, it is important to note that, as can be seen, eight institutions in France are in the top 20. It should be noted that the average overall transfer for the EU-27 countries for the period 1999–2009 was 5%. There are 10 institutions above 5%, and it should be remembered that the entire series is studied here, from 1999 to 2019, where transfer in the last 10 years was low until the technology or research is adopted by the industry.

The Institut National de Recherche en Agriculture, Alimentation et Environnement (INRAE) in France has a transfer rate of 33%, with a level of specialization of its publications of 38%. The case of Université Paris-Saclay (France) is also noteworthy, with a trans-fer rate of 12% despite the low level of specialisation of its publications (6%); the same can be seen with the Institut National de la Santé et de la Recherche Médicale (France), with a transfer of 11% and a specialization of less than 5%, and, finally, Institut Pasteur Paris, with 12% transfer rates versus 11% specialization. A curious situation is that of the two institutions that are mentioned as highly specialized, but have a low level of transferL Swedish University of Agricultural Sciences (2%), and Wageningen University & Research (4%).

Regarding international collaboration, three institutions stand out in particular, United States Department of Agriculture (308), Harvard University (258), and University of Oxford (207).

4.3. Top Journals Used for the Publications Cited in Patents

Table 2 lists the top 20 journals in which these patent-cited works have been published. These 20 journals account for 14,217 articles out of the total 27,917, which is half of the publications (50.93%). The mega-journal PLos ONE stands out in terms of the number of publications with 3379 articles. In 2014 Binfield [43] defined the four main criteria for a mega-journal: a very broad thematic scope, scientific solvency of the article, open access generally through article processing charges (APC) and a broad editorial board of academic publishers. Under these four criteria, PLos ONE appeared in 2006. Since its launch, its number of publication increased until it reached its maximum in 2013 with 32,055 documents indexed in Scopus, from this moment on the number of documents indexed in Scopus has decreased, reaching 16,316 in 2019. Categorized in both SJR and JCR as Multidisciplinary, it is positioned in the first quartile in SJR while moving to the second quartile in JCR.

Taking SJR as a reference, all the journals are positioned in at least one of their categories in the first quartile. However, if positioning in JCR is analyzed, of the Top 20 journals studied, three do not reach a position in the first quartile. To the already mentioned PLos ONE, one must add European Food Research and Technology and International Journal of Systematic and Evolutionary Microbiology.

The dominant categories in SJR are Plant Science and Genetics, with seven journals indexed in these categories, followed by Food Science and Medicine (miscellaneous), with six journals in each category. In JCR, the Plant Science category, eight journals are indexed, and in Food Science and Technology, six journals are indexed. From an editing perspec-tive, nine of the Top 20 Journals were published in the United States, and the re-maining eleven were published in European countries: United Kingdom, Netherlands, and Germany.

SciVal employs the All Science Journal Classification (ASJC) categories to classify Scopus sources, i.e., journals. Note that the same journal can be assigned one or more categories of the ASJC classification. The following field names are classified under the subject area Agricultural and Biological Sciences:

- Agricultural and Biological Sciences (all)
- Agricultural and Biological Sciences (miscellaneous)
- Animal Science and Zoology
- Agronomy and Crop Science
- Aquatic Science
- Ecology, Evolution, Behavior, and Systematics
- Food Science
- Forestry
- Horticulture
- Insect Science
- Plant Science
- Soil Science

Table 2. Top 20 journals and their metrics. (Data 2019).

Journal	N	SJR Category. Rank SJR	SJR Indicator	CiteScore Scopus	JCR Category. Rank JCR	JCR Impact Factor	Impact Factor (5 Years) JCR
PLoS ONE	3379	Multidisciplinary. 10/145-Q1	1.023	5.2	Multidisciplinary Sciences. 27/71-Q2	2.740	3.227
Journal of Virology	1885	Insect Science. 2/145-Q1 Immunology. 31/225-Q1 Microbiology. 19/158-Q1 Virology. 9/71-Q1	2.406	7.9	Virology. 8/37-Q1	4.501	4.288
Applied and Environmental Microbiology	1628	Food Science. 11/327-Q1 Biotechnology. 33/324-Q1 Ecology. 30/391-Q1 Applied Microbiology and Biotechnology. 8/119-Q1	1.594	7.1	Microbiology. 39/136-Q2 Biotechnology & Applied Microbiology. 37/156-Q1	4.016	4.597
Journal of Agricultural and Food Chemistry	1427	Agricultural and Biological Sciences (miscellaneous). 33/298-Q1 Chemistry (miscellaneous). 61/463-Q1	1.086	6.1	Agriculture, Multidisciplinary. 4/58-Q1 Chemistry, Applied. 15/71-Q1 Food Science & Technology. 21/139-Q1	4.192	4.290
Plant Physiology	758	Plant Science. 13/483-Q1 Genetics. 21/346-Q1 Physiology. 8/186-Q1	3.616	12.5	Plant Sciences. 10/234-Q1	6.902	7.520
Plant Journal	655	Plant Science. 16/483-Q1 Cell Biology. 31/300-Q1 Genetics. 28/346-Q1	3.161	9.8	Plant Sciences 13/234-Q1	6.141	6.629
Food Chemistry	576	Food Science. 10/327-Q1 Analytical Chemistry. 8/126-Q1 Medicine (miscellaneous). 185/2754-Q1	1.775	10.7	Chemistry, Applied. 5/71-Q1 Food Science & Technology. 6/139-Q1 Nutrition & Dietetics. 10/89-Q1	6.306	6.219
Plant Cell	510	Plant Science. 6/483-Q1 Cell Biology. 20/300-Q1	5.399	14.1	Plant Sciences. 6/234-Q1 Biochemistry & Molecular Biology. 23/297-Q1 Cell Biology. 23/195-Q1	9.618	10.144

Table 2. *Cont.*

Journal	N	SJR Category. Rank SJR	SJR Indicator	CiteScore Scopus	JCR Category. Rank JCR	JCR Impact Factor	Impact Factor (5 Years) JCR
Journal of Experimental Botany	343	Plant Science. 19/483-Q1 Physiology. 15/186-Q1	2.647	9.8	Plant Sciences. 14/234-Q1	5.908	7.011
International Journal of Food Microbiology	327	Food Science. 22/327-Q1 Safety, Risk, Reliability and Quality. 13/394-Q1 Microbiology. 37/158-Q1 Medicine (miscellaneous). 298/2754-Q1	1.364	7.4	Microbiology. 35/136-Q2 Food Science & Technology. 23/139-Q1	4.187	4.226
Phytochemistry	323	Horticulture. 9/90-Q1 Plant Science. 106/483-Q1 Biochemistry. 208/456-Q2 Molecular Biology. 255/414-Q3 Medicine (miscellaneous). 821/2754-Q2	0.763	4.9	Plant Sciences. 47/234-Q1 Biochemistry & Molecular Biology. 155/297-Q3	3.044	3.374
Plant Molecular Biology	308	Agronomy and Crop Science. 11/363-Q1 Plant Science. 27/483-Q1 Genetics. 66/346-Q1 Medicine (miscellaneous). 191/2754-Q1	1.730	7.6	Plant Sciences. 42/234-Q1 Biochemistry & Molecular Biology. 138/297-Q2	3.302	4.065
Current Biology	301	Agricultural and Biological Sciences (miscellaneous). 4/298-Q1 Biochemistry, Genetics and Molecular Biology (miscellaneous). 17/271-Q1 Neuroscience (miscellaneous). 0/151-Q1	3.958	13.8	Biology: 3/93-Q1 Biochemistry & Molecular Biology. 24/297-Q1 Cell Biology. 24/195-Q1	9.601	10.174

Table 2. *Cont.*

Journal	N	SJR Category. Rank SJR	SJR Indicator	CiteScore Scopus	JCR Category. Rank JCR	JCR Impact Factor	Impact Factor (5 Years) JCR
Theoretical and Applied Genetics	290	Agronomy and Crop Science. 3/363-Q1 Biotechnology. 23/324-Q1 Genetics. 54/346-Q1 Medicine (miscellaneous). 154/2754-Q1	1.968	7.2	Agronomy. 5/91-Q1 Plant Sciences. 18/234-Q1 Genetics & Heredity. 37/178-Q1 Horticulture. 2/36-Q1	4.439	4.603
Journal of Dairy Science	287	Animal Science and Zoology. 10/429-Q1 Food Science. 17/327-Q1 Genetics. 88/346-Q2	1.440	5.4	Agriculture, Dairy & Animal Science. 5/63-Q1 Food Science & Technology. 37/139-Q1	3.333	3.432
Journal of Food Engineering	276	Food Science. 23/327-Q1	1.338	7.5	Engineering, Chemical. 28/143-Q1 Food Science & Technology. 16/139-Q1	4.499	4.332
European Food Research and Technology	264	Food Science. 88/327-Q2 Biochemistry. 237/456-Q3 Biotechnology. 107/324-Q2 Chemistry (miscellaneous). 123/463-Q2 Industrial and Manufacturing Engineering. 85/484-Q1	0.654	3.8	Food Science & Technology. 58/139-Q2	2.366	2.341
Planta	253	Plant Science. 50/483-Q1 Genetics. 107/346-Q2	1.259	5.4	Plant Sciences. 41/234-Q1	3.390	3.687
PLoS Genetics	223	Ecology, Evolution, Behavior and Systematics. 15/663-Q1 Cancer Research. 17/214-Q1 Genetics. 19/346-Q1 Molecular Biology. 29/414-Q1 Genetics (clinical). 7/99-Q1	3.744	9.0	Genetics & Heredity. 26/178-Q1	7.528	8.555
International Journal of Systematic and Evolutionary Microbiology	204	Ecology, Evolution, Behavior and Systematics. 122/663-Q1 Microbiology. 56/158-Q2 Medicine (miscellaneous). 504/2754-Q1	1.020	4.2	Microbiology. 86/136-Q3	2.415	2.415

Using the above classification, it is possible to establish which field names the stated publications were classified under. Note that the indexing of articles in the scientific categories is done by the indexing category of the journal. This information is provided direct-ly by Scopus; see Figure 5. In this case, Scopus indexes the work into the scientific catego-ries as the journal it is published in is indexed. Three different groups can be clearly seen: the three that are around 20% (Food Science, Plant Science, Agricultural and Biolog-ical Sciences (all)), those that are 5–10% (Agronomy and Crop Science, Insect Science, Ecology, Evolution, Behavior and Systematics, Animal Science and Zoology) and those that are below 3% (Horticulture, Aquatic Science, Soil Science, Forestry, Agricultural and Biological Sciences (miscellaneous)). Therefore, the transfer in patents is mainly led by the field of food science, followed by plant science. The first three categories together account for almost 60% of all these publications (59.3%).

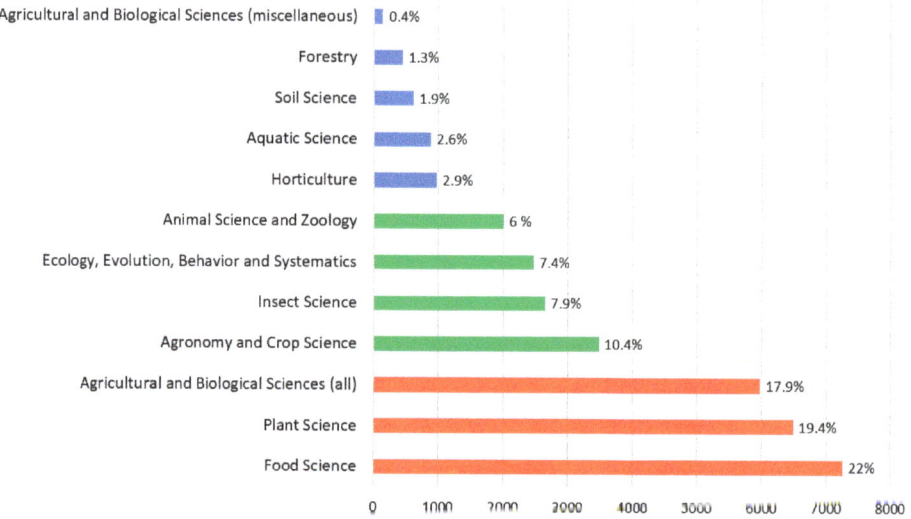

Figure 5. Field names for ASJC on Agricultural and Biological Sciences.

Of all the papers published in journals, 2020 of them were review papers, forming 7.3% of the total. Note that the review works are only 3.4% of the total scientific production of this category. This means that they are very important studies in the patent field, as they reflect the state-of-the-art in a particular field and provide a context for the patent. Finally, it should be mentioned that all these publications have an average of six authors. This should, therefore, be the number considered as the average number of authors for papers in this scientific field.

4.4. The Quality of the Articles

The journal's quality criteria do not measure the quality of individual articles published in that journal. A journal can publish articles of excellent quality that may be overlaid by others of lesser quality, resulting in an overall count that determines the final quality of the journal. The Field-Weighted Citation Impact (FWCI) allows the quality of an article to be measured, so that if its value equals or exceeds the value 1, the article has exceeded the citation expectation for that article.

This section only analyses data from articles cited in patents (N-AGRI-CP). Figure 6 shows how, in four of the Top 20 of journals with the highest transference, more than 90% of the published articles equaled or exceeded the FWCI's benchmark of 1. This means that more than 90% of the articles have had a higher-than-expected citation level for the year of publication, the type of publication and the discipline in which they are categorized.

Plant Cell stands out, with 98.2% of its articles with a value equal to or greater than 1. Five jour-nals have a value equal to or greater than the benchmark for between 80 and 89% of their articles. Seven do so for 70–79% of their articles. Of the top 20, the lowest value is 50.8% of the articles in the European Food Research and Technology journal.

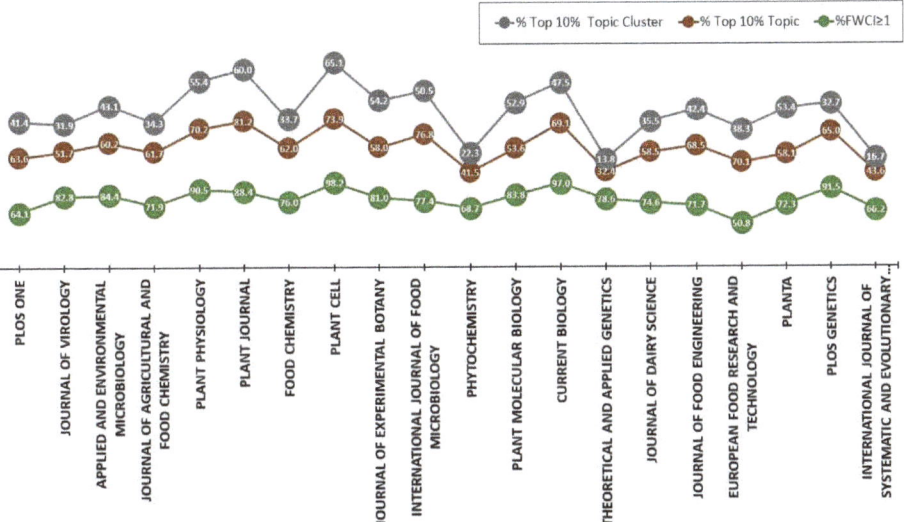

Figure 6. Percentage of articles Field-Weighted Citation Index (FWCI) ≥ 1, Top 10% Topic y Topic Cluster.

Together with the FWCI, Figure 4 also shows the percentage of articles in the Top20 journals that are in the Topic and Topic Cluster's Top 10%. These values are obtained from analysis of the Topic Prominence Percentile and the Topic Cluster Prominence Per-centile, showing the percentage of publications with a percentile equal to or greater than 90% (first decile).

If the analysis is focused on the Top 10% Topic, the highest value is reached by Plant Journal, with 81.2% of its articles placing in the Top 10%, followed by International Journal of Food Microbiology (76.8%) and Plant Cell (73.9%). The lowest value is seen for Theoretical and Applied Genetics, with 32.4%. If the Topic Clusters are considered, in the top 10% for the three highest values is Plant Cell, with 65.1% of its publications, Plant Journal, with 60, and Plant Physiology, with 55.4%. The lowest value is again found in Theoretical and Applied Genetics, with 13.8%.

4.5. The Open Access and European Funding Agencies

In this section an analysis is made of the publications that have been funded by European programs and of those that are open access, always within the field of study.

There are different types of open access, commonly referred to as open access "routes" or "pathways". Gold Open Access allows free access to the final article, as published, and can be used in accordance with the conditions established by the license of use. The second option is Green Open Access, where the final reader will also have access to the final article. The difference between these two types is that through the first option (Gold Open Access) the deposit and therefore access to the article is made through an open access journal with peer review and generally upon a fee for Article Publishing Charge (APC). In the second way (Green Open Access), the author deposits the article, once accepted (postprint) or an unreviewed article (preprint), in a website or digital resource repository, without having to pay an APC, although a period of embargo is usually imposed by the journal in which the full text cannot be accessed, a period of time that can oscillate between 6 and

24 months. In addition to these two routes, there is a third route, Bronze Open Access, in which full text articles are accessible from the editor's website but cannot be reused as authors do not have a license to do so. There is also a fourth type, which we can call hybrid (Hybrid Open Access), which refers to hybrid open access journals, in which there are both subscription and open access arti-cles; in this case, the author pays for open access publication. Finally, there is a fifth way, the diamond route, which generally comprises journals from government institutions or scientific associatio ns, which publish open access without payment by the author.

In the Agricultural and Biological Sciences category, 3288 publications were found, funded by both EU and member country research programs. This is less than 12% of the total. Of these, 548 publications appear to be funded by the EU, through its various research programs discussed above, i.e., 17% of those funded through some form of research program. In summary, EU-funded research accounts for 2% of all published work.

An analysis of the papers in OA shows that, among the 548 papers funded by the EU, 399 are not in OA, i.e., 73%. Of these, for those that are OA, i.e., 149, 23 are in OA Gold, 93 in OA Green, 24 in OA Bronze, and 9 in OA Hybrid.

This section highlights the low impact on the number of scientific publications that the EU's research programs have had in the Agricultural and Biological Sciences category, in relation to being cited in patents, as they have formed 2% of the total number of published papers. Only 27% of the funded papers have been in some form of OA.

4.6. Topics of the Publications Cited in Patents

The topics covered for all these publications can be summarized in two fields: Topic Cluster name, and Topic name. Table 3 shows the first 20 Topic Cluster names and Topic names.

Table 3. Top 20 Topic Cluster names and Topic names.

Topic Cluster Name	N
Arabidopsis, Plants, Genes	2464
Cheeses, Caseins, Milk	905
Metagenome, Probiotics, Bacteria	858
Breads, Starch, Glutens	574
Viruses, Mosaic Viruses, Phytoplasma	423
Tea, Polyphenols, Anthocyanins	388
HIV-1, HIV, HIV Infections	368
Wines, Vitis, Grapes	347
Cellulose, Lignin, Cellulases	344
Salmonella, Escherichia Coli, Listeria Monocytogenes	313
Shoots, Explants, Callus	289
Ethylenes, Apples, Fruit	284
Olea, Oils, Oils and Fats	278
Drying, Moisture Determination, Thermal Processing (Foods)	275
Broiler Chickens, Laying Hens, Swine	269
Spermatozoa, Semen, Oocytes	263
Plants, Rhizosphere, Rhizobium	253
Adenoviridae, Neoplasms, Dependovirus	251
Hepacivirus, Hepatitis B Virus, Hepatitis C	251
Photosystem II Protein Complex, Photosynthesis, Chlorophyll	244

There are many genetic issues in the main topic cluster names. Gene-expression analysis is increasingly important in biological research related to plant breeding. It is therefore not surprising that the most relevant topic cluster name is, Arabidopsis, Plants, Genes. *Arabidopsis thaliana* is a small weed of the cruciferous family that has become one of the most important systems for the study of many aspects of plant biology [78]. Its unique characteristics offer several advantages when considering it as a research model. Firstly, it is a true diploid with a very short life cycle (6–8 weeks), of self-pollination, and produces numerous seeds that remain viable for many years [79]. Its rapid growth allows the analysis of many individuals in a minimum space and therefore, the consequent rapid amplification of the genotypes useful for later studies [80]. Secondly, its compact genome with relatively few repeated sequences and a low DNA content [81], makes it by far the smallest known genome higher plant, and therefore an ideal system for genetic and molecular studies. Thirdly, it can be transformed by Agrobacterium tumefaciens and through the Ti plasmid it is possible to introduce genes of interest and keep them stable [82].

The second relevant topic cluster name related to genetics is Metagenome, Probiotics, Bacteria. Metagenomics is a set of techniques used to determine the microbial population that can be found in each environment, studied in the community context [83].

It is interesting to note the large number of topic cluster names related to food and nutritional properties: Cheeses-Caseins- Milk; Breads-Starch-Glutens; Tea-Polyphenols-Anthocyanins; or Olea- Oils-Oils and Fats. The consumer is increasingly demanding and directly influences the supply and demand for dairy products, demanding higher quality products. They choose between the lipid and protein components of milk and those present in cheese, such as fatty acids, caseins, and whey proteins. The Food Industry usually seeks to increase milk protein, especially casein, which is considered to be the best quality [84]. Likewise, the growing demand for gluten-free products has encouraged the design of many gluten-free bakery products [85]. Regarding to polyphenols, phenolic compounds are mainly considered to be responsible for the main organoleptic features of foods and beverages of plant origin, particulary their color and taste properties. They also contribute to health and are associated with the consumption of diets high in fruit and vegetables or drinks of vegetable origin such as wine or tea [86]. Much research highlights the beneficial health effects of the Mediterranean diet, which is distinguished by the consumption of virgin olive oil as the main source of dietary fat [87], of course this is linked to the olive orchard (*Olea europaea*).

Another of the Topic Cluster names related to food is that of Wines, Vitis, Grapes. It is not surprising that the organoleptic qualities of wine are the subject of major studies given the high economic value of this industry. The final taste of wine is influenced by many factors, but perhaps the most decisive ones are on the one hand the variety of grape used as raw material, and in this regard, there is a market trend towards monovarietal wines, and on the other hand the species of wine yeast used, as each species of wine yeast performs a specific metabolic activity, and therefore determines the final concentrations of flavor compounds in the final wine. Of the studies cited in patents, it is worth highlighting the one related to the quantitative determination of the odorants of fifty-two young red wines from different grape varieties: Garnacha, Tempranillo, Cabernet Sauvignon and Merlot [88]. Another important study is related to the function of yeast species and strains in wine flavor [89].

Finally, another food-related topic cluster name is Drying, Moisture Determination, Thermal Processing (Foods). Of the most cited papers in this field, two are reviews. The first is related to the phenomenon of shrinkage of foodstuffs observed during different dehydration processes [90], and the other to with thermal pasteurization, which is known to be used to reduce microbial populations in foods, but which has the disadvantage of destroying heat-sensitive nutrients and food qualities such as taste, color, and texture [91]. However, research papers in this field highlight studies in food processing and the preservation of ultrasound techniques [92], and those related to the mentioned technique and the interesting compounds of the grape (bioactive substances such as anthocyanins) [93].

The topic names are more specific and, therefore, less numerous in terms of their ap-pearance, but it is interesting to indicate to which Topic Cluster name they belong, as shown in Table 4. It can be verified that among the 20 most important topic names, seven are from the Topic Cluster name of Arabidopsis, Plants, Genes. On the other hand, two are from the second most important Topic Cluster name, "Cheeses, Caseins, Milk" and an-other two from the third "Metagenome, Probiotics, Bacteria".

Table 4. Top 20 Topic names.

Topic Name	N	Topic Cluster Name
Cinnamyl Alcohol Dehydrogenase, Lignification, 4-Coumarate-Coa Ligase	123	Arabidopsis, Plants, Genes
Virgin Olive Oil, Oleuropein, Elenolic Acid	121	Olea, Oils, Oils and Fats
Nicotiana Benthamiana, Taliglucerase Alfa, Molecular Farming	107	Viruses, Mosaic Viruses, Phytoplasma
Hepatitis C Virus, Virus Internalization, RNA Replication	102	Hepacivirus, Hepatitis B Virus, Hepatitis C
Lactobacillus Amylovorus, Bifidobacterium Animalis, Probiotic Agent	95	Metagenome, Probiotics, Bacteria
Endoreduplication, Arabidopsis, Leaf Growth	89	Arabidopsis, Plants, Genes
Immunologic Receptors, Passalora Fulva, Plant Immunity	84	Arabidopsis, Plants, Genes
Anthocyanins, Chalcone Isomerase, Dihydroflavanol 4-Reductase	83	Arabidopsis, Plants, Genes
Rennet, Milk Protein Concentrate, Caseins	79	Cheeses, Caseins, Milk
Glucose-1-Phosphate Adenylyltransferase, Starch Synthase, Endosperm	73	Breads, Starch, Glutens
Glucosinolates, Neoglucobrassicin, Glucoerucin	72	Glucosinolates, NF-E2-Related Factor 2, Brassica
Coffee Beans, Coffea Arabica, Melanoidins	71	Coffee, Caffeine, Energy Drinks
Bacteriocins, Lactobacillales, Biopreservatives	68	Metagenome, Probiotics, Bacteria
Gynoecium, Flowering, Carpels	68	Arabidopsis, Plants, Genes
Strigolactones, Orobanche, Striga Hermonthica	67	Arabidopsis, Plants, Genes
Neutralizing Antibodies, Human Immunodeficiency Virus Vaccine, GP 140	65	HIV-1, HIV, HIV Infections
Adenoviridae, Adenovirus Receptor, Human Adenoviruses	64	Adenoviridae, Neoplasms, Dependovirus
Peptidyl-Dipeptidase A, Protein Hydrolysates, Antihypertensive Effect	64	Cheeses, Caseins, Milk
Pulsed Electric Fields, Pasteurization, Heat Inactivation	64	Drying, Moisture Determination, Thermal Processing (Foods)
Systemic Acquired Resistance, S-Methyl Benzo(1,2,3)Thiadiazole-7-Carbothioate, Salicylic Acids	64	Arabidopsis, Plants, Genes

If an analysis is made by the individual words of the Topic Cluster name and Topic name, Table 5 is obtained. The topic clusters include those related to genetics or molecular biology, such as Genes, Arabidopsis, Metagenome, Genome. Additionally, there are related to specific foods such as Cheeses, Milk, Breads or Oils. The third group can be understood as covering food related constituents such as Caseins, Probiotics, Glutens, or Starch. It is noteworthy that there is a Topic Cluster name of specific animals, i.e., swine. Regarding the Topic names, food issues predominate, especially those related to dairy products such as Probiotic Agent, Lactobacillales, Rennet, Pasteurization, or those related to cereals such as Dough or Glutens.

Table 5. Main words for the top 20 Topic Cluster names and Topic names.

Topic Cluster Name	N	Topic Name	N
Genes	2907	Arabidopsis	500
Plants	2841	Probiotic Agent	267
Arabidopsis	2464	Lactobacillales	204
Neoplasms	1504	Nicotiana Benthamiana	166
Bacteria	1027	Virus Internalization	142
Caseins	905	Rennet	138
Cheeses	905	Dough	135
Milk	905	Hepatitis C Virus	135
Metagenome	858	Carotenoids	132
Probiotics	858	Endosperm	129
Genome	722	Anthocyanins	127
Viruses	672	4-Coumarate-Coa Ligase	125
Glutens	620	Pasteurization	125
Breads	574	Cinnamyl Alcohol Dehydrogenase	123
Starch	574	Lignification	123
Escherichia Coli	562	Elenolic Acid	121
Oils	482	Virgin Olive Oil	121
Swine	481	Agrobacterium	121
Mosaic Viruses	423	Plant Immunity	120
Phytoplasma	423	Glutens	119

5. Conclusions

This paper provides a comprehensive analysis of the current approach to research in the agricultural and biological sciences from the perspective of technology innovation transfer, using patent citation of scientific output as an indicator. This type of approach is encompassed within the Triple Helix concept, where the efforts of academia, industry and governments are brought together.

The great challenge of agriculture, as an economic activity, and of agronomy, as a science, is to provide food for the world's population. The European Union is a geographically densely inhabited area with a long tradition of agricultural research. In the 1999–2019 period, almost one million papers were published by the EU-27 countries in Agri-cultural and Biological Sciences category. Since 2013, these publications stabilized at around 650,000 per year. Only 2.8% of these publications have been cited by patents. That is about 1700 per year, decreasing in the last 10 years; this is the estimated period of the impact of scientific production on patents. These papers have had an average of six au-thors. Review articles have accounted for 7%, when, in this scientific field as a whole, they account for 3.4%.

The systematic benchmarking of results is necessary to help take steps towards improving one's own scientific activity, in order to collect information and to develop a framework for the future. In addition, this allows the concepts on which the evaluation of academic performance or publications is based, i.e., benchmarking based on indicators, to identify best practices for the improvement of the initial situation. Therefore, for further benchmarking purposes, the main results are shown below as an initial framework.

The results validate the relevance of applying bibliometric indicators to a patent. Forty percent of this research was carried out in collaboration with 130 countries outside the EU-27. This certainly shows the great collaboration that exists between the EU-27 coun-tries and the rest of the world. The top five countries in this regard are Germany, France, Spain, Italy, and the Netherlands. The institutions that lead the research cited in patents are the central research institutions of the countries mentioned above: CNRS (France), INRAE (Italy), or CSIC (Spain). This is probably due to the large volume of scientific pro-duction that these institutions have. If attention is paid to the degree of specialization of the institutions, understood as the percentage of articles in the Agricultural and Biological Sciences category in relation to the total number of published works, there are three institutions with more than 30%; these are the Swedish University of Agricultural Sciences (58%), AgroParisTech (52%), Wageningen University & Research (48%), and INRAE (38%).

The journals used for this scientific production are mainly indexed in the SJR Plant Science, and Genetics categories, followed by Food Science. According to the JCR classification, they would also be classified under Plant Science, and Food Science & Technology. A total of 90% of the published articles equaled or exceeded the FWCI's benchmark 1; this means that the articles have had a higher-than-expected citation level for the year of pub-lication, the type of publication and the discipline in which they are categorized. If the analysis is focused on the top 10% Topic, the highest value is reached by Plant Journal, with 81.2% of its articles placed in the top 10%, followed by International Journal of Food Microbiology (76.8%) and Plant Cell (73.9%).

This manuscript highlights the low impact that the EU's research programs have had on the number of scientific publications in the Agricultural and Biological Sciences category, in relation to being cited in patents, as they have formed 2% of the total number of published papers. Only 27% of the funded papers were in some form of OA.

The top three Topic Cluster names were: "Arabidopsis, Plants, Genes", "Cheeses, Caseins, Milk", and "Metagenome, Probiotics, Bacteria". The top three Topic names were: "Cinnamyl Alcohol Dehydrogenase, Lignification, 4-Coumarate-Coa Ligase", "Virgin Olive Oil, Oleuropein, Elenolic Acid", and "Nicotiana Benthamiana, Taliglucerase Alfa, Molecular Farming".

In summary, the research topics most reflected in patents are those related to genetics (Arabidopsis, Metagenome, Genome), to major food issues (Cheeses, Milk, Breads or Oils and to food and beverage products that are of great concern at present (Caseins, Probiotics, Glutens, or Starch).

The use of patents for decision-making is not yet a widespread tool on all innovative research fronts; this work can be a benchmark for future policy decisions regarding the directions research institutions should take in their future development. The results provide evidence of the potential of the methodology developed and the metrics obtained to represent the patent transfer contributions of national science systems as an indicator of technological innovation.

From this point of view, the current strategic research plan of both the EU-27 and its member countries' systems should seek to enhance the development of the science base for an industry based on the transfer to industry. Transfer to patents has proven to be long-term, and university rankings and demands on researchers are short-term. Trying to link the two issues would improve the search for innovations for industry itself, which, in the end, would translate into an improvement in the quality of life of citizens.

Author Contributions: M.C. and A.A. conceived and wrote the article; A.A. and E.S.-M. analyzed the data; M.C., A.A. and F.M.-A. wrote the paper. A.A. and F.M.-A. supervised the research. E.S.-M. and F.M.-A. revised the manuscript. They share the structure and aims of the manuscript, paper drafting, editing and review. All authors have read and agreed to the published version of the manuscript.

Funding: This research received no external funding.

Institutional Review Board Statement: Not applicable.

Informed Consent Statement: Not applicable.

Data Availability Statement: Data retrieved from Scopus, SciVal, JCR, and SJR databases.

Acknowledgments: The authors would like to thank to the CIAIMBITAL (University of Almería, CeiA3) for its support.

Conflicts of Interest: The authors declare no conflict of interest.

References

1. Doré, T.; Makowski, D.; Malézieux, E.; Munier-Jolain, N.; Tchamitchian, M.; Tittonell, P. Facing up to the paradigm of ecological intensification in agronomy: Revisiting methods, concepts and knowledge. *Eur. J. Agron.* **2011**, *34*, 197–210. [CrossRef]
2. Ericksen, P. Conceptualizing food systems for global environmental change research. *Glob. Environ. Chang.* **2008**, *18*, 234–245. [CrossRef]
3. Gattullo, C.E.; Mezzapesa, G.N.; Stellacci, A.M.; Ferrara, G.; Occhiogrosso, G.; Petrelli, G.; Castellini, M.; Spagnuolo, M. Cover Crop for a Sustainable Viticulture: Effects on Soil Properties and Table Grape Production. *Agronomy* **2020**, *10*, 1334. [CrossRef]
4. Tilman, D.; Cassman, K.G.; Matson, P.A.; Naylor, R.; Polasky, S. Agricultural sustainability and intensive production practices. *Nature* **2002**, *418*, 671–677. [CrossRef]
5. Ahmad, N.; Mukhtar, Z. Genetic manipulations in crops: Challenges and opportunities. *Genomics* **2017**, *109*, 494–505. [CrossRef] [PubMed]
6. Poudel, S.; Poudel, B.; Acharya, B.; Poudel, P. Pesticide use and its impacts on human health and environment. *Environ. Ecosyst. Sci.* **2020**, *4*, 47–51. [CrossRef]
7. Laraus, J. The problems of sustainable water use in the Mediterranean and research requirements for agriculture. *Ann. Appl. Biol.* **2004**, *144*, 259–272. [CrossRef]
8. Baranski, M.R. Wide adaptation of Green Revolution wheat: International roots and the Indian context of a new plant breeding ideal, 1960–1970. *Stud. Hist. Philos. Sci. Part C* **2015**, *50*, 41–50. [CrossRef]
9. Roesch-McNally, G.E.; Basche, A.D.; Arbuckle, J.; Tyndall, J.C.; Miguez, F.E.; Bowman, T.; Clay, R. The trouble with cover crops: Farmers' experiences with overcoming barriers to adoption. *Renew. Agric. Food Syst.* **2018**, *33*, 322–333. [CrossRef]
10. Zapata-Sierra, A.J.; Manzano-Agugliaro, F. Controlled deficit irrigation for orange trees in Mediterranean countries. *J. Clean. Prod.* **2017**, *162*, 130–140. [CrossRef]
11. La Malfa, G.; Leonardi, C. Crop practices and techniques: Trends and needs. *Acta Hortic.* **2001**, *1*, 31–42. [CrossRef]
12. Pertot, I.; Caffi, T.; Rossi, V.; Mugnai, L.; Hoffmann, C.; Grando, M.S.; Gary, C.; Lafond, D.; Duso, C.; Thiery, D.; et al. A critical review of plant protection tools for reducing pesticide use on grapevine and new perspectives for the implementation of IPM in viticulture. *Crop. Prot.* **2017**, *97*, 70–84. [CrossRef]
13. Berk, P.; Hocevar, M.; Stajnko, D.; Belsak, A. Development of alternative plant protection product application techniques in orchards, based on measurement sensing systems: A review. *Comput. Electron. Agric.* **2016**, *124*, 273–288. [CrossRef]
14. Sanders, K. Orange Harvesting Systems Review. *Biosyst. Eng.* **2005**, *90*, 115–125. [CrossRef]

15. Méndez, V.; Pérez-Romero, A.; Sola-Guirado, R.R.; Miranda-Fuentes, A.; Manzano-Agugliaro, F.; Zapata-Sierra, A.; Rodríguez-Lizana, A. In-Field Estimation of Orange Number and Size by 3D Laser Scanning. *Agronomy* **2019**, *9*, 885. [CrossRef]
16. Novas, N.; Alvarez-Bermejo, J.; Valenzuela, J.L.; Gázquez, J.; Manzano-Agugliaro, F. Development of a smartphone application for assessment of chilling injuries in zucchini. *Biosyst. Eng.* **2019**, *181*, 114–127. [CrossRef]
17. El Khaled, D.; Novas, N.; Gazquez, J.A.; Garcia, R.M.; Manzano-Agugliaro, F. Fruit and Vegetable Quality Assessment via Dielectric Sensing. *Sensors* **2015**, *15*, 15363–15397. [CrossRef]
18. Manzano-Agugliaro, F.; García-Cruz, A.; Fernández-Sánchez, J.S. Women's labour and mechanization in mediter-ranean greenhouse farming. *Outlook Agric.* **2013**, *42*, 249–254.
19. Welch, R.M.; Graham, R.D. A new paradigm for world agriculture: Meeting human needs: Productive, sustainable, nutritious. *Field Crop. Res.* **1999**, *60*, 1–10. [CrossRef]
20. Ruiz-Real, J.L.; Uribe-Toril, J.; Arriaza, J.A.T.; Valenciano, J.D.P. A Look at the Past, Present and Future Research Trends of Artificial Intelligence in Agriculture. *Agronomy* **2020**, *10*, 1839. [CrossRef]
21. Olesen, J.; Trnka, M.; Kersebaum, K.; Skjelvåg, A.; Seguin, B.; Peltonensainio, P.; Rossi, F.; Kozyra, J.; Micale, F. Impacts and adaptation of European crop production systems to climate change. *Eur. J. Agron.* **2011**, *34*, 96–112. [CrossRef]
22. Pretty, J. Intensification for redesigned and sustainable agricultural systems. *Science* **2018**, *362*, eaav0294. [CrossRef] [PubMed]
23. Van Etten, J.; Beza, E.; Calderer, L.; Van Duijvendijk, K.; Fadda, C.; Fantahun, B.; Kidane, Y.G.; Van De Gevel, J.; Gupta, A.; Mengistu, D.K.; et al. First experiences with a novel farmer citizen science approach: Crowdsourcing participatory variety selection through on-farm triadic comparisons of technologies (tricot). *Exp. Agric.* **2016**, *55*, 275–296. [CrossRef]
24. Li, X.; Gagliardi, D.; Miles, I. Variety in the innovation process of UK research and development service firms. *RD Manag.* **2019**, *50*, 173–187. [CrossRef]
25. Van Norman, G.A.; Eisenkot, R. Technology transfer: From the research bench to commercialization: Part 1: Intel-lectual property rights—Basics of patents and copyrights. *JACC Basic Transl. Sci.* **2017**, *2*, 85–97. [CrossRef]
26. Blind, K. The influence of regulations on innovation: A quantitative assessment for OECD countries. *Res. Policy* **2012**, *41*, 391–400. [CrossRef]
27. Di Cataldo, V. From the European patent to a community patent. *Colum. J. Eur. Law* **2002**, *8*, 19.
28. Schäfers, A. The Luxembourg Patent Convention, The Best Option for the Internal Market. *JCMS J. Common. Mark. Stud.* **1987**, *25*, 193–207. [CrossRef]
29. Deng, Y. The effects of patent regime changes: A case study of the European patent office. *Int. J. Ind. Organ.* **2007**, *25*, 121–138. [CrossRef]
30. Heikkilä, J.T.; Verba, M.A. The role of utility models in patent filing strategies: Evidence from European countries. *Science* **2018**, *116*, 689–719. [CrossRef]
31. Brack, H.P. Utility Models and Their Comparison with Patents and Implications for the US Intellectual Property Law System. Boston College Intellectual Property and Technology Forum. 2009, pp. 1–15. Available online: http://bciptf.org/wp-content/uploads/2011/07/13-iptf-Brack.pdf (accessed on 29 January 2021).
32. Moser, P. How do patent laws influence innovation? Evidence from nineteenth-century world's fairs. *Am. Econ. Rev.* **2005**, *95*, 1214–1236. [CrossRef]
33. Guimón, J. Promoting university-industry collaboration in developing countries. *World Bank* **2013**, *3*, 12–48.
34. Galvez-Behar, G. The 1883 Paris Convention and the Impossible Unification of Industrial Property. In *Patent Cultures*; Cambridge University Press (CUP): Cambridge, UK, 2021; pp. 38–68.
35. Blakeney, M. Patenting of plant varieties and plant breeding methods. *J. Exp. Bot.* **2012**, *63*, 1069–1074. [CrossRef]
36. Enghardt, F.; Hekker, F. The Trade Mark Reform Package and the classification of goods and services. *J. Intellect. Prop. Law Pract.* **2016**, *11*, 822–825. [CrossRef]
37. Kiewiet, B. Plant variety protection in the European Community. *World Pat. Inf.* **2005**, *27*, 319–327. [CrossRef]
38. Pisoschi, A.M.; Pisoschi, C.G. Is open access the solution to increase the impact of scientific journals? *Science* **2016**, *109*, 1075–1095. [CrossRef]
39. Leydesdorff, L.; Meyer, M. The decline of university patenting and the end of the Bayh–Dole effect. *Science* **2010**, *83*, 355–362. [CrossRef]
40. Watanabe, C.; Tsuji, Y.S.; Griffy-Brown, C. Patent statistics: Deciphering a 'real'versus a 'pseudo'proxy of innovation. *Technovation* **2001**, *21*, 783–790. [CrossRef]
41. Zhang, T.; Chen, J.; Jia, X. Identification of the Key Fields and Their Key Technical Points of Oncology by Patent Analysis. *PLoS ONE* **2015**, *10*, e0143573. [CrossRef]
42. Kuan, C.-H.; Huang, M.-H.; Chen, D.-Z. Ranking patent assignee performance by h-index and shape descriptors. *J. Inf.* **2011**, *5*, 303–312. [CrossRef]
43. Luan, C.; Zhou, C.; Liu, A. Patent strategy in Chinese universities: A comparative perspective. *Science* **2010**, *84*, 53–63. [CrossRef]
44. Kang, K.; Sohn, S.Y. Evaluating the patenting activities of pharmaceutical research organizations based on new technology indices. *J. Inf.* **2016**, *10*, 74–81. [CrossRef]
45. Bornmann, L.; Mutz, R.; Daniel, H.-D. Are there better indices for evaluation purposes than the h index? A comparison of nine different variants of the h index using data from biomedicine. *J. Am. Soc. Inf. Sci. Technol.* **2008**, *59*, 830–837. [CrossRef]

46. Diaz Perez, M.; Rivero Amador, S.; de Moya-Anegon, F. Latin American technological production of greatest in-ternational visibility. 1996–2007. A case study: Brasil. *Rev. Española Doc. Científica* **2010**, *33*, 34–62.
47. Makhoba, X.; Pouris, A. A patentometric assessment of selected R&D priority areas in South Africa, a comparison with other BRICS countries. *World Pat. Inf.* **2019**, *56*, 20–28. [CrossRef]
48. Díaz-Pérez, M.; De-Moya-Anegón, F. El análisis de patentes como estrategia para la toma de decisiones innovadoras, El Profesional de la Información. *Prof. Inf.* **2008**, *17*, 293–302.
49. Noruzi, A.; Abdekhoda, M. Google Patents: The global patent search engine. *Webology* **2014**, *11*, 1–12.
50. Moskovkin, V.M.; Shigorina, N.A.; Popov, D. The possibility of using the Google Patents search tool in patentometric analysis (based on the example of the world's largest innovative companies). *Sci. Tech. Inf. Process.* **2012**, *39*, 107–112. [CrossRef]
51. Huang, M.H.; Dong, H.R.; Chen, D.Z. Globalization of collaborative creativity through cross-border patent activities. *J. Informetr.* **2012**, *6*, 226–236. [CrossRef]
52. Huang, M.-H.; Chen, S.-H.; Lin, C.-Y.; Chen, D.-Z. Exploring temporal relationships between scientific and technical fronts: A case of biotechnology field. *Science* **2013**, *98*, 1085–1100. [CrossRef]
53. Motta, G.D.S.; Quintella, R.H. Assessment of Non-Financial Criteria in the Selection of Investment Projects for Seed Capital Funding: The Contribution of Scientometrics and Patentometrics. *J. Technol. Manag. Innov.* **2012**, *7*, 172–197. [CrossRef]
54. Park, H.W.; Hong, H.D.; Leydesdorff, L. A comparison of the knowledge-based innovation systems in the economies of South Korea and the Netherlands using Triple Helix indicators. *Science* **2005**, *65*, 3–27. [CrossRef]
55. Etzkowitz, H.; Leydesdorff, L. The dynamics of innovation: From National Systems and "Mode 2" to a Triple Helix of university–industry–government relations. *Res. Policy* **2000**, *29*, 109–123. [CrossRef]
56. Meyer, M.S.; Siniläinen, T.; Utecht, J.T. Towards hybrid Triple Helix indicators: A study of university-related patents and a survey of academic inventors. *Science* **2003**, *58*, 321–350. [CrossRef]
57. Baldini, N. University patenting and licensing activity: A review of the literature. *Res. Eval.* **2006**, *15*, 197–207. [CrossRef]
58. Kim, H.; Huang, M.; Jin, F.; Bodoff, D.; Moon, J.; Choe, Y.C. Triple helix in the agricultural sector of Northeast Asian countries: A comparative study between Korea and China. *Science* **2011**, *90*, 101–120. [CrossRef]
59. Chihib, M.; Salmerón-Manzano, E.; Novas, N.; Manzano-Agugliaro, F. Bibliometric Maps of BIM and BIM in Uni-versities: A Comparative Analysis. *Sustainability* **2019**, *11*, 4398. [CrossRef]
60. Garrido-Cardenas, J.A.; Esteban-García, B.; Agüera, A.; Pérez, J.A.S.; Manzano-Agugliaro, F. Wastewater Treatment by Advanced Oxidation Process and Their Worldwide Research Trends. *Int. J. Environ. Res. Public Health* **2019**, *17*, 170. [CrossRef]
61. Garrido-Cardenas, J.A.; González-Cerón, L.; Manzano-Agugliaro, F.; Mesa-Valle, C. Plasmodium genomics: An approach for learning about and ending human malaria. *Parasitol. Res.* **2019**, *118*, 1–27. [CrossRef]
62. Garrido-Cardenas, J.A.; Mesa-Valle, C.; Manzano-Agugliaro, F. Human parasitology worldwide research. *Parasitology* **2017**, *145*, 699–712. [CrossRef]
63. Binfield, P. Novel Scholarly Journal Concepts. In *Opening Science. The Evolving Guide on How the Internet Is Changing Research, Collaboration and Scholarly Publishing*; Bartling, S., Friesike, S., Eds.; Springer: Cham, Switzerland, 2014; pp. 155–163. ISBN 978 3 319 00026 8.
64. Colledge, L. Output and outcome metrics. In *Snowball Metrics Recipe Book*; Colledge, L., Ed.; Elsevier: Amsterdam, The Netherlands, 2017; p. 87. Available online: www.snowballmetrics.com/ (accessed on 21 January 2021).
65. Li, B.; Ch'Ng, E.; Chong, A.Y.-L.; Bao, H. Predicting online e-marketplace sales performances: A big data approach. *Comput. Ind. Eng.* **2016**, *101*, 565–571. [CrossRef]
66. Li, R.-T.; Khor, K.A.; Yu, L. Identifying Indicators of Progress in Thermal Spray Research Using Bibliometrics Analysis. *J. Therm. Spray Technol.* **2016**, *25*, 1526–1533. [CrossRef]
67. Yu, M.-C.; Wu, Y.J.; Alhalabi, W.; Kao, H.-Y.; Wu, W.-H. ResearchGate: An effective altmetric indicator for active researchers? *Comput. Hum. Behav.* **2016**, *55*, 1001–1006. [CrossRef]
68. Ekpo, E.; Hogg, P.; McEntee, M. A Review of Individual and Institutional Publication Productivity in Medical Radiation Science. *J. Med. Imaging Radiat. Sci.* **2016**, *47*, 13–20. [CrossRef]
69. Avanesova, A.A.; Shamliyan, T.A. Comparative trends in research performance of the Russian universities. *Science* **2018**, *116*, 2019–2052. [CrossRef]
70. Liu, X.; Sun, R.; Wang, S.; Wu, Y.J. The research landscape of big data: A bibliometric analysis. *Libr. Hi Tech.* **2019**, *38*, 367–384. [CrossRef]
71. Colledge, L. SciVal. Usage and Patent Metrics. Guidebook. 2019. Available online: https://p.widencdn.net/1ldn6j/ACAD_SV_EB_SciValUsageandPatentGuide_WEB (accessed on 21 January 2021).
72. Zijlstra, H.; McCullough, R. *CiteScore: A New Metric to Help You Track Journal Performance and Make Decisions*; Elsevier: Amsterdam, The Netherlands, 2016. Available online: https://www.elsevier.com/editors-update/story/journal-metrics/citescore-a-new-metric-to-help-you-choose-the-right-journal (accessed on 21 January 2021).
73. Garfield, E. Citation Indexes for Science: A New Dimension in Documentation through Association of Ideas. *Science* **1955**, *122*, 108–111. [CrossRef]
74. Larivière, V.; Sugimoto, C.R. The Journal Impact Factor: A Brief History, Critique, and Discussion of Adverse Effects. In *Springer Handbook of Science and Technology Indicators*; Moed, H., Schmoch, U., Thelwall, E., Glänzel, W., Eds.; Springer: Berlin/Heidelberg, Germany, 2019.

75. González-Pereira, B.; Guerrero-Bote, V.; Moya-Anegón, F. A new approach to the metric of journals' scientific prestige: The SJR indicator. *J. Inf.* **2010**, *4*, 379–391. [CrossRef]
76. Bergstrom, C.T.; West, J. Comparing Impact Factor and Scopus CiteScore. Eigenfactor.org. 2016. Available online: http://eigenfactor.org/projects/posts/citescore.php (accessed on 21 January 2021).
77. Elsevier. Research Metrics GuidebookResearch Intelligence. 2019. Available online: https://p.widencdn.net/5pyfuk/ACAD_RL_EB_ElsevierResearchMetricsBook_WEB (accessed on 21 January 2021).
78. Meyerowitz, E.M. Arabidopsis thaliana. *Annu. Rev. Genet.* **1987**, *21*, 93–111.
79. Czechowski, T.; Stitt, M.; Altmann, T.; Udvardi, M.K.; Scheible, W. Genome-Wide Identification and Testing of Superior Reference Genes for Transcript Normalization in Arabidopsis. *Plant Physiol.* **2005**, *139*, 5–17. [CrossRef]
80. Schmid, M.; Davison, T.S.; Henz, S.R.; Pape, U.J.; Demar, M.; Vingron, M.; Schölkopf, B.; Weigel, D.; Lohmann, J.U. A gene expression map of Arabidopsis thaliana development. *Nat. Genet.* **2005**, *37*, 501–506. [CrossRef] [PubMed]
81. Leutwiler, L.S.; Hough-Evans, B.R.; Meyerowitz, E.M. The DNA of Arabidopsis thaliana. *Mol. Genet. Genom.* **1984**, *194*, 15–23. [CrossRef]
82. Lloyd, A.M.; Barnason, A.R.; Rogers, S.G.; Byrne, M.C.; Fraley, R.T.; Horsch, R.B. Transformation of Arabidopsis thaliana with Agrobacterium tumefaciens. *Science* **1986**, *234*, 464–466. [CrossRef] [PubMed]
83. Garrido-Cardenas, J.A.; Manzano-Agugliaro, F. The metagenomics worldwide research. *Curr. Genet.* **2017**, *63*, 819–829. [CrossRef] [PubMed]
84. Boukria, O.; El Hadrami, E.M.; Boudalia, S.; Safarov, J.; Leriche, F.; Aït-Kaddour, A. The Effect of Mixing Milk of Different Species on Chemical, Physicochemical, and Sensory Features of Cheeses: A Review. *Foods* **2020**, *9*, 1309. [CrossRef]
85. Horstmann, S.W.; Lynch, K.M.; Arendt, E.K. Starch Characteristics Linked to Gluten-Free Products. *Foods* **2017**, *6*, 29. [CrossRef]
86. Cheynier, V. Polyphenols in foods are more complex than often thought. *Am. J. Clin. Nutr.* **2005**, *81*, 223S–229S. [CrossRef]
87. Ghanbari, R.; Anwar, F.; Alkharfy, K.M.; Gilani, A.; Li, M. Valuable Nutrients and Functional Bioactives in Different Parts of Olive (*Olea europaea* L.)—A Review. *Int. J. Mol. Sci.* **2012**, *13*, 3291–3340. [CrossRef]
88. Ferreira, V.; Lopez, R.; Cacho, J.F. Quantitative determination of the odorants of young red wines from different grape varieties. *J. Sci. Food Agric.* **2000**, *80*, 1659–1667.
89. Romano, P.; Fiore, C.; Paraggio, M.; Caruso, M.; Capece, A. Function of yeast species and strains in wine flavour. *Int. J. Food Microbiol.* **2003**, *86*, 169–180.
90. Mayor, L.; Sereno, A. Modelling shrinkage during convective drying of food materials: A review. *J. Food Eng.* **2004**, *61*, 373–386. [CrossRef]
91. Garcia-Gonzalez, L.; Geeraerd, A.; Spilimbergo, S.; Elst, K.; Van Ginneken, L.; Debevere, J.; Van Impe, J.; Devlieghere, F. High pressure carbon dioxide inactivation of microorganisms in foods: The past, the present and the future. *Int. J. Food Microbiol.* **2007**, *117*, 1–28. [CrossRef] [PubMed]
92. Knorr, D.; Zenker, M.; Heinz, V.; Lee, D.-U. Applications and potential of ultrasonics in food processing. *Trends Food Sci. Technol.* **2004**, *15*, 261–266. [CrossRef]
93. Corrales, M.; Toepfl, S.; Butz, P.; Knorr, D.; Tauscher, B. Extraction of anthocyanins from grape by-products assisted by ultrasonics, high hydrostatic pressure or pulsed electric fields: A comparison. *Innov. Food Sci. Emerg. Technol.* **2008**, *9*, 85–91. [CrossRef]

Article

Identification and Analysis of Strawberries' Consumer Opinions on Twitter for Marketing Purposes

Juan D. Borrero [1,*] and Alberto Zabalo [2]

[1] Department of Business and Marketing, Agricultural Economic Research Group Huelva University, 21071 Huelva, Spain
[2] Department of Agroforestry Sciences, Huelva University, 21071 Huelva, Spain; alberto.zabalo@dcaf.uhu.es
* Correspondence: jdiego@uhu.es; Tel.: +34-607-514-200

Abstract: Data are currently characterized as the world's most valuable resource and agriculture is responding to this global trend. The challenge in that particular field of study is to create a Digital Agriculture that help the agri-food sector grow in a fair, competitive environment. As automated machine learning techniques and big data are global research trends in agronomy, this paper aims at comparing different marketing techniques based on Content Analysis to determine the feasibility of using Twitter to design marketing strategies and to determine which techniques are more effective, in particular, for the strawberry industry. A total of 2249 hashtags were subjected to Content Analysis using the Word-count technique, Grounded Theory Method (GTM), and Network Analysis (NA). Findings confirm the results of previous studies regarding Twitter's potential as a useful source of information due to its lower execution and analysis costs. In general, NA is more effective, cheaper, and faster for Content Analysis than that based both on GTM and automated Word-count. This paper reveals the potential of strawberry-related Twitter data for conducting berry consumer studies, useful in increasing the competitiveness of the berry sector and filling an important gap in the literature by providing guidance on the challenge of data science in agronomy.

Keywords: twitter; content analysis; Network Analysis (NA); Grounded Theory Method (GTM); berry growers

Citation: Borrero, J.D.; Zabalo, A. Identification and Analysis of Strawberries' Consumer Opinions on Twitter for Marketing Purposes. *Agronomy* **2021**, *11*, 809. https://doi.org/10.3390/agronomy11040809

Academic Editors: Francisco Manzano-Agugliaro and Esther Salmerón-Manzano

Received: 11 March 2021
Accepted: 16 April 2021
Published: 20 April 2021

Publisher's Note: MDPI stays neutral with regard to jurisdictional claims in published maps and institutional affiliations.

Copyright: © 2021 by the authors. Licensee MDPI, Basel, Switzerland. This article is an open access article distributed under the terms and conditions of the Creative Commons Attribution (CC BY) license (https://creativecommons.org/licenses/by/4.0/).

1. Introduction

Data are currently characterized as the world's most valuable resource, or the oil of the digital era [1]. Agriculture is responding to the changing environment. It is trying to create digitization strategies that will enable and catalyze a Digital Agriculture and that help the agri-food sector grow in a fair competitive environment. Many companies recognize the need to incorporate a social network strategy as part of their overall marketing efforts [2–4]. In fact, 90% of marketing specialists consider Social Networks Sites (SNSs) to be important for their marketing strategy [5], because of their becoming an important channel for communications with consumers due to the large volume of users and the possibility of collecting data directly from them.

Social media marketing is an inexpensive alternative to traditional methods of involving consumers [3,6–8]. This is especially significant for small- and medium-sized enterprises since their resources, including marketing budgets, are usually smaller than those of their larger counterparts [9]. Nevertheless, the agribusiness literature still lags in this field [3,10].

The current work focuses on Twitter [11] because it is one of the most popular SNSs on which messages, called tweets, circulate openly, becoming important for both individuals and organizations to broadcast and discuss opinions in real-time [12].

The public characteristic of Twitter allowed the obtaining of data that, duly processed, contributed to the analysis with a solid quantitative foundation in which topics of interest could be identified.

Some authors have analyzed tweets in areas such as consumer food preferences or habits [13–15] or communication of Corporate Social Responsibility (CSR) of agri-food companies [16], but on agri-food research it has not been explored as appropriate [14] and, to date, we did not find any research that investigates the usefulness of SNSs media marketing to increase in the competitiveness of the agri-producer sector.

In the context of SNSs, we can understand that the food-related communication of consumers through Twitter can very well reflect their interests and, therefore, serve as a basis for defining the production and marketing strategies of the agri-producer sector.

Besides, for farmers—the lowest power in the agri-food value chain [17]—knowing the interests of consumers is vital in their production strategy.

The European Union (EU) considers digitization of agriculture a key strategy to bring a number of benefits to farmers, such as increased profitability and access to new markets [18] and it is also an excellent lever to accelerate the transition towards a climate-neutral, circular, and more resilient economy [19,20].

One of the most dynamic sectors in the agri-food market is that of fresh products since, in addition to being perishable, they are beneficial for human health.

As it was established in many researches, berries gain more consumers' attention [21–23] as they look out for healthier dietary options [24–27].

Spain and USA are the main global fresh berry exporters (and producers) followed by Mexico, Chile, and Peru, with USA, Canada, and UK being the largest importers (and consumers) [28,29]. Among all those fresh berries, strawberries are the most consumed by volume [30], and they led the global organic berries market share in 2019 [31].

In addition, as main berry producing and consuming countries are English and Spanish speakers, we analyze the Twitter behavior of these two profiles.

To date, there is still a lack of research' aspects into the social media marketing that would help berry growers to successfully develop their businesses [32], our paper sheds light on three important issues that are critical for berry firms preparing to start using social marketing strategies. First, our study complements prior research suggesting evidence of the value relevance of engaging in a social media strategy. As long as berry firms are not able to evaluate the consequences of social media strategies on their value, they cannot effectively align such initiatives with their organizational goals. Second, as a communication platform, Twitter may be used to foster relational bonds with berry customers, thus leading to long-term relationships and reliable repeat business, which is consistent with the basic principles of relationship marketing. Finally, while considerable academic research has been carried out to explore social networks, few empirical studies have examined or attempted to compare in term of costs the effectiveness of different marketing research techniques when a SNS is used as a source of information. Thus, the final purpose of this research is to assess the usefulness of consumer-generated content (CGC) from Twitter for berry firms.

Still, research suggests that understanding users' motives can provide useful insights into how berry consumer behavior. In this context, we develop a theoretical framework for theory building by using and comparing three different techniques used for Content Analysis: Word-count, Grounded Theory Method (GTM), and Network Analysis (NA).

Hence, with the overall aim of using different techniques to discover the main topics included in tweets that included the hashtags #strawberries or #fresas (strawberries in Spanish), and assess the potential of Twitter data for consumer marketing research, we focus on the following research questions: RQ1: "Does the content of hashtags associated with the search criteria reflect the interests of the strawberry consumer?"; RQ2: "Do non-explicit relationships among consumers' hashtags reflect different and more-or-less relevant topics of interest for berry-industry?"; and RQ3: "Which hashtag Content Analysis technique has been more effective?"

The evidence in relation to these assumptions will shed light on the potential of Twitter data to elucidate berry fruit consumer behavior and, in consequence, to assess the utility for fresh food industry.

2. Literature Review

While most daily decisions are determined by emotional and spontaneous processes [33], current practices in consumer marketing based on direct questions require consumers to reflect before answering [34], which could lead to biases in the answers that compromise the validity of the data obtained [35,36].

Additionally, although consumption studies conducted in supermarkets have been strongly recommended and remain the most common practice [37,38], they have the great inconvenience of their cost in terms of time and money. Resorting to social media analysis methods to study consumer behavior is increasingly frequent since, in addition to increasing efficiency and being cheaper, they are closer to consumer thinking than more traditional techniques such as surveys [39].

Among SNSs, Twitter has become one of the most popular microblogging services and is attracting the interest of marketing and consumer science researchers [40,41] because provides access to instinctive consumer information obtained in real-life situations. Thus, its primary purpose is to allow people to share their immediate thoughts, but it also has the potential to be an important data source regarded consumer behavior related to agri-food products.

Word-count analysis has dominated [40,42] the research with Twitter data related to agricultural food. Manual content analysis is still one of the core methods used in food-related Twitter research [15]. However, in our digitized media environment, Automated Content Analysis (ACA) has gained importance and popularity [43]. Recently, quantitative techniques for extracting intelligence from food-related tweets as sentiment analysis [44,45] using Partition Around Medoids (PAM) and clustering algorithms [46]; or text analysis using Machine Learning (ML) such as Support Vector Machine (SVM) and hierarchical clustering [47], or n-gram [14] are being used.

ACA offers a wide range of text capturing, ML and Natural Language Processing (NLP) techniques for mining intelligence from SNSs [48] that can then be utilized for analysis of keywords, summarization of text or clustering by employing the above techniques [49].

Content analysis approaches can summarize large volumes of text into closely grouped themes [50]. They have also been used in conjunction with NA [51], enabling studies to be conducted whereby visual graphs based on co-occurrence of keywords can be developed [52]. Using such parameters, it may be possible to develop theories surrounding network level attributes [53].

However, it was found that food-related SNS research remains fragmented and, although in the domain of agribusiness is in a preliminary stage, it has lots of potential in terms of theoretical, mathematical, and empirical research. Although rarely used to date with berry fruit-related Twitter data, Word-count, GTM [47,54]—complemented with automated content analysis- and NA [55], will be analyzed in this field of research.

3. Material and Methods

In this study, social tags (hashtags in Twitter) are analyzed, which serve as mechanisms for the semantic unification of concepts within a social network [56,57], as Twitter is based on short messages—less than 280 characters.

We developed a five-stage of theory building based on the following big data-driven research by [58] and text mining model by [59]:

1. Data acquisition: Automatic data acquisition from social media;
2. Data processing: Transformation and cleaning with text meaning;
3. Data understanding: Factor identification with Word-count (term frequency analysis) technique;
4. Theory development: To analyze keywords using GTM to identify association rules among them and major emerging themes;

5. Data Insights: Automated content analysis through NA (community detection and modularity analysis) and visualization techniques to generate deep insights from the textual data.

3.1. Automatic Data Acquisition

To identify which terms to use as a seed for extraction, a previous analysis of the entries on the internet was made. Although the first intention was to use the term 'berries' it was decided to use the words 'strawberries' and 'fresas' as search criteria since they were the terms that contained the most entries in Google. Thus, the criterion for obtaining tweets was particularized for those tweets containing the hashtags #strawberries or #fresas (strawberries in Spanish).

Data were recovered using R software [60] through the Twitter package [61] that provides communication with Twitter Application Programming Interface (API), searches for and collect tweets with specific keywords, and thereby collects the data and stores in a database. This package, which has been previously used to explore consumer perception of different products [62,63], mines data from the information contained in the social network associated with the hashtag.

The extraction did not specify the language of the tweets or the specific geographical location from which they were published, so any tweet (available after authentication on the server) containing the keyword could be retrieved from its public location.

As the number of tweets recovered in previous research has varied considerably, from a few to millions [64,65], a request was made in 30 June 2019 to extract a total of 9999 tweets (excluding both replies and retweets). The automatic data extraction was made twice, with #strawberries and with #fresas, based on all tweets who contained the searched hashtag within them.

3.2. Data Processing: Text Cleaning, Tokenization, and Data Loading

Not all primary datasets may be useful unless the collection is conducted appropriately [66,67]. For example, if tweets are extracted based on hashtags as #strawberries, how does one identify the rest of the keywords objectively which may help to get concepts or categories of concepts?

The processing phase consists of filtering and manipulating tweets to clean and remove a large part of data which do not meaningfully contribute to the research question as retweets [58], or removing terms that do not contain content such as stop words, numbers, and punctuation marks, or converting them to lowercase to eliminate ambiguity.

Analyzing Twitter large data without properly handling social bots has serious implications. For the purpose of this study, we used volume and frequency as criteria to categorize them [68].

Tokenization refers to divide the tweet content into minimal units with their own meaning, that is, words, which for this analysis will be hashtags. Prior to the division of the text, the elements under study were tweets and each was in a row, thus fulfilling the condition of an observation as a record. When performing the tokenization, the element to study becomes each token (hashtag), but several hashtags can be found in the same tweet. To resolve this point, each token list must be unnested, doubling the number of records as many times as hashtags come in the same tweet.

The entire cleaning and tokenization process was automated by designing a function that was implemented in the R script.

This information, after cleaning and filtering elements, was saved in a Comma-separated Values (.csv) format.

3.3. Data Understanding with Word-Count

Text cleaning and processing are necessary in order to extract intelligence from the unstructured texts (tweets) extracted.

In this study, an automatic text processing method based in tokenization was performed to know hashtag frequencies. The procedure of tokenization splits sentences into words and extracts hashtags that form the basic units of our analysis.

Individual word analysis is often applied to Twitter data analysis, but it is potentially problematic because it ignores the broader context of the tweet. However, such an analysis has the potential to quickly summarize large volumes of data. In this case, because they are words written intentionally to reinforce their meaning, a count of hashtags contained in tweets was made.

3.4. Themes with Grounded Theory Method (GTM)

To improve the transparency of the research, the grounded-theory approach was performed following the [69] methodology.

Following an interpretive, inductive approach, hashtags automatically retrieved were constantly analyzed and manually compared in an iterative process in which hashtags (initial codes)—words with a low level of abstraction—were classified into similar groups by assigning conceptual labels (focused codes) to the units of meaning according to the coding procedure until the discovery of emerging categories and themes with high level abstraction [69–72].

In our case, we highlight a top-down approach, where two coders, who speak English and Spanish fluently and with more than two years of experience in consumer behavior research, working together in the process of constant comparison of data until reaching a consensus on the establishment of the final topics. They followed a manual process to identify emerging themes through reading each hashtag in the context of the others and their frequency and manually categorizing it through applying a conceptual framework [71].

Constant comparative analysis entails that coders need to make comparisons between empirical data and conceptual labels, between conceptual labels and themes, among data, among conceptual labels, and among different 'slices of data' in order to reach higher levels of abstraction and advance with the conceptualization [73]. During the analysis, coders had to be sensitive to data analysis that guided them to what to do next.

Data reduction was achieved by limiting the analysis to those aspects that were relevant with a view to the research questions [74].

Credibility (truth value), transferability (applicability), and dependability (consistency) [75] were used to evaluate the trustworthiness of the grounded theory. In order to increase the credibility of the findings and diminishing the encoder bias, we employed triangulation through interviews with producers and consumers. A total of 20 interviews were conducted with consumers chosen from among university staff (five administrative staff, five professors, five students, and five Erasmus students), and five berry managers from the area were interviewed. To facilitate transferability, authors provided to the coders background data to establish the context of the study to allow comparisons to be made [76]. Finally, consistency was achieved via automatic data extraction.

Therefore, through an inductive and iterative process, using constant comparison of hashtags, emergent categories were discovered [70,72], the topics in the data [77].

3.5. Insights with Network Analysis (NA)

In recent years, social media research has begun to overcome the quantitative perspective to explore other aspects through network theory.

This study utilizes the overall network structures in order to identify key hashtags and similarities among strawberry-related hashtags. The two network-structural attributes were established as indicators of information flow characteristics: centralization, or contribution of a node according to its location in the network; and modularity, or the division of a network's force between clusters.

Measurements of centrality have been widely used to capture patterns of information flow in a network [78,79]. Using degree centrality, nodes with more connections are considered more important. Twitter networks are 2-mode, meaning that each link has a direction, from user to hashtag. To analyze hashtags relations, the 2-mode networks were transformed into 1-mode.

The process of identifying the underlying structure of the data in terms of grouping the most similar elements is called clustering. Elements included in the same cluster should be similar, and elements included in different clusters should be dissimilar. The concept of similarity or dissimilarity will depend on some kind of metric. One of the most well-known algorithms for community detection was proposed by [80]. This method for measuring the modularity was modified trying to reduce the computational demands significantly through several new approaches [81,82].

The algorithm selected in this work to choose the most appropriate method for better identifying communities talking on a particular topic (strawberries in this case) was the Louvain modularity [83]. This is a bottom-up algorithm, similar to the earlier method by [81] where initially every vertex belongs to a separate community, and vertices are moved between communities iteratively in a way that maximizes the vertices local contribution to the overall modularity. The algorithm stops when it is not possible to increase this modularity.

This methodological level begins with the .csv file import into the open-source network analysis and visualization software called Gephi [84], written in Java on the NetBeans platform. The 2-mode Twitter network with directed ties indicating links between users and hashtags was transformed into 1-mode network with undirected ties among hashtags.

The Fruchterman–Reingold algorithm, followed by the Force Atlas 2 algorithm were chosen because they allow for the attraction of the most central nodes and separation of those least central. The first has the function of arranging the nodes from the attraction–repulsion relationship of the gravitational force created by the algorithm itself, and the second serves to disperse the groups, creating space for the larger nodes.

Finally, the nodes in the graphs were colored according to the communities to which they belonged calculated according to the Louvain modularity [83], and the size of the node was represented in the graph proportionally to the number of links (degree).

4. Results

We made two extractions with a total of 9999 tweets in 30 June 2019. The script execution first collected the last 9999 tweets that included the hashtag #strawberries and then the other 9999 tweets that included #fresas. In the first case, 2184 tweets contained hashtags other than strawberries and in the second, 1596 tweets contained hashtags other than fresas were obtained. The time period covered by the extraction was about seven months, from 2 December 2018 for #strawberries, and 12 December 2018 for #fresas. Most of the strawberries world production from the main producers is concentrated in this period.

In total, 11,150 hashtags were collected and treated, of which 2249 hashtags were valid for the Content Analysis.

After a filtering, tokenization, and debugging process, a total of 579 unique hashtags for the term 'strawberries' and 163 for 'fresas' were obtained.

Figure 1 and Table 1 show the extraction procedure as well as some basic statistics about the number of tokens analyzed.

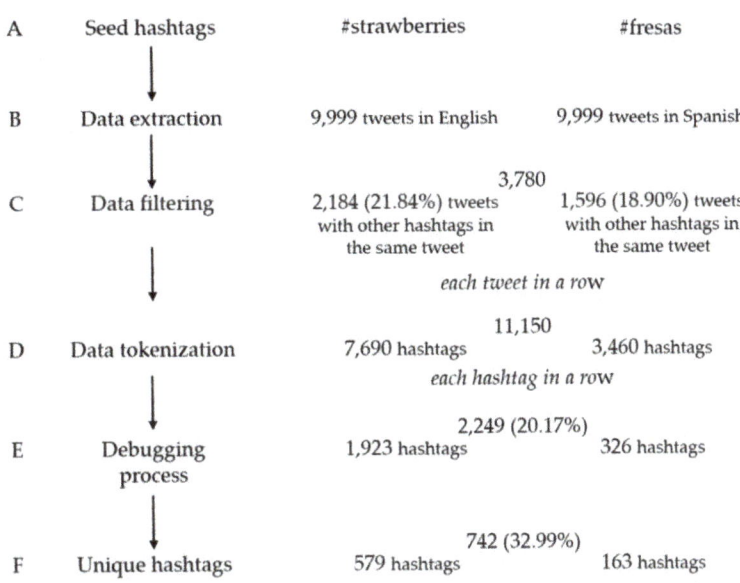

Figure 1. Data collection procedure.

Table 1. Descriptive statistics.

	#Strawberries	%	#Fresas	%
Number of tweets with a single token	259	11.88%	274	17.19%
Number of tweets with two tokens	539	24.69%	324	20.31%
Number of tweets with three tokens	205	9.38%	324	20.31%
Number of tweets with four tokens	198	9.06%	224	14.06%
Number of tweets with five tokens	225	10.31%	150	9.38%
Number of tweets with six tokens	239	10.94%	150	9.38%
Number of tweets with seven tokens	123	5.63%	25	1.56%
Number of tweets with eight tokens	198	9.06%	25	1.56%
Number of tweets with nine tokens	96	4.38%	50	3.13%
Number of tweets with ten or more tokens	102	4.69%	50	3.13%
Total of tweets	2184		1596	
Number of tokens	7690		3460	
Mean	3.52		2.17	
Standard deviation	3.09		2.54	

To summarize this large amount of heterogeneous data, only the 50 most frequent hashtags were considered (Figures 2 and 3).

Hashtag frequencies using the word-count technique allowed to discover emerging categories. In that regard, 'Anuga' (the Cologne food trade fair) was the word most used in fresas-related tweets, confirming that there are users who use the term in a professional context. There were also cases in which productive practices were described, such as 'urbanfarming' and other terms associated with farms ('seeds', 'cucumbers', 'peppers', 'tomatoes', 'spinach', 'germinate', 'substrate'). The most common words in the tweets related to the moment of the occasion were 'dessert' and 'breakfast.'

'Chocolate' was also among the most frequent words, indicating the association with 'sweets', although there were also others ('vitamins', green juices, healthy life, and healthy lives) more related to 'health'. The words 'love' and 'lovers', which also appeared, indicate their association with impulse-buying situations. Without reading tweets in their entirety, the high frequency of appearance of the words 'fideua' or 'paella' seemed absurd. However, these words reflect how Twitter is often used to list daily activities.

Several of the most used words were related to the specific context of the feeding situations. For example, words like 'Sunday' were found with a relatively high frequency, which indicates that people talked about the time of the meal in their tweets. We also found words that refer to special occasions or places (for example, 'Acapulco', the city in Mexico), as well as words related to other people involved in the occasion of eating ('family' or 'bestfriends'). Words related to specific agri-food products, such as other fruit (especially 'blueberries' and 'apples'), were also identified.

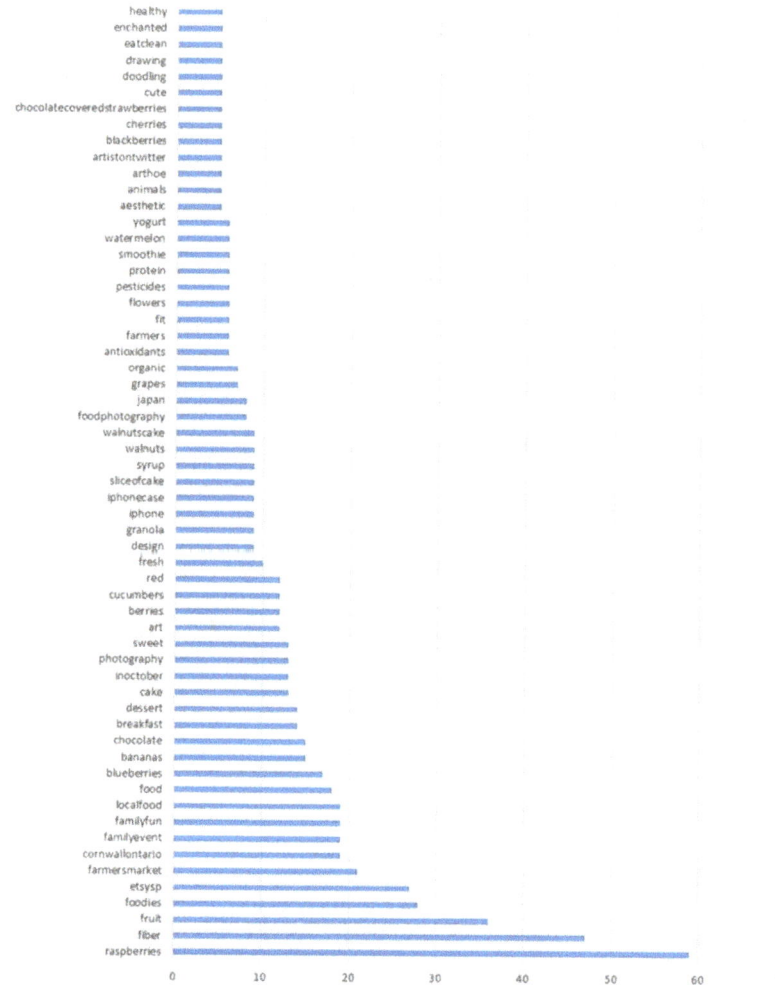

Figure 2. Top 50 English hashtags frequency. Extracted from Tweets that include the hashtag #strawberries.

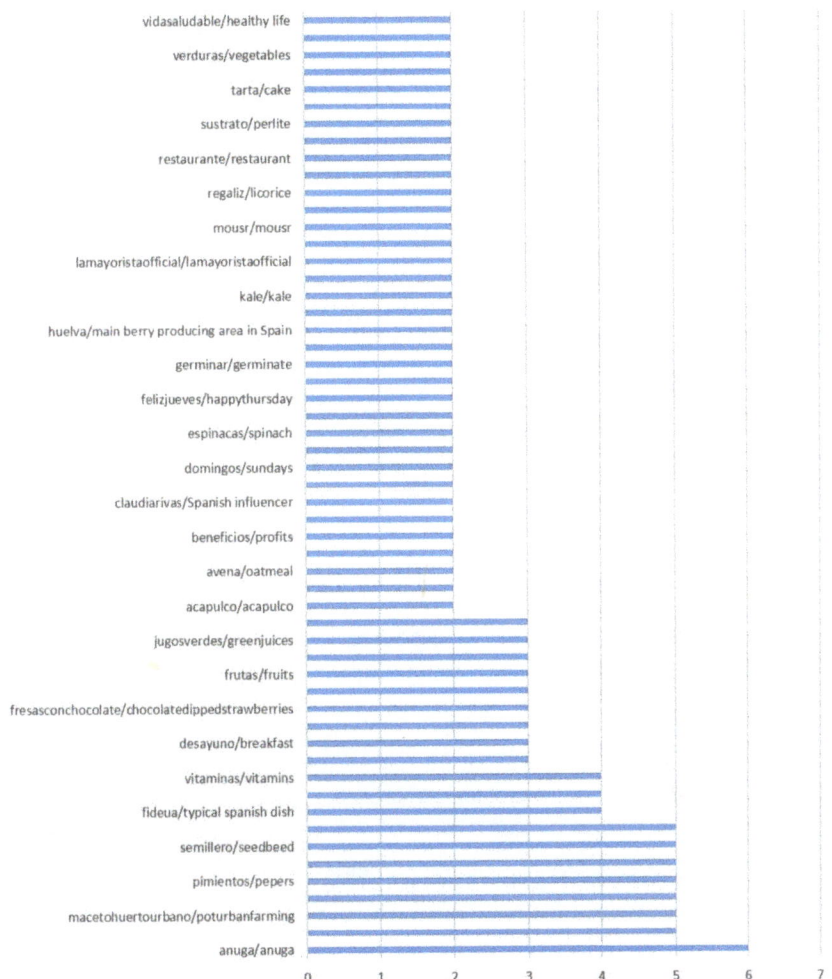

Figure 3. Top 50 Spanish hashtags frequency. Extracted from Tweets that include the hashtag #fresas (strawberries).

On the other hand, the hashtags frequency analysis of 'strawberries' highlighted the association with other fruit (but this time with 'raspberries' in addition to 'blueberries' and 'bananas', and to a lesser extent with 'blackberries', 'grapes', and 'watermelon'), health (but this time as 'fiber' and with many more terms like 'organic', 'antioxidants', 'fit', 'eatclean', and 'healthy'), sweets ('chocolate', 'cake', 'sweet'), and acquaintances or family ('familyevent' and 'familyfun').

Finally, there are mentions of times and modes of consumption ('breakfast' and 'dessert'). A total of three new associations appear, one referring to local consumption ('localfood' and 'farmersmarket'), another to art ('photography', 'art', 'design'), and the third to consumption mode ('smoothie' and 'yogurt'). The reference to love appears to a lesser extent ('flowers') and urban production does not appear.

In total, 13 focused codes (categories) were built (as numbers). In the Spanish data analysis, categories 10–12 do not exist, and in the analysis of strawberries, the category seven was not built (Table 2).

Table 2. Inductive construction of categories and emerging themes (above #strawberries; below #fresas).

Themes	Focused Codes	Initial Codes for #Strawberries
1. Fruits	1	raspberries, fruits, blueberries, banana, berries, grapes, watermelon, blackberries
	2	breakfast, dessert, food
2. Context	3	familyevent, familyfun
	6	farmersmarket
	8	flowers
3. Consumption	4	chocolate, cake, syrup, sliceofcake, granola, walnutcake, chocolatecoveredstrawberries
		sweet
		smoothie, yogurt
4. Healthy lifestyle	5	fiber, organic, antioxidants, eatclean, healthy, fresh
		fit
	9	localfood
5. Production	11	farmers, pesticides
	12	incotober
6. Art	10	photography, design, foodphotography, red, cute

Themes	Focused Codes	Initial Codes for #fresas (strawberries)
1. Fruits	1	frutas (fruits), frutosrojos (berries), manzana (apple), arándanos (blueberries)
2. Context	2	desayuno (breakfast), postres (dessert)
	3	domingos (Sundays), felizdomingo (happysunday), felizjueves (happythursday), lafamilia (thefamily), mejoresamigos (bestfriends)
	6	fideua (typical spanish dish), paella (typical spanish dish), bar(bar), restaurante (restaurant)
	8	amor (love), rosas(roses)
		acapulco (beautiful beach in Mexico)
	13	Anuga (the leading food fair in the world)
3. Consumption	4	Chocolate (chocolate), fresasconchocolate (chocolatedippedstrawberries), fresascubiertasdechocolate (chocolatecoveredstrawberries), avena (oatmeal), perversodechocolate (perversechocolate)
		Dulces (sweets), tarta (cake), tusdulces (yoursweets)
4. Healthy lifestyle	5	Jugosverdes (greenjuices)
		Vitaminas (vitamins)
		Vidamassana (healthierlife), vidasaludable (healthylife)
5. Production	7	Macetahuertourbano (urban farm), semillero (seedbed), pepinos (cucumbers), pimientos (peppers), tomates (tomatoes), germinar (germinate), sustrato (substrate), verduras (vegetables)
	9	Huelva (localcity)
6. Art		

The manual coding process based on GTM identified five common themes and a sixth just for the strawberry dataset. The manual coding process based on GTM identified five common themes, and a sixth just for the strawberry dataset: substitute fruits (1), context of consumption (2), food consumption (3), lifestyle (4), associations with production (5), and associations with art (6). (Table 2).

The first theme is the comparison the strawberries with other products, mainly fruits.

The context of the activities related to agri-food products was a frequent topic in tweets analyzed. People described where, when, and with whom they were talking about strawberries. The most frequently mentioned places of consumption were restaurants (in the Spanish case) and farmersmarket (in English). In general, the 'home' location was mentioned less frequently than elsewhere, suggesting that people do not tend to make explicit reference to their homes when they tweet about feeding situations. Instead, they seem to refer to places considered different or special (e.g., Acapulco). Some feeding situations were motivated by a special occasion, such as special days (happy Sunday) or events (familyevent). The hashtags also contained information about a specific situation in time (breakfast, dessert).

Epicurean attitudes toward food such as 'chocolate perverso' were found in tweets. Additionally included were references to emotions (love, cute, happy Sunday, happy holidays).

References about healthy and unhealthy strawberries' aspects, such as vitamins, healthy life, fiber, organic, antioxidants, eat clean, and healthy are also regularly presented in tweets.

Finally, Figure 4 shows the network obtained with the application of these spatialization criteria, observing how the position of the hashtags in the 'fresas' network is much more defined than in 'strawberries' network.

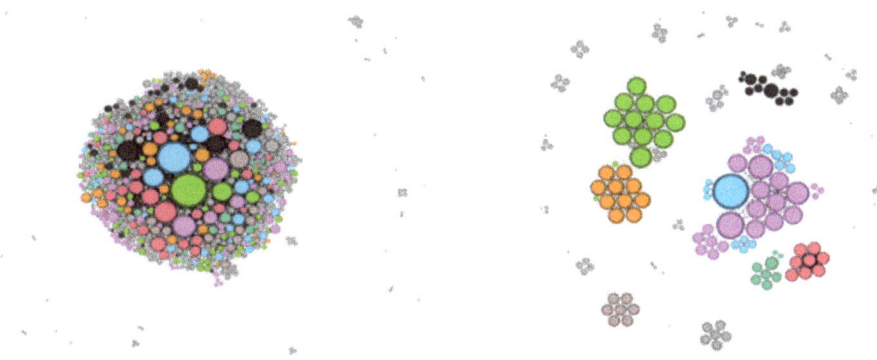

Figure 4. Spatial network distribution after application of the algorithms (**left**, 'strawberries'; **right**, 'fresas'). The colors represent the cluster and the separation among hashtags is based on their relative positions.

The final procedure was the detection of communities by calculating the modularity [83] (Figures 5 and 6). Overall, five communities appear on the #strawberries network and seven on the #fresas network.

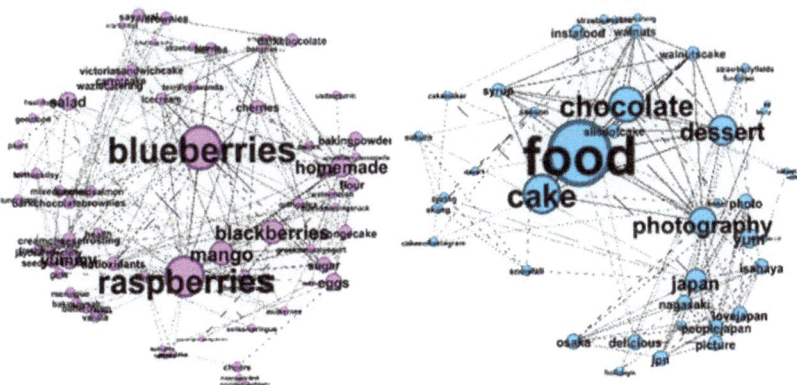

Figure 5. *Cont.*

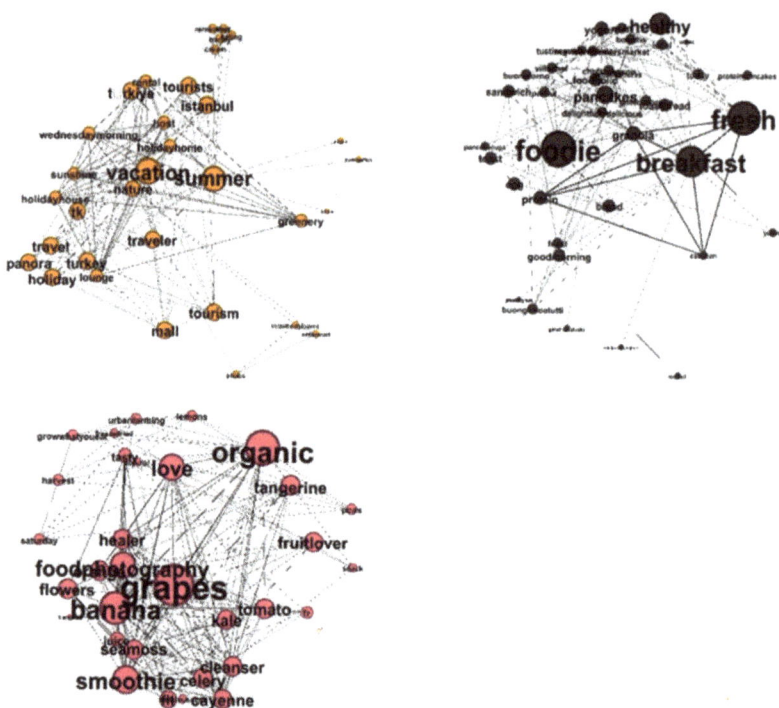

Figure 5. Communities in the English keyword network. Separate analysis of the communities with visualization of the words that compose it. The size of the node represents the degree or importance within the group.

Figure 6. *Cont.*

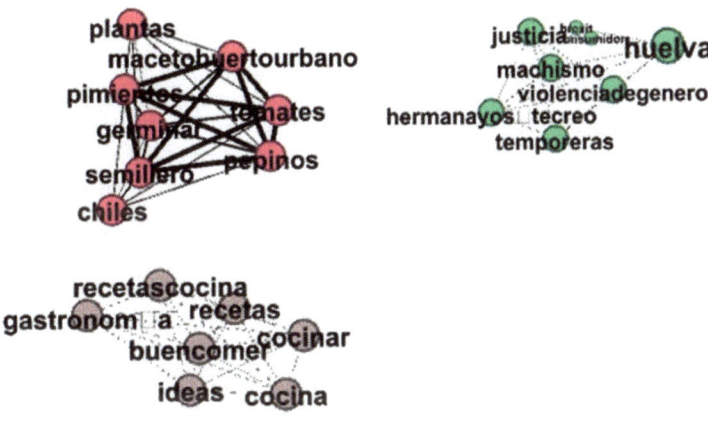

Figure 6. Communities in the Spanish keyword network. Separate analysis of the communities with visualization of the words that compose it. The size of the node represents the degree or importance within the group.

5. Discussion

The research discovered that people in their tweets described associations with topics that interest them. This confirmed our aim that these tweet content analysis techniques can serve as a low-cost marketing tools.

Sometimes, people tweeted about a craving for certain agri-food products, in line with other research [85,86].

The context of eating strawberries was one of the main issues that emerged from the content analysis. This accords with the perception that context is recognized as a key variable in the choice of agri-food product [87–89].

The data contained in our study shown that people tweeted mainly on strawberry consumption situations in a positive emotional state. According to [90], people use positive words when describing or remembering eating experiences due to a positive disposition toward food. Our results also reflect that meals are positively remembered when they involve family and friends [91,92].

Twitter was also used when making plans to eat, either with family or friends, which could be related to a social activity [93]. Related with this aspect, some strawberry consumption situations were motivated by a special occasion, such as special days (happy Sunday) or events (familyevent) [89,94].

Tweets that describe feeding situations in restaurants, or 'farmersmarket' could be related to the growing fruit consumption outside home, even on the street [95]. This suggests that Twitter could offer researchers the opportunity to recover impulsively generated and real-life data. In this regard, it should be pointed out that consumers increasingly use smartphones to access social networks, which allows data to be collected in any situation. Worldwide, 52% of total web traffic originates from mobile phones, representing 74% of the traffic on social networks [96].

In light of previous research on agri-food choices, patterns coincided with expectations [85,86,97] as it was common for tweets to include content about the specific fruits that people were eating, buying, or preparing.

Epicurean attitudes toward strawberries [98] were found in tweets. Associations included positive words such as 'chocolate perverso'. There were also references to emotions (love, cute, happy Sunday, happy holidays). This is in accordance with the fact that eating has been widely reported to be associated with emotional factors [99,100].

Tweets also give importance to healthy food as authors point out [101–103]. Thus, the ecofriendly labels and standards can play a significant role in influencing consumer purchase decisions [104]. However, results seem to indicate that users are not interested in communicate about sustainability in contrary to [42].

Manual content analysis using GTM revealed some differences between users who talk about 'fresas' and those who talk about 'strawberries'. While the former give more relevance to the situation of eating out (restaurant, bar) and production (urban, farming, or professional fair), the latter pay more attention to the healthy lifestyle, the product's connections with art, and food security (pesticides).

However, automated content analysis through NA, in addition to reflecting the differences between the two user profiles, also discovers new issues that did not emerge with frequency analysis or GTM methodology. Regarding the differences between user groups, it was observed that 'fresas' users relate chocolate consumption (sweets) with special days (diadelosenamorados) while those using 'strawberries' associate this consumption with art (photography). The 'fresas' user also comments on eating out (restaurant, tapas, bar), while the community analysis detects a new context environment in the 'strawberries' group most related to holidays (vacations, summer, tourism). Finally, both groups of users talk about health, but the coding by GTM classifies certain fruits (apples) within the theme 'fruits' and the communities include them within the 'health' theme (apples, spinach).

Last but not least, in addition to the relationships between topics, NA detected new communities that did not emerge with the GTM analysis or Word-count technique: in relation to the hashtag 'strawberries', a group related to leisure (vacations, summer, tourism), and for 'fresas', one community related to recipes (recipes, cooking) and another related to exploitation of working people (Huelva, seasonal workers, justice, sexism, gender violence).

Finally, to use Twitter data for marketing research, we must be aware of its limitations and how to address them: related to the information analyzed from Twitter, hashtags are minimal units, but while they do not report on everything that a tweet can express (emotions, context, feelings), they are key words in the tweet that have been intentionally written by the user. Another limitation is that Twitter users are a non-representative sample of the general population [105] and, although men and women are equally represented, the distribution is largely skewed toward younger and better educated people. However, such systematic bias of the sample may decrease over time as more people become active users.

Due to the fact that the Spanish hashtag frequencies are much lower than the English ones (see Table 1 and Figure 3), the use of this analyses for the intended marketing purposes in this study is more limited in the Spanish case than in the English one. Nevertheless, a bigger selection of tweets in the automatic data extraction would resolve this problem.

Taking into account the aforementioned Twitter limitations, there are several directions for future research.

It would be interesting to expand the automatic data extraction to others hashtags (i.e., blueberries and raspberries) or to see the distribution or frequency of the number of hashtags in the tweets in addition to strawberries (strawberries) to find differences in this regard when considering these different sub-networks.

Moreover, the network analysis could be expanded taking into account other centrality metrics (i.e., intermediation) or clustering (i.e., PAM).

This study could be expanded by investigating consumer perceptions that result in the identification of the activities most desired by consumers. In addition, future research should try to examine the effectiveness of the different SNSs in various outcome measures, such as commitment, purchase intention, and brand affinity. It would also be interested to compare GTM and NA with other ML-based content analysis or apply them to other terms.

6. Conclusions

By accepting that social networks have transformed and will continue to transform the way in which companies and consumers communicate, berry industry must use social media data as part of their overall marketing efforts, as data are numerous, valid, and cheap to obtain.

It is true that tweets represent user opinions. We have tried to argue (although not to demonstrate) that hashtags contained in tweets are units with meaning that, analyzed as a whole, can lead us to obtain a representation of user' interests, although it is also true that they do not represent the entire Strawberry consumer population. However, we have obtained several interesting conclusions.

The results of this study provide a very necessary first step to providing such guidance considering Twitter as a useful source of information for berry-consumer marketing research. Firstly, Automated Content Analysis (ACA) demonstrates which hashtags represent the main user interests.

Secondly, NA found out non-explicit relationships among consumers' hashtags that reflect relevant topics for marketers of berry industry.

Thirdly, our study contributes to explore a global research trends that is the agri-food data from social networks. Using #strawberries and #fresas as search criteria, it was found that simple analysis based on word counting yields less information than the results of the other techniques used, highlighting two obstacles:

(i) inclusion of non-relevant hashtags; and
(ii) no identification of underlying issues or relationships. Even though content analysis using GTM provided much deeper information, it took much more time. NA, in addition to being faster, proved to be more efficient and allowed the discovery of new underlying themes and relationships.

In addition to the comparison of the three content analysis techniques, the analysis of the two separate datasets provided wealthy information to understand the differences between these markets.

Thus, in word-count analysis, it appears that English-speaking users have an orientation towards organic food rather than to vitamins of Spanish-speaking users. The first profile seems to be concerned about pesticides, while the second more about production. Regarding the differences in modes of consumption, the first group seems to describe more types of consumption such as yogurt and smoothies, while the Spanish-speaking market tweets more of consuming in more impulsive situations. Beauty and art are connotations that appear only in the first group. GTM reinforces these ideas, especially (and important) that of pesticides vs. farming, organic vs. vitamins, and art. Finally, NA brought to light more defined groups in the Spanish network and new topics, which, with the previous techniques, could not be intuited, such as social justice relationships in production (sexism, racism, immigration) and recipes in the Spanish dataset and holidays (vacation, summer) in the English dataset.

Although some relationships are incoherent with the consumption of strawberries, making it necessary to refine or to analyze in depth, connections have been found that a conventional market study can establish, such as places, situations, or feelings in which consumption is favored, but to a lower cost.

In this type of analysis, it is possible to determine consumer groups and to know their motivations and desires, which allows offering the product in formats with more acceptance or with elaborations (i.e., strawberries and chocolate), as well as segmenting the marketing campaigns.

On the other hand, it might be a challenge to evaluate whether the presence of a specific hashtag in a tweet is random, i.e., to reveal whether there is a pattern reflecting a trend in the consumer 'opinion.

In conclusion, ACA, and specifically NA, can imply opportunities so that, in a simple and cheap way, the berry producing sector can better understand consumer behavior.

To maximize its benefits, this agri-food sector could strategically build a technological knowledge base of social media analytics, and strategically manage and support its use by facilitating IT-marketing and IT-organization alignments [106].

Despite the stated limitations, results confirm the potential of strawberry-related hashtags from Twitter for conducting berry consumer studies, useful in increasing the competitiveness of the berry sector and filling an important gap in the literature by providing guidance on the challenge of data science in agronomy.

Author Contributions: Conceptualization, methodology, software, validation and formal analysis and writing—original draft preparation, J.D.B.; investigation, resources, data curation, writing—review and editing, J.D.B. and A.Z.; funding acquisition, A.Z. All authors have read and agreed to the published version of the manuscript.

Funding: This research was funded by Department of Agroforestry Science-University of Huelva.

Conflicts of Interest: The authors declare no conflict of interest.

References

1. The Economist. The World's Most Valuable Resource is No Longer Oil, but Data. Available online: https://www.economist.com/leaders/2017/05/06/the-worlds-most-valuable-resource-is-no-longer-oil-but-data (accessed on 20 February 2021).
2. Appel, G.; Grewal, L.; Hadi, R.; Stephen, A.T. The future of social media in marketing. *J. Acad. Mark. Sci.* **2020**, *48*, 79–95. [CrossRef]
3. Galati, A.; Crescimanno, M.; Tinervia, S. Social media as a strategic marketing tool in the Sicilian wine industry: Evidence from Facebook. *Wine Econ. Policy* **2017**, *6*, 40–47. [CrossRef]
4. Li, F.; Larimo, J.; Leonidou, L.C. Social media marketing strategy: Definition, conceptualization, taxonomy, validation, and future agenda. *J. Acad. Mark. Sci.* **2021**, *49*, 51–70. [CrossRef]
5. Stelzner, M. 2016 Social Media Marketing Industry Report: How Marketers Are Using Social Media to Grow Their Business. Available online: http://www.socialmediaexaminer.com (accessed on 20 September 2020).
6. Kaplan, A.M.; Haenlein, M. Users of the world, unite! The challenges and opportunities of social media. *Bus. Horiz.* **2010**, *53*, 59–68. [CrossRef]
7. Nobre, H.; Silva, D. Social network marketing strategy and SME strategy benefits. *J. Transnatl. Manag.* **2014**, *19*, 138–151. [CrossRef]
8. Stebner, S.; Baker, L.M.; Peterson, H.H.; Boyer, C.R. Marketing with more: An in-depth look at relationship marketing with new media in the green industry. *J. Appl. Commun.* **2017**, *101*. [CrossRef]
9. Atanassova, I.; Clark, L. Social media practices in SME marketing activities: A theoretical framework and research agenda. *J. Cust. Behav.* **2015**, *14*, 163–183. [CrossRef]
10. Yao, B.; Shanoyan, A.; Peterson, H.H.; Boyer, C.; Baker, L. The use of new-media marketing in the green industry: Analysis of social media use and impact on sales. *Agribusiness* **2019**, *35*, 281–297. [CrossRef]
11. Twitter. About. Available online: https://about.twitter.com/company (accessed on 20 September 2020).
12. Zanini, M.; Lima, V.; Migueles, C.; Reis, I.; Carbone, D.; Lourenco, C. Soccer and twitter: Virtual brand community engagement practices. *Mark. Intell. Plan.* **2019**, *37*, 791–805. [CrossRef]
13. Guèvremont, A. Improving consumers' eating habits: What if a brand could make a difference? *J. Consum. Mark.* **2019**, *36*, 885–900. [CrossRef]
14. Moreno-Sandoval, L.G.; Sánchez-Barriga, C.; Buitrago, K.E.; Pomares-Quimbaya, A.; Garcia, J.C. Spanish Twitter Data Used as a Source of Information About Consumer Food Choice. In *Machine Learning and Knowledge Extraction, Lecture Notes in Computer Science, Proceedings of the CD-MAKE 2018, Hamburg, Germany, 27–30 August 2018*; Holzinger, A., Kieseberg, P., Tjoa, A., Weippl, E., Eds.; Springer: Cham, Switzerland, 2018; Volume 11015, p. 11015.
15. Vidal, L.; Ares, G.; Machín, L.; Jaege, S.R. Using Twitter data for food-related consumer research: A case study on "what people say when tweeting about different eating situations". *Food Qual. Prefer.* **2015**, *45*, 58–69. [CrossRef]
16. Araujo, T.; Kollat, J. Communicating effectively about CSR on Twitter. *Internet Res.* **2018**, *28*, 419–431. [CrossRef]
17. Cucagna, M.E.; Goldsmith, P.D. Value adding in the agri-food value chain. *Int. Food Agribus. Manag. Rev.* **2018**, *21*, 293–316. [CrossRef]
18. Ciampi, K.; Cavicchi, A. *Dynamics of Smart Specialisation Agri-food Trans-Regional Cooperation, JRC Technical Reports, S3 Policy Brief Series*; Publications Office of the European Union: Luxembrug, 2017. [CrossRef]
19. European Commission. The European Green Deal. Available online: https://ec.europa.eu/info/sites/info/files/european-green-deal-communication_en.pdf (accessed on 30 March 2021).
20. Council of European Union. Draft Council Conclusions on Digitalisation for the Benefit of the Environment. Available online: https://data.consilium.europa.eu/doc/document/ST-13957-2020-INIT/en/pdf (accessed on 30 March 2021).
21. De Cicco, A. The Fruit and Vegetable Sector in the EU—A statistical Overview. Eurostat. 2020. Available online: https://ec.europa.eu/eurostat/statistics-explained/index.php/The_fruit_and_vegetable_sector_in_the_EU_-_a_statistical_overview (accessed on 30 March 2021).
22. Willer, H.; Schaak, D.; Lernoud, J. Organic farming and market development in Europe and the European union. *Org. Int. World Org. Agric.* **2018**, 217–250.
23. Mezzetti, B. EUBerry: The Sustainable Improvement of European Berry Production. Quality, and Nutritional Value in a Changing Environment. *Int. J. Fruit Sci.* **2013**, *13*, 60–66. [CrossRef]
24. Baby, B.; Antony, P.; Vijayan, R. Antioxidant and anticancer properties of berries. *Crit. Rev. Food Sci. Nutr.* **2018**, *58*, 1–17. [CrossRef]
25. Bhat, R.; Geppert, J.; Funken, E.; Stamminger, R. Consumers Perceptions and Preference for Strawberries—A Case Study from Germany. *Int. J. Fruit Sci.* **2015**, *15*, 405–424. [CrossRef]

26. Castro, D.; Teodoro, A. Anticancer properties of bioactive compounds of berry fruits—A review. *Br. J. Med. Med. Res.* **2015**, *6*, 771–794. [CrossRef]
27. Skrovankova, S.; Sumczynski, D.; Mlcek, J.; Jurikova, T.; Sochor, J. Bioactive Compounds and Antioxidant Activity in Different Types of Berries. *Int. J. Mol. Sci.* **2015**, *16*, 24673. [CrossRef] [PubMed]
28. COMTRADE. Available online: https://comtrade.un.org/db/mr/rfCommoditiesList.aspx?px=S1&cc=0519 (accessed on 30 March 2021).
29. FAO. Available online: http://www.fao.org/faostat/en/#data/QC (accessed on 30 March 2021).
30. Sobekova, K.; Thomsen, M.R.; Ahrendsen, B.L. Market trends and consumer demand for fresh berries. *Appl. Stud. Agribus. Commer.* **2013**, *7*, 11–14. [CrossRef]
31. Fortune Business Insights. Organic Berries Market. Markek Research Report. Summary. Available online: https://www.fortunebusinessinsights.com/organic-berries-market-103191 (accessed on 30 March 2021).
32. Wang, J.; Yue, C.; Gallardo, K.; McCracken, V.; Luby, J.; McFerson, J. What Consumers Are Looking for in Strawberries: Implications from Market Segmentation Analysis. *Agribusiness* **2017**, *33*, 56–69. [CrossRef]
33. Kahneman, D. Maps of bounded rationality: Psychology for behavioral economics. *Am. Econ. Rev.* **2003**, *93*, 1449–1475. [CrossRef]
34. Decker, R.; Trusov, M. Estimating aggregate consumer preferences from online product reviews. *Int. J. Res. Mark.* **2010**, *27*, 293–307. [CrossRef]
35. Köster, E.P. The psychology of food choice: Some often encountered fallacies. *Food Qual. Prefer.* **2003**, *14*, 359–373. [CrossRef]
36. Podsakoff, P.M.; MacKenzie, S.B.; Lee, J.Y.; Podsakoff, N.P. Common method biases in behavioral research: A critical review of the literature and recommended remedies. *J. Appl. Psychol.* **2003**, *88*, 879–903. [CrossRef]
37. Lawless, H.T.; Heymann, H. Sensory evaluation of food. In *Principles and Practices*, 2nd ed.; Springer: New York, NY, USA, 2010.
38. Meiselman, H.L. The future in sensory/consumer research: Evolving to a better science. *Food Qual. Prefer.* **2013**, *27*, 208–214. [CrossRef]
39. Chamlertwat, W.; Bhattarakosol, P.; Rungkasiri, T.; Haruechaiyasak, C. Discovering Consumer Insight from Twitter via Sentiment Analysis. *J. UCS* **2012**, *18*, 973–992.
40. Carr, J.; Decreton, L.; Qin, W.; Rojas, B.; Rossochacki, T.; Wen Yang, Y. Social media in product development. *Food Qual. Prefer.* **2015**, *40*, 354–364. [CrossRef]
41. Gong, S.; Juanjuan, Z.; Ping, Z.; Xuping, J. Tweeting as a Marketing Tool: A Field Experiment in the TV Industry. *J. Mark. Res.* **2017**, *54*, 833–850. [CrossRef]
42. Ruggeri, A.; Samoggia, A. Twitter communication of agri-food chain actors on palm oil environmental, socio-economic, and health sustainability. *J. Consum. Behav.* **2018**, *17*, 75–93. [CrossRef]
43. Boumans, J.W.; Trilling, D. Taking stock of the toolkit: An overview of relevant automated content analysis approaches and techniques for digital journalism scholars. *Digit. J.* **2016**, *4*, 8–23. [CrossRef]
44. Mattila, M.; Salman, H. Analysing Social Media Marketing on Twitter using Sentiment Analysis. Bachelor's Thesis, KTH Royal Institute of Technology, Stockholm, Sweden, 2018.
45. Mishra, N.; Singh, A. Use of twitter data for waste minimisation in beef supply chain. *Ann. Oper. Res.* **2018**, *270*, 337–359. [CrossRef]
46. Mostafa, M.M. Clustering halal food consumers: A Twitter sentiment analysis. *Int. J. Market. Res.* **2019**, *61*, 320–337. [CrossRef]
47. Singh, A.; Shuklab, N.; Mishra, N. Social media data analytics to improve supply chain management in food industries. *Transp. Res. Part E Logist. Transp. Rev.* **2018**, *114*, 398–415. [CrossRef]
48. Chau, M.; Xu, J. Business Intelligence in Blogs: Understanding Consumer Interactions and Communities. *MIS Q.* **2012**, *36*, 1189–1216. [CrossRef]
49. Yanai, K.; Kawano, Y. Twitter Food Photo Mining and Analysis for One Hundred Kinds of Foods. In *Advances in Multimedia Information Processing—Proceedings of the PCM 2014: 15th Pacific-Rim Conference on Multimedia, Kuching, Malaysia, 1–4 December 2014*; Ooi, W.T., Snoek, C., Tan, H.K., Ho, C.K., Huet, B., Ngo, C.-W., Eds.; Springer: Cham, Switzerland, 2014; pp. 22–32.
50. Hannigan, T.R.; Haans, R.F.; Vakili, K.; Tchalian, H.; Glaser, V.L.; Wang, M.S.; Kaplan, S.; Jennings, P.D. Topic modeling in management research: Rendering new theory from textual data. *Acad. Manag. Ann.* **2019**, *13*, 586–632. [CrossRef]
51. Angelopoulos, S.; Merali, Y. Sometimes a cigar is not just a cigar: Unfolding the transcendence of boundaries across the digital and physical. In Proceedings of the ICIS International Conference in Information Systems, Seoul, Korea, 10–13 December 2017.
52. Barabási, A.L. *Network Science*; Cambridge University Press: Cambridge, UK, 2016.
53. Zuo, M.Z.; Angelopoulos, S.A.; Ou, C.X.; Carol, X.C.; Liu, H.L.; Liang, Z.L. Identifying Dynamic Competition in Online Marketplaces Through Consumers. Clickstream Data. 2020. Available online: https://doi.org/10.2139/ssrn.3598889 (accessed on 22 April 2020).
54. Anninou, I.; Foxall, G.R. Consumer decision-making for functional foods: Insights from a qualitative study. *J. Consum. Mark.* **2017**, *34*, 552–565. [CrossRef]
55. Eskandari, F.; Lake, A.A.; Weeks, G.; Butler, M. Twitter conversations about food poverty: An analysis supplemented with Google Trends analysis. *Lancet* **2019**, *394*. [CrossRef]
56. Nam, H.; Joshi, Y.V.; Kannan, P.K. Harvesting Brand Information from Social Tags. *J. Mark.* **2017**. [CrossRef]
57. Tsur, O.; Rappoport, A. What's in a hashtag? Content based prediction of the spread of ideas in microblogging communities. In Proceedings of the Fifth ACM International Conference on Web Search and Data Mining, Seattle, WA, USA, 8–12 February 2012; ACM: New York, NY, USA, 2012; pp. 643–652.

58. Kar, A.K.; Dwivedi, Y.K. Theory building with big data-driven research—Moving away from the "What" towards the "Why". *Int. J. Inf. Manag.* **2020**, *54*, 1–10. [CrossRef]
59. Zaki, M.; McColl-Kennedy, J.R. Text mining analysis roadmap (TMAR) for service research. *J. Serv. Mark.* **2020**, *34*, 30–47. [CrossRef]
60. R Core Team. *R: A Language and Environment for Statistical Computing*; R Foundation for Statistical Computing: Vienna, Austria, 2013.
61. Gentry, J. Package 'twitteR'. 2014. Available online: http://cran.r-project.org/web/packages/twitteR/twitteR.pdf (accessed on 21 October 2020).
62. Breen, J. R by Example: Mining Twitter for Consumer Attitudes Towards Airlines. Cambridge Aviation Research. 2011. Available online: http://es.slideshare.net/jeffreybreen/r-by-example-mining-twitter-for (accessed on 20 October 2020).
63. Worch, T. What should you know about analysing social media data using twitteR: The experience of a practitioner. In Proceedings of the 6th European Conference on Sensory and Consumer Research, Copenhagen, Denmark, 7–10 September 2014.
64. Fried, D.; Surdeanu, M.; Kodbourov, S.; Hingle, M.; Bell, D. Analyzing the language of food on social media. In Proceedings of the 2014 IEEE International Conference on Big Data, Washington, DC, USA, 27–30 October 2014.
65. Linvill, D.L.; McGee, S.E.; Hicks, L.K. Colleges' and universities' use of Twitter: A content analysis. *Public Relat. Rev.* **2012**, *38*, 636–638. [CrossRef]
66. George, G.; Osinga, E.C.; Lavie, D.; Scott, B.A. Big data and data science methods for management research. *Acad. Manag. J.* **2016**, *59*, 1493–1507. [CrossRef]
67. Tirunillai, S.; Tellis, G.J. Does chatter really matter? Dynamics of user-generated content and stock performance. *Mark. Sci.* **2012**, *31*, 198–215. [CrossRef]
68. Liu, X. A big data approach to examining social bots on Twitter. *J. Serv. Mark.* **2019**, *33*, 369–379. [CrossRef]
69. Forkmann, S.; Henneberg, S.C.; Witell, L.; Kindström, D. Driver configurations for successful service infusion. *J. Serv. Res.* **2017**, *20*, 275–291. [CrossRef]
70. Charmaz, K. *Constructing Grounded Theory: A Practical Guide through Qualitative Analysis*; SAGE Publications Ltd.: London, UK, 2006.
71. Humphreys, A.; Rebecca, J.W. Automated text analysis for consumer research. *J. Consum. Res.* **2017**, *44*, 1274–1306. [CrossRef]
72. Glaser, B.G. *Basics of Grounded Theory Analysis*; Sociology Press: Mill Valley, CA, USA, 1992.
73. Gregory, R.W. Design science research and the grounded theory method: Characteristics, differences, and complementary uses. In Proceedings of the 18th European Conference on Information Systems (ECIS 2010), Pretoria, South Africa, 7–9 June 2010.
74. Schreier, M. *Qualitative Content Analysis in Practice*; Sage: Thousand Oaks, CA, USA, 2012.
75. Guba, E.G. Criteria for assessing the trustworthiness of naturalistic inquiries. *Educ. Commun. Technol. J. Theory Res. Dev.* **1981**, *29*, 75–91.
76. Shenton, A.K. Strategies for ensuring trustworthiness in qualitative research projects. *Educ. Inf.* **2004**, *22*, 63–75. [CrossRef]
77. Strauss, A.; Corbin, J. *Basics of Qualitative Research*; Sage: Thousand Oaks, CA, USA, 1990.
78. Borgatti, S.P. Centrality and network flow. *Soc. Netw.* **2005**, *27*, 55–71. [CrossRef]
79. Freeman, L.C. Centrality in social networks: Conceptual clarification. *Soc. Netw.* **1979**, *1*, 215–239. [CrossRef]
80. Girvan, M.; Newman, M.E.J. Community structure in social and biological networks. *Proc. Natl. Acad. Sci. USA* **2002**, *99*, 7821–7826. [CrossRef] [PubMed]
81. Clauset, A.; Newman, M.E.J.; Moore, C. Finding community structure in very large networks. *Phys. Rev. E* **2004**, *70*, 66111. [CrossRef] [PubMed]
82. Clauset, A. Finding local community structure in networks. *Phys. Rev. E* **2005**, *72*, 26132. [CrossRef] [PubMed]
83. Blondel, V.D.; Guillaume, J.L.; Lambiotte, R.; Lefebvre, E. Fast unfolding of communities in large networks. *J. Stat. Mech. Theory Exp.* **2008**, *10*, P10008. [CrossRef]
84. Bastian, M.; Heymann, S.; Jacomy, M. Gephi: An Open Source Software for Exploring and Manipulating Networks. Association for the Advancement of Artificial Intelligence (www.aaai.org). 2009. Available online: https://gephi.org/publications/gephi-bastian-feb09.pdf (accessed on 23 February 2021).
85. Jaeger, S.R.; Bava, C.M.; Worch, T.; Dawson, J.; Marshall, D.W. The food choice kaleidoscope. A framework for structured description of product, place and person as sources of variation in food choices. *Appetite* **2011**, *56*, 412–423. [CrossRef]
86. Kyutoku, Y.; Minami, Y.; Koizumi, K.; Okamoto, M.; Kusakabe, Y.; Dan, I. Conceptualization of food choice motives and consumption among Japanese in light of meal, gender, and age effects. *Food Qual. Prefer.* **2012**, *24*, 213–217. [CrossRef]
87. Köster, E.P. Diversity in the determinants of food choice: A psychological perspective. *Food Qual. Prefer.* **2009**, *20*, 70–82. [CrossRef]
88. Köster, E.P.; Mojet, J. Theories of food choice development. In *Understanding Consumers of Food Products*; Frewer, L., van Trijp, H.C.M., Eds.; Woodhead Publishing: Cambridge, UK, 2006; pp. 93–124.
89. Meiselman, H.L. Experiencing food products within a physical and social context. In *Product Experience*; Schifferstein, H.N.J., Hekkert, P., Eds.; Elsevier: Oxford, UK, 2008; pp. 559–580.
90. Desmet, P.M.A.; Schifferstein, H.N.J. Sources of positive and negative emotions in food experience. *Appetite* **2008**, *50*, 290–301. [CrossRef]
91. Piqueras-Fiszman, B.; Jaeger, S.R. What makes meals 'memorable'? A consumer-centric exploration. *Food Res. Int.* **2014**. [CrossRef]
92. Piqueras-Fiszman, B.; Jaeger, S.R. Emotions associated to mealtimes: Memorable meals and typical evening meals. *Food Res. Int.* **2014**. [CrossRef]
93. Rappoport, L.; Downey, R.G.; Huff-Corzine, L. Conceptual differences between meals. *Food Qual. Prefer.* **2001**, *13*, 489–495. [CrossRef]

94. Bisogni, C.A.; Winter Falk, L.; Madore, E.; Blake, C.E.; Jastran, M.; Sobal, J.; Devine, C.M. Dimensions of everyday eating and drinking episodes. *Appetite* **2007**, 218–231. [CrossRef]
95. Liu, M.; Kasteridis, P.; Yen, S.T. Breakfast, lunch, and dinner expenditures away from home in the United States. *Food Policy* **2013**, *38*, 156–164. [CrossRef]
96. ComScore. The Global Mobile Report. comScore Inc. 2017. Available online: https://www.comscore.com/Insights/Presentations-and-Whitepapers/2017/The-Global-Mobile-Report (accessed on 30 June 2020).
97. Yates, L.; Warde, A. The evolving content of meals in Great Britain. Results of a survey in 2012 in comparison with the 1950s. *Appetite* **2014**, *84*, 299–308. [CrossRef]
98. Vehkalahti, K.; Tahvonen, R.; Tuorila, H. Hedonic responses and individual definitions of an ideal apple as predictors of choice. *J. Sens. Stud.* **2013**, *28*, 346–357.
99. Canetti, L.; Bachar, E.; Berry, E.M. Food and emotion. *Behav. Process.* **2002**, *60*, 157–164. [CrossRef]
100. Macht, M. How emotions affect eating: A five-way model. *Appetite* **2008**, *50*, 1–11. [CrossRef] [PubMed]
101. Aprile, M.C.; Caputo, V.; Nayga, R.M., Jr. Consumers' valuation of food quality labels: The case of the European geographic indication and organic farming labels. *Int. J. Consum. Stud.* **2012**, *36*, 158–165. [CrossRef]
102. Eldesouky, A.; Mesias, F.J.; Escribano, M. Perception of Spanish consumers towards environmentally friendly labelling in food. *Int. J. Consum. Stud.* **2020**, *44*, 64–76. [CrossRef]
103. Samoggia, A.; Bertazzoli, A.; Ruggeri, A. Food retailing marketing management: Social media communication for healthy food. *Int. J. Retail Distrib. Manag.* **2019**, *47*, 928–956. [CrossRef]
104. McEachern, M.G.; Warnaby, G. Exploring the relationship between consumer knowledge and purchase behaviour of value-based labels. *Int. J. Consum. Stud.* **2008**, *32*, 414–426. [CrossRef]
105. Mellon, J.; Prosser, C. Twitter and Facebook are not Representative of the General Population: Political Attitudes and Demographics of British Social Media users. *Res. Politics* **2017**. [CrossRef]
106. Wang, Y.; Deng, Q.; Rod, M.; Shaobo, J. A thematic exploration of social media analytics in marketing research and an agenda for future inquirí. *J. Strateg. Mark.* **2020**. [CrossRef]

Article

Global Trends in Coffee Agronomy Research

Héctor Madrid-Casaca [1], Guido Salazar-Sepúlveda [2], Nicolás Contreras-Barraza [3], Miseldra Gil-Marín [4] and Alejandro Vega-Muñoz [5,*]

Citation: Madrid-Casaca, H.; Salazar-Sepúlveda, G.; Contreras-Barraza, N.; Gil-Marín, M.; Vega-Muñoz, A. Global Trends in Coffee Agronomy Research. *Agronomy* **2021**, *11*, 1471. https://doi.org/10.3390/agronomy11081471

Academic Editors: Francisco Manzano Agugliaro and Esther Salmerón-Manzano

Received: 30 June 2021
Accepted: 23 July 2021
Published: 24 July 2021

Publisher's Note: MDPI stays neutral with regard to jurisdictional claims in published maps and institutional affiliations.

Copyright: © 2021 by the authors. Licensee MDPI, Basel, Switzerland. This article is an open access article distributed under the terms and conditions of the Creative Commons Attribution (CC BY) license (https://creativecommons.org/licenses/by/4.0/).

[1] Facultad de Ciencias Económicas, Administrativas y Contables, Universidad Nacional Autónoma de Honduras, Tegucigalpa 11101, Honduras; hector.madrid@unah.edu.hn
[2] Departamento de Ingeniería Industrial, Facultad de Ingeniería, Universidad Católica de la Santísima Concepción, Concepción 4090541, Chile; gsalazar@ucsc.cl
[3] Facultad de Economía y Negocios, Universidad Andres Bello, Santiago 8370035, Chile; nicolas.contreras@unab.cl
[4] Facultad de Administración y Negocios, Universidad Autónoma de Chile, Santiago 7500912, Chile; miseldra.gil@uautonoma.cl
[5] Public Policy Observatory, Universidad Autónoma de Chile, Santiago 7500912, Chile
* Correspondence: alejandro.vega@uautonoma.cl

Abstract: This article empirically provides a scientific production trends overview of coffee agronomy at the global level, allowing us to understand the structure of the epistemic community on this topic. The knowledge contributions documented are examined using a bibliometric approach (spatial, productive, and relational) based on data from 1618 records stored in the Web of Science (JCR and ESCI) between 1963 and May 2021, applying traditional bibliometric laws and using VOSviewer for the massive treatment of data and metadata. At the results level, there was an exponential increase in scientific production in the last six decades, with a concentration on only 15 specific journals; the insertion of new investigative peripheral and semiperipheral countries and organizations in worldwide relevance coauthorship networks, an evolution of almost 60 years in relevant thematic issues; and a co-occurring concentration in three large blocks: environmental sustainability of forestry, biological growth variables of coffee, and biotechnology of coffee species; topic blocks that, although in interaction, constitute three specific communities of knowledge production that have been delineated over time.

Keywords: agroforestry; bibliometrics; coffee biology; coffee biotechnology; coffee industry; coffee species; environmental sustainability; global research; scientific documentation

1. Introduction

This article empirically analyzes the global trends of research in coffee agronomy in terms of its evolution over time, the sources of documentation of scientific production, the geography of knowledge generation (national and organizational), and the topics under study. Research that has been marked by the sustainability agenda in coffee agroforestry has promoted interest in research on organic production systems, the preservation of local cultures and knowledge, biodiversity conservation, and agroecological principles to reconcile sustainable agriculture, considering socioeconomic and cultural contexts that vary at the local level, in order to devise economic development models to further improve the benefits and family budget for family and collective agriculture given its remarketing [1–12]. As a result, socioenvironmental standards and certifications have experienced a strong development in the coffee sector during the last decade [13–17].

Due to the Kyoto Protocol in the face of climate change, coffee agroforestry in general and organic farms have gained increased attention as a strategy for carbon sequestration (C), synergistically conserving biodiversity and reducing greenhouse gas (GHG) emissions [18–23]. Thus, to mitigate global climate change, researchers have proposed new production mechanisms, including studies of flowering phenology, the measurement of transpiration and water potential in different microclimatic conditions, and the periodic pruning of shade trees to increase the addition of organic matter and the return of nutrients to soil [24–29].

In addition, with increasing patterns of climate variability, water resources for agriculture may become more unpredictable and scarcer [30–33]. For this reason, the presence of shade trees (adequate pruning), the reuse of secondary treated wastewater (with fertilizer management and adequate nutritional conditions), irrigation performance and management (depths and technologies), and groundwater balance seek to reduce soil evaporation and coffee transpiration as measures to preserve water within the agroecosystem [2,34–40].

On the other hand, irrigation systems have become a common technique to improve coffee yields because they provide a more controlled production environment and avoid production losses due to water deficits, this subject being of interest to several researchers [1,2,41–46]. It has been pointed out that soil water deficit is one of the main factors affecting the vegetative development and productivity of coffee, so improving irrigation systems is an important area for researchers to demonstrate its effects [44]. On the other hand, climate change has increased the presence of coffee leaf rust (CLR) (*H. vastatrix*), which is one of the main diseases that strongly affect production [47–49], with an important influence on its costs [50,51]. Given the importance of the management of this disease, it is necessary to have a thorough knowledge of the species and its localities, as the level of incidence and intensity is determined by the microclimates depending on the geographic zones. It is for the same reason that strategies and systems must be designed to predict this disease [52]; however, an important task is to evaluate the pesticide efficiency [53] and determine the tolerance of different coffee genotypes [54]. Thus, drip irrigation techniques can be used to provide nutrients during the growth cycle of the plant based on the plant's nutrient absorption rate, where fertigation improves nutrient use efficiency by gradually providing nutrients and according to their absorption rate [1].

With respect to contaminants in coffee, the use of organic fertilizers and nonselective herbicides (*glyphosate*) has been found to have transitory effects that could result in irreversible and prolonged damage to crop growth and drying conditions (in *Coffea arabica* L.), in addition to Ochratoxin A (OTA), the main mycotoxin found in coffee [55–58]. In terms of pest treatment, the use of yeasts in dual culture with filamentous fungi, the effect of the citrus mealybug *Planococcus citri* (Risso, 1813), and the study of 144 microorganisms previously isolated from the fruit of *Coffea arabica* to evaluate their proteolytic activities have been discussed [59–61]. There is also the use of shade trees (leguminous and nonleguminous) and organic inputs, particularly in areas where coffee berry disease is prevalent [62]

Scientific studies on coffee agronomy and the effect of micronutrients refer to the application of foliar spraying of Zn sulfate on crops and their yield in Arabian coffee ("Mundo Novo"), obtaining a positive response to increasing concentrations of ZnSO4 applied in oil to the leaves. On the other hand, the use of zinc (Zn) in treatments of acid clay soils in the southeastern region of Brazil had a positive result on soil attributes (chemical fertility, micronutrients, organic matter, and acidity), causing an improvement in pest management and soil recovery [63–66].

Harvesting and subsequent drying are two of the most important operations in coffee production systems, for which the drying technology must be adjusted under different parameters such as mathematical evaluations of the drying curves for different coffee species, thermal losses in the coffee dryer, rotation times, and energy efficiency in a fixed-bed coffee dryer for (*Coffea arabica* L.) [67,68], as well as a continuous rotator and its humidity percentages [69–72].

Finally, the conservation of biodiversity is an important challenge to maintain the richness of native, productive, and mitigation species in agricultural activity, the habitat being a conservation factor of great relevance for the sustainable development of coffee plantations; therefore, the subject is of great interest for several researchers [73–77]. It has been demonstrated that the rural development of geographic areas with population and production has been adapted to the local environment by following the farming methods and knowledge of the local people, protecting habitats, forestation areas, and the different forms of life in the ecosystem [78]. These landscapes have a relatively high level of agrobiodiversity compared to conventional (monoculture) agriculture [74]. Thus, coffee farms, as habitat fragments, can act as buffer zones and biological corridors between protected forests and other areas [79]. The use of shade trees in agroforestry can also offer an effective coping mechanism to implement in agricultural areas that suffer from extreme climates [25]. This includes crop diversification, coffee marketing activity (e.g., certification of coffee production and postharvest coffee processing), and migrant labor schemes [13,80,81].

2. Materials and Methods

We used a set of articles as a homogeneous basis for citation, counting the main collection of Web of Science (WoS) [82], by selecting articles published in WoS-indexed journals in the Science Citation Index (WoS-SCI), Social Science Citation Index (WoS-SSCI), and Emerging Science Citation Index (WoS-ESCI) based on a search vector [83] about coffee (TS = coffee) restricted to the WoS Agronomy category (WC = agronomy) and with unrestricted time parameters, performing the extraction on 22 May 2021.

The resulting set of articles was analyzed bibliometrically in terms of their exponential growth to ensure a critical mass of documented scientific production that ensures interest in the international scientific community [84,85], determining the time median and its contemporary and obsolete periods. In terms of concentrations, Bradford's law of concentrations was applied to the journals, fragmented into thirds of articles, avoiding the exponential decrease in decreasing performance by expanding the search of references in scientific journals peripheral to the topic under study [86–91]. Lotka's law about authors was applied to identify the most prolific group of authors and study them in isolation from the majority of authors with a smaller number of articles based on the unequally distributed scientific production among authors [92]. The Hirsch index or h-index was used for articles based on the set of articles most cited by the scientific community and the citations they have received in other publications of the WoS core collection, established as the "n" documents cited "n" times or more [93,94]. Zipf's law on words was applied to empirically determine words with the highest frequency of occurrence in the set of articles studied [95]. Information processing and the visualization of spatiality, coauthorship, and cooccurrence [96–99] were processed with VOSviewer, using fragmentation analysis with thematic and time trend visualization outputs [100–108].

3. Results

The results show that there is an exponential growth of documented mainstream research in coffee agronomy between the 1960s and the year 2020, with a scientific production that reached 120 articles in the last recorded year and half-periods of production (median years of publication) located in 2009, as shown in Figure 1. This evidence of compliance with Price's law allows us to consistently give way to other types of bibliometric analysis.

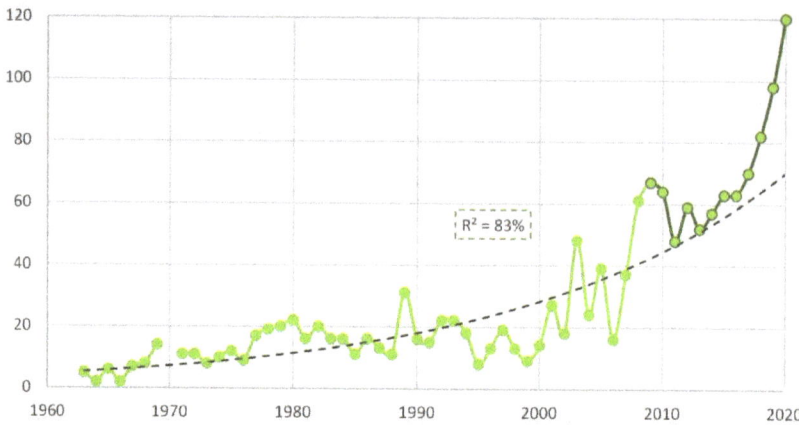

Figure 1. Temporary trend of publications on coffee agronomy (1960s–2020s).

3.1. Scientific Production Environments

The first trend that we identified shows the places or environments of production, including both the bibliographic space of publication sources and the geographic space where this knowledge is generated.

3.1.1. Bibliographic Environments of Scientific Production

In terms of Bradford's law, publications on coffee agronomy between 1963 and 2020 are concentrated in a core of 3 journals out of 125 in total that constitute Bradford's core or main third of article concentration in a small number of journals. In the nucleus zone or first third of articles, 30% are covered, concentrated in three journals; in Zone 2 or the second third of articles, 37% of articles are covered (to complete the 2/3 of articles), concentrated in 12 semiperipheral journals; and finally, Zone 3 or last third covers the remaining articles (34%), dispersed in 110 peripheral journals. As for the Bradford multiplier calculated at 6.6 (average growth rate in the number of journals from one zone to the next), it allows us to calculate the theoretical series of journals that should be found in each zone so that the dispersion of journals is 22.2% lower than the situation we could theoretically have found in Zones 2 and 3, as detailed in Table 1.

Table 1. Publications on coffee agronomy by Bradford's zones between 1963 and 2020.

Zone	Number of Articles in Thirds (%)		Journals (%)		Bradford Multipliers	Journals (Theoretical Serie (S_{SB}))	
Nucleus	482	(30%)	3	(2%)		$3 \times (n^0)$	3
Zone 1	592	(37%)	12	(10%)	4.0	$3 \times (n^1)$	20
Zone 2	544	(34%)	110	(88%)	9.2	$3 \times (n^2)$	130
Total	1618	(100%)	125 *	(100%)	n = 6.6		153 *
						% error (ε_p) =	−22.2%

* Real and theoretical value, incorporated for percentage error calculation.

The Bradford zone calculation is reported, as indicated in Table 1. Given a core zone $a = 3$ and a mean multiplier $n = 6.6$, Equation (1) for the geometric series summation of Bradford (S_{SB}) is:

$$S_{SB} = \sum_{i=1}^{3} \left(a * n^{i-1} \right) = 3 + 20 + 130 = 153 \qquad (1)$$

with an error percentual margin (ε_p) in Equation (2):

$$\varepsilon_p = \left(\frac{(Real - Estimated)}{Real}\right) * 100 = \left(\frac{(125-153)}{125}\right) * 100 = -22.2\% \qquad (2)$$

Figure 2 shows the evolution of articles published in the 15 journals of the nucleus and Zone 1, in the contemporary half-period from 2009 to 2020, which are mainly published by major worldwide editor companies or by Brazilian institutions (whose characteristics are detailed in Appendix A). As can be seen, the behavior is not homogeneous among the journals, and not all journals show increasing trends.

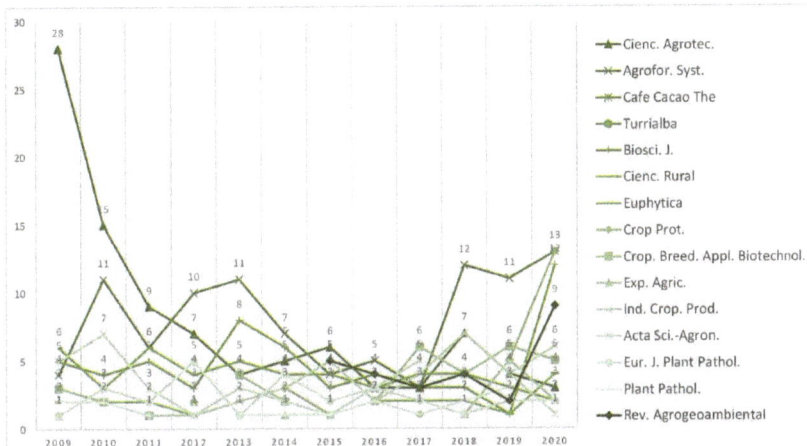

Figure 2. Publications on coffee agronomy in journals from major Bradford zones (2009–2020).

Table 2 below shows that the journals in the WoS core collection (included in the Journal Citation Reports of Clarivate™ version 2020) with an increasing trend of publications on coffee agronomy are Crop Prot. (Q2), Crop. Breed. Appl. Biotechnol. (Q3), Ind. Crop. Prod. (Q1), and Agrogeoambiental Rev. (Emerging Sources Citation Index (ESCI)).

Table 2. Journals from major Bradford zones (1960–2020 and 2009–2020).

Journals	1960–2020	2009–2011	2012–2014	2015–2017	2018–2020	Trends	JIF: JCR-WoS
Cienc. Agrotec.	197	52	16	12	14	↓—	1.390; Q2
Agrofor. Syst.	153	21	28	12	36	→	2.549; Q2
Cafe Cacao The	132	0	0	0	0	0	0; N/A
Turrialba	127	0	0	0	0	0	0; N/A
Biosci. J.	61	14	17	10	16	→	0.347; Q4
Cienc. Rural	57	15	13	11	9	↓—	0.843; Q4
Euphytica	52	6	3	5	7	→	1.895; Q2
Crop Prot.	45	3	5	2	18	↑+	2.571; Q2
Crop. Breed. Appl. Biotechnol.	43	6	6	8	15	↑+	1.282; Q3
Exp. Agric.	41	1	3	1	8	→	2.118; Q2
Ind. Crop. Prod.	37	2	6	11	17	↑+	5.645; Q1
Acta Sci.-Agron.	36	15	6	9	5	↓—	2.042; Q2
Eur. J. Plant Pathol.	35	6	10	7	5	→	1.907; Q2
Plant Pathol.	31	5	2	4	5	→	2.590; Q2
Rev. Agrogeoambiental	27	0	0	12	15	↑+	ESCI *
Total	1074	179	168	196	300	↑+	—

* Emerging Sources Citation Index, journal without journal impact factor calculation (JIF). N/A: not available, discontinued calculation.

3.1.2. Geographical Environments of Scientific Production

Regarding the geography of knowledge production on coffee agronomy, the set of extracted articles shows 89 countries of authorial affiliation (See Figure 3). Brazil has the largest number of contributions, participating in the coauthorship of 655 articles. Followed at a distance by France (150 articles) and USA (113 articles), all other countries have contributions of less than 100 articles.

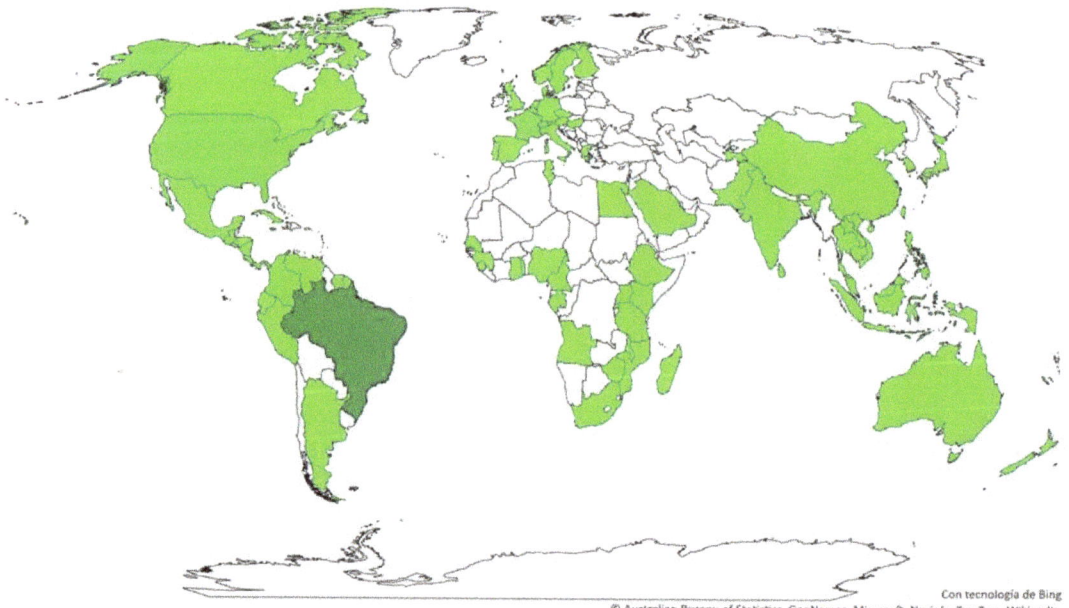

Figure 3. Geography of documented scientific production.

As shown in Figure 4, among these 89 countries, there is a high degree of association between geographically distributed coauthors, although some countries participate in producing knowledge on this topic in isolation: Greece, Hungary, Sierra Leone, Sri Lanka, and Tunisia. The greater width of the lines represents a stronger coauthorship connection between countries and the colors the average number of years of publication; thus, countries with purple nodes have a higher average publication age, and those with a reddish color indicate a lower average publication age.

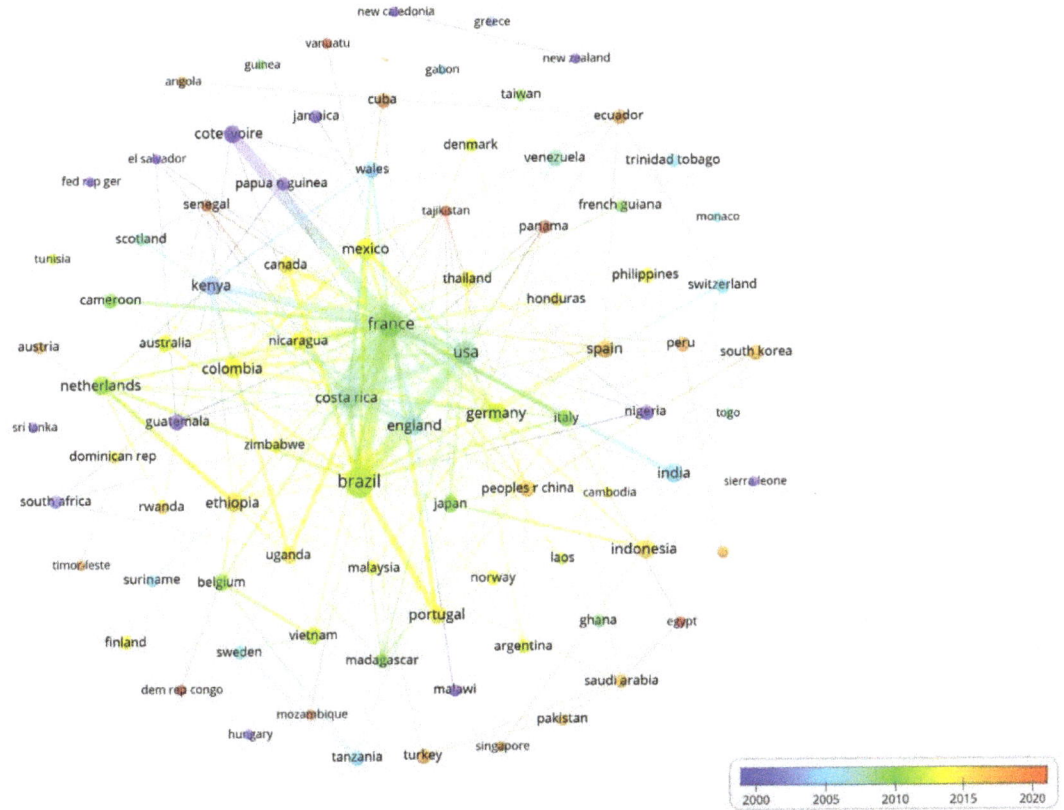

Figure 4. Coffee agronomy research coauthorship network: countries level.

For the graph of coauthorship in Figure 4, the calculation of the "total link strength" was obtained based on the relationships with other countries and the number of joint collaborations using VOSviewer, from which the 14 highest positions are presented in Table 3 (indicator out of 20) and show the best-connected countries within the group of world knowledge production in coffee agronomy. The high contribution of Brazil to the total number of articles is noteworthy (40%), followed at a distance by France (9%). In addition, the articles with contributions from Brazil exceed 4500 citations from other publications indexed in the WoS core collection and the citations received for articles with contributions from France are close to 3000.

Although all the data and metadata analyzed in this article are arranged in English by WoS, this geographical distribution is reflected idiomatically in the articles. In total, 69% of the articles are published in English (contemporary is 77%), followed by articles in Portuguese, French, and Spanish, among others, as shown in Table 4.

Table 3. Relevant countries in coffee agronomy research.

Rank	Country	Published Articles	Contribution at 1618	Citations Received by WoS Core	Total Link Strength
1.	France	150	9%	2898	200
2.	Costa Rica	80	5%	2136	112
3.	Brazil	655	40%	4537	94
4.	USA	113	7%	2378	85
5.	United Kingdom	35	2%	641	42
6.	Kenya	32	2%	392	39
7.	Germany	29	2%	368	36
8.	Mexico	45	3%	670	34
9.	Nicaragua	12	1%	193	31
10.	Netherlands	28	2%	453	30
11.	Ethiopia	27	2%	285	28
12.	Colombia	41	3%	276	27
13.	Canada	11	1%	65	22
14.	Portugal	25	2%	484	20
15.	Japan	16	1%	220	20
16.	Uganda	12	1%	222	20

Table 4. Publication languages in coffee agronomy research.

Language	Articles (1960–2020)	% of 1618	Articles (2009–2020)	% of 846	Avg. Cit. per Article (2009–2020)
English	1120	69%	652	77%	5898/652 = 9.05
Portuguese	269	17%	135	16%	577/135 = 4.27
French	126	8%	21	2%	76/21 = 3.62
Spanish	95	6%	36	4%	32/36 = 0.89
German	3	0%	0	0%	0
Japanese	2	0%	0	0%	0
Indonesian	2	0%	2	0%	1/2 = 0.50
Hungarian	1	0%	0	0%	0
Total	1618	100%	846	100%	6584/846 = 7.78

3.2. Actors of Scientific Production in Coffee Agronomy

Among these actors, we identified authors and their affiliation organizations in search of trends in research on coffee agronomy.

3.2.1. Author Affiliation Organizations Network

To reduce, in terms of relevance, the number of author-affiliated organizations, the Hirsch index or h-index was used, and therefore, only the 52 documents cited 52 times or more (for a resulting h-index of 54 citations) were considered, all published in English (in contrast, 297 articles did not present citations, and there are 191 with only one citation). Thus, the 1242 author-affiliated organizations present in the 1618 articles under study were reduced to 129 organizations. This set of high citation (impact) articles was published between 1986 and 2015, and among the organizations contributing to this production are the Federal University of Lavras (with two affiliations: "univ fed lavras" and "univ fed lavras ufla") and the Federal University of Viçosa ("univ fed vicosa") and most other universities in Brazil. Another highlight is the high average number of citations received by the Tropical Agricultural Research and Teaching Center (CATIE). Table 5 shows the top 10 organizations in terms of coauthorship contributions in published articles and Figure 5 shows the coauthorship network among the 129 organizations.

Table 5. Relevant author affiliation organizations in coffee agronomy research.

Organization	Documents (A)	Citations (B)	Avg. Cit. (C = B/A)	Links	Total Link Strength	Avg. Pub. Year
Univ Fed Lavras (Federal University of Lavras)	172	1113	6	20	87	2011
Univ Fed Vicosa (Federal University of Viçosa)	122	1030	8	22	74	2012
Univ Fed Lavras UFLA (Federal University of Lavras)	99	492	5	12	52	2012
CIRAD [1]	64	855	13	26	78	2012
Univ Sao Paulo (University of Sao Paulo)	53	456	9	22	36	2005
CATIE [2]	43	1410	33	24	61	2012
EPAMIG [3]	29	178	6	8	35	2008
Univ Fed Espirito Santo (Federal University of Espirito Santo)	29	191	7	12	29	2014
Univ Fed Uberlandia (Federal University of Uberlandia)	25	78	3	6	13	2014
Univ Estadual Paulista (Sao Paulo State University)	23	246	11	7	15	2014

[1] Centre de Coopération Internationale en Recherche Agronomique pour le Développement/Center for International Cooperation in Agricultural Research for Development (CIRAD, París and Montpellier, France), [2] Centro Agronómico Tropical de Investigación y Enseñanza/Tropical Agricultural Research and Teaching Center (CATIE, Costa Rica and other countries), and [3] Empresa de Pesquisa Agropecuária de Minas Gerais/Agricultural Research Company of Minas Gerais (EPAMIG, Brasil).

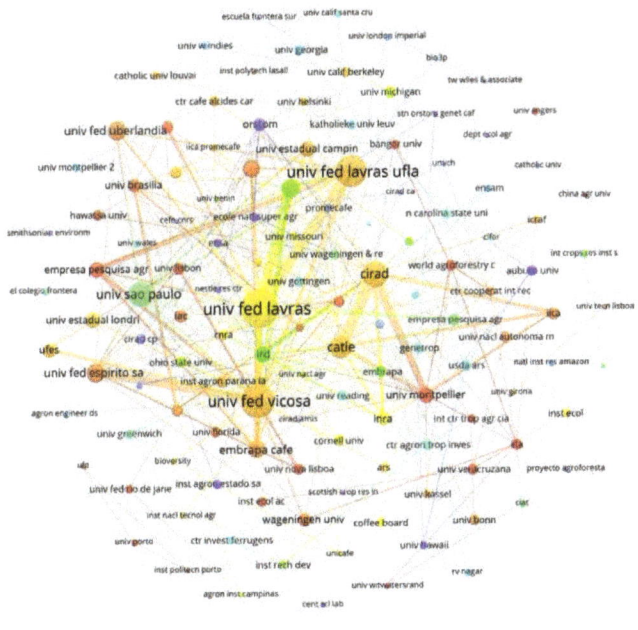

Figure 5. Coffee agronomy research coauthorship network, organizations level.

3.2.2. Prolific Coauthors Network

For the total 4670 authors contributing to the 1618 articles, 68 authors were estimated to be prolific (root square = 4670), and 57 authors with at least 7 publications were chosen, which, as shown in Figure 6, constitute 5 clusters, detailed in Appendix B.

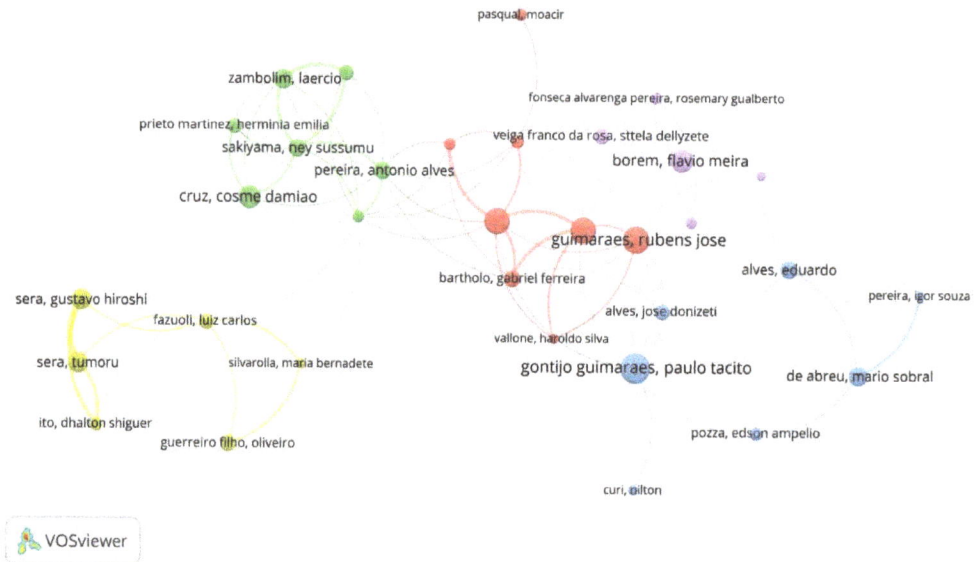

Figure 6. Coffee agronomy research coauthorship network, authors level.

As for the number of publications, Paulo Tácito Gontijo Guimaraes, PhD in Agronomy, with an emphasis in Fertilization and Soil Fertilization, and Coordinator of the Plant and Soil Nutrition Laboratory of EPAMIG Sul (Lavras, Brazil), has conducted research on topics related to the fertilization, quality, and seedlings of coffee. He is the author with the most publications, with 20 articles being cited 128 times in the WoS core collection. A second relevant author identified in this study is Philippe Lashermes, a researcher at the Institut de Recherche pour le Développement (IRD, France) and codirector of the international initiative that sequenced the coffee genome, who, in the present study, records 18 publications and 897 citations in the WoS core collection. Some other relevant authors on this topic are Rubens José Guimaraes (18 publications), Antonio Nazareno Guimaraes Mendes (17 articles), and Gladyston Rodrigues Carvalho (17 articles) (see Table 6).

Table 6. Relevant authors in coffee agronomy research.

	Authors	Articles	Citations	Total Link Strength
1	Paulo Tácito Gontijo Guimaraes	20	128	9
2	Philippe Lashermes	18	897	16
3	Rubens José Guimaraes	18	96	35
4	Antonio Nazareno Guimaraes Mendes	17	82	27
5	Gladyston Rodrigues Carvalho	17	66	36

3.3. Subjects of Scientific Production in Coffee Agronomy

Through text data mining, 5142 keywords were identified (author keywords and keywords plus) and approximately 72 outstanding keywords, 66 being chosen as outstanding keywords with an occurrence of 15 or more times, which present an average age that covers the last decade. Among the outstanding keywords with a more recent average age, in yellow to red colors, the following concepts stand out: climate change, organic matter, growth, shade trees, etc. (see Figure 7). This tendency to the proliferation of new research topics is inserted within three major research areas that are identified in Figure 8 by establishing fragmented clusters with the relevant keywords: the theme of environmental sustainability in forestry (in green), another with respect to the variables of biological growth of coffee (in blue), and finally oriented towards the biotechnology of coffee species (in red).

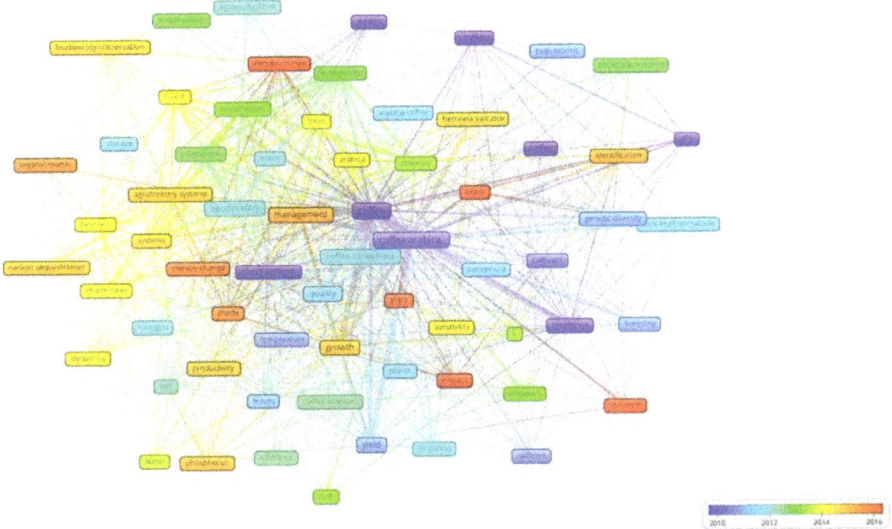

Figure 7. Coffee agronomy research keywords co-occurrence network, temporary visualization.

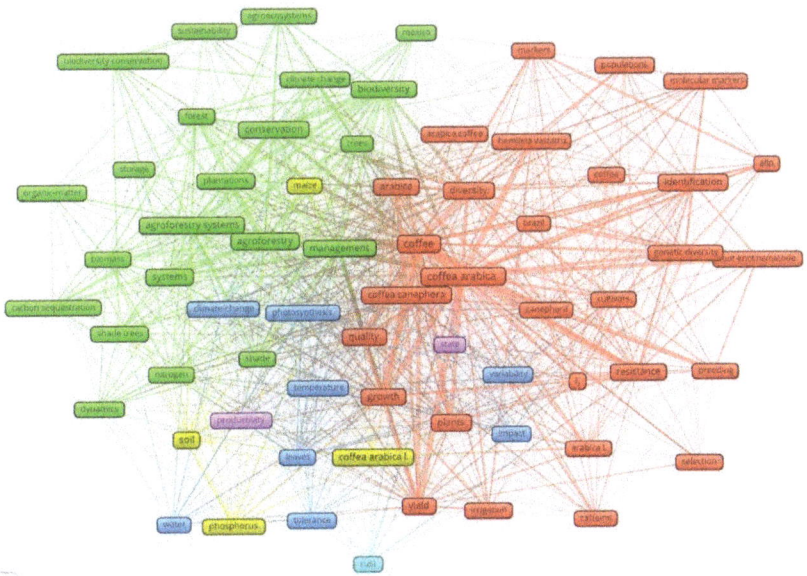

Figure 8. Coffee agronomy research keywords co-occurrence network, thematic visualization.

4. Discussion

This article empirically contributes to establishing a general overview of the trends in the scientific production of coffee agronomy at the global level, which allows us to understand the structure of the epistemic community on this specific agronomic topic, managing to identify three main thematic areas of research of the coffee, a product of the various research agendas worldwide. Thematic coffee research areas, including the environmental sustainability of forestry, biological growth variables of coffee, and biotechnology of coffee species, display marked differences from a panoramic perspective of analysis. Although there are studies of literature reviews related to the areas of our findings, such as "Reductions in water, soil and nutrient losses and pesticide pollution in agroforestry practices: a review of evidence and processes" [109] and "Effects of shade trees on robusta coffee growth, yield and quality. A meta-analysis" [30]. These stand out for their use of multiple word combinations connected by Boolean operators but not an enveloping search vector [83], and the use of a selection method (e.g., PRISMA), which allowed them to systematize the selection of articles and to gain depth in their analysis, but their tendency to reduce the number of articles analyzed (only 83 and 30 articles, respectively) gives our study an advantage in terms of coverage by using bibliometrics as a meta-analytical method that is not reductive [110].

On the other hand, there is also literature review research such as "Remodeling agro-industrial and food wastes into value-added bioactives and biopolymers" [111] and "Challenges of organic agriculture to produce composts and vermicompost to produce medicinal plants—a socioeconomic demand" [109], which contribute to the topics related to the lines of study but possess less systematic rigorous methodology. In cases such as these, our research proposes a bibliometric methodology, defining a search vector, using homogeneous and structured databases, and incorporating a large sample size (1618 articles). Thus, in the extensive literature reviewed, no other mainstream articles have been identified that can provide a meta-analytical coverage as broad as ours, and no other sources have been identified that, to date, allow us to account for the tendency patterns that the global epistemic community of research on coffee agronomy has adopted over time. In the following conclusions, we will report on the diverse findings that are identified as results and their implications.

5. Conclusions

This article bibliometrically shows the existence of an exponentially growing trend of publications in this research topic, with an adjustment of over 80%, achieving a critical mass of documented scientific production in mainstream articles that show the interest on the part of the international scientific community for research in coffee agronomy. This growth rate of the published knowledge on coffee agronomy allows determining the average time with a period of current technical obsolescence, exceeding 12 years after the publication of a document, except for articles with citations above the historical average and determined as classics in this theme. Thus, in the period of contemporary production (2009 to date), three journals (JCR-WoS) are identified with a growing tendency to publish articles on this topic.

In turn, the article identifies a trend of three main journals that are concentrated in the first third of Bradford or core zone articles, with 30% of the total number of articles (482 of 1618) partly generated by the completion of registration of the journals Café Cacao The and Turrialba in the Journal Citation Report (JCR) of WoS. Thus, it is in the journals Ciencia e Agrotecnologia (published by the Federal University of Lavras), Agroforestry Systems (published by Springer, Kluwer Academic Publishers), and Cafe Cacao The (published by CIRAD—Cultures Perennes, only until 1994), where there is a broad and deep discussion on the topic under study. It should be identified that outside this Bradford core (Zones 1 and 2), there is an exponential diminishing of decreasing performance when trying to expand the search for references on coffee agronomy, as for this specific topic, it would be about peripheral scientific journals.

Geographically, the recent generation of knowledge presents a tendency to a concentration in Brazil as an emerging pole of knowledge production on coffee agronomy, which contributes with 655 articles out of a total of 1618 (40%). On the other hand, the Federal University of Lavras stands out both in the number of documents and citations. Regarding the level of authorship, according to Lotka's law, the conformation of five research groups stand out, where not only prolific authors (high production) but also prominent authors (high production and high citation) stand out, such as Paulo Tácito Gontijo Guimaraes and Philippe Lashermes. It is of future interest to be able to study them in isolation from the "other" authors with a smaller number of articles and establish in depth the origin of their unequal level of scientific production in comparison with the common authors on this topic. The scientific production that has been generated in languages other than English (23%) is also of interest to analyze in the future, especially the degree of international collaboration, citation, and use as input for other publications that transcend the base language.

As we have pointed out in our study, we also used Zipf's law to empirically determine the words with the highest frequency of occurrence (keywords and keyterms) in the set of articles studied. Thus, using fragmentation analysis through VOSviewer, thematic and time trend visualization outputs were analyzed. The thematic trends that have evolved in these six decades are identified to strengthen three major research areas: environmental sustainability of forestry, biological growth variables of coffee, and biotechnology of coffee species.

The limitations of this study are due to the wide coverage of articles reviewed (1618), which affects the degree of depth and specificity of the analyzes, and the results should be understood at the level of trends and meta-analytic behaviors. However, this opening of 5142 keywords also generates possibilities for greater segmentation at the level of systematic reviews, such as those mentioned in the discussion, in search of greater depth in specific topics related to coffee agronomy, and the panoramic character of the bibliometric meta-analysis escapes. Another limitation to be considered is related to the way in which both the authors, the journal, and WoS (Clarivate, London, UK) register data and metadata of the articles under study, especially due to the lack of uniformity in the terms used by the authors in the keywords, titles, and abstracts of their manuscripts. In view of this, the requirements of concentration and high occurrence imposed by bibliometric methods make it possible to generate error filters, assuming that errors in data and metadata should occur with low occurrence.

In terms of future research challenges, specific bibliometric and systematic review analyses in the three areas identified should be carried out as lines of future coffee agronomy research (silvicultural environmental sustainability, biological growth variables, and biotechnology of species). The strong relationship between coffee production, contribution in published articles, and local editions of magazines (JCR-WoS) in Brazil make it an interesting national case to study in greater depth and establish explanations of its evolution from coffee agronomic production to the production of knowledge on agronomic coffee.

Supplementary Materials: The following are available online at https://www.mdpi.com/article/10.3390/agronomy11081471/s1, Table S1: Agronomy_Coffee_1618.txt.

Author Contributions: Conceptualization, H.M.-C. and A.V.-M.; methodology, A.V.-M.; software, M.G.-M. and A.V.-M.; validation, G.S.-S.; formal analysis, A.V.-M.; writing—original draft preparation, H.M.-C., N.C.-B. and G.S.-S.; writing—review and editing, N.C.-B., G.S.-S. and A.V.-M.; visualization, A.V.-M.; project administration, A.V.-M.; funding acquisition, N.C.-B., G.S.-S. and A.V.-M. All authors have read and agreed to the published version of the manuscript.

Funding: The article processing charge (APC) was partially funded by Universidad Católica de la Santísima Concepción.

Data Availability Statement: The analyzed dataset has been included as Supplementary Materials.

Conflicts of Interest: The authors declare no conflict of interest.

Appendix A. Relevant Journals in Coffee Agronomy

This appendix details the 15 journals from major Bradford zones (2009–2020) shown in Figure 2 and Table 2. In Table A1, the journals are presented, detailing standardized abbreviated name, full name, International Standard Serial Number (ISSN), publisher, articles published in the Web of Science (WoS) database indexed between 1960–2020 and 2009–2020 (contemporary semiperiod), and the WoS categories to which the journal is attached.

Table A1. Details of journals from major Bradford zones (2009–2020).

Journals	Full Name	ISSN	Publisher	1960–2020	2009–2020	WoS Category
Cienc. Agrotec.	Ciencia e Agrotecnologia	1413-7054	Univ Fed Lavras	197	94	Agriculture, Multidisciplinary; Agronomy
Agrofor. Syst.	Agroforestry Systems	0167-4366	Springer	153	97	Agronomy; Forestry
Cafe Cacao The	Cafe Cacao The	0007-9510	CIRAD-Cultures Perennes	132	0 *	Agronomy
Turrialba	Turrialba	0041-4360	Inter-Amer Inst Cooperat Agric	127	0 **	Agronomy
Biosci. J.	Bioscience Journal	1516-3725	Univ Fed Uberlandia	61	57	Agriculture, Multidisciplinary; Agronomy; Biology
Cienc. Rural	Ciencia Rural	0103-8478	Univ Fed Santa Maria	57	48	Agronomy
Euphytica	Euphytica	0014-2336	Springer	52	21	Agronomy; Plant Sciences; Horticulture
Crop Prot.	Crop Protection	0261-2194	Elsevier	45	28	Agronomy
Crop. Breed. Appl. Biotechnol.	Crop Breeding and Applied Biotechnology	1984-7033	Brazilian Soc Plant Breeding	43	35	Agronomy; Biotechnology & Applied Microbiology
Exp. Agric.	Experimental Agriculture	0014-4797	Cambridge Univ Press	41	13	Agronomy
Ind. Crop. Prod.	Industrial Crops and Products	0926-6690	Elsevier	37	36	Agricultural Engineering; Agronomy
Acta Sci.-Agron.	Acta Scientiarum-Agronomy	1807-8621	Univ Estadual Maringa	36	35	Agronomy

Table A1. *Cont.*

Journals	Full Name	ISSN	Publisher	1960–2020	2009–2020	WoS Category
Eur. J. Plant Pathol.	European Journal of Plant Pathology	0929-1873	Springer	35	28	Agronomy; Plant Sciences; Horticulture
Plant Pathol.	Plant Pathology	0032-0862	Wiley	31	16	Agronomy; Plant Sciences
Rev. Agrogeoambiental ***	Revista Agrogeoambiental	1984-428X	Inst Fed Sul Minas Gerais	27	27	Agronomy

* In zero since 1995. ** In zero since 1991. *** Emerging Sources Citation Index, journal without impact factor calculation (IF). N/A: currently not available in the Journal Citation Report.

Appendix B. Prolific Authors by Cluster

This appendix details the five clusters of prolific coauthors network shown in Figure 6 (see Table A2).

Table A2. Cluster of prolific coauthors network.

Cluster	Authors
Cluster 1	Bartholo, Gabriel Ferreira Botelho, Cesar Elias Carvalho, Gladyston Rodrigues De Rezende, Juliana Costa Guimaraes Mendes, Antonio Nazareno Guimaraes, Rubens Jose Pasqual, Moacir Vallone, Haroldo Silva
Cluster 2	Baiao De Oliveira, Antonio Carlos Caixeta, Eveline Teixeira Cruz, Cosme Damiao Pereira, Antonio Alves Prieto Martinez, Herminia Emilia Sakiyama, Ney Sussumu Zambolim, Laercio
Cluster 3	Alves, Eduardo Alves, Jose Donizeti Curi, Nilton De Abreu, Mario Sobral Gontijo Guimaraes, Paulo Tacito Pereira, Igor Souza Pozza, Edson Ampelio
Cluster 4	Fazuoli, Luiz Carlos Guerreiro Filho, Oliveiro Ito, Dhalton Shiguer Sera, Gustavo Hiroshi Sera, Tumoru Silvarolla, Maria Bernadete
Cluster 5	Borem, Flavio Meira Da Silva, Fabio Moreira Fonseca Alvarenga Pereira, Rosemary Gualberto Malta, Marcelo Ribeiro Veiga Franco Da Rosa, Sttela Dellyzete

References

1. Byrareddy, V.; Kouadio, L.; Kath, J.; Mushtaq, S.; Rafiei, V.; Scobie, M.; Stone, R. Win-win: Improved irrigation management saves water and increases yield for robusta coffee farms in Vietnam. *Agric. Water Manag.* **2020**, *241*, 106350. [CrossRef]
2. Venturin, A.Z.; Guimarafes, C.M.; de Sousa, E.F.; Machado, F.; Josa, A.; Rodrigues, W.P.; Serrazine Ãcaro, D.; Bressan-Smith, R.; Marciano, C.R.; Campostrini, E. Using a crop water stress index based on a sap flow method to estimate water status in conilon coffee plants. *Agric. Water Manag.* **2020**, *241*, 106343. [CrossRef]
3. Lyngbaek, A.E.; Muschler, R.G.; Sinclair, F.L. Productivity and profitability of multistrata organic versus conventional coffee farms in Costa Rica. *Agrofor. Syst.* **2001**, *53*, 205–213. [CrossRef]
4. De Souza, H.N.; De Graaff, J.; Pulleman, M.M. Strategies and economics of farming systems with coffee in the Atlantic Rainforest Biome. *Agrofor. Syst.* **2011**, *84*, 227–242. [CrossRef]
5. Sibelet, N.; Ba, S.N. Strategies of Ugandan farmers facing coffee wilt disease. *Cah. Agric.* **2012**, *21*, 258–268. [CrossRef]
6. Bertrand, B.; Montagnon, C.; Georget, F.; Charmetant, P.; Etienne, H. Creation and dissemination of Arabica coffee varieties: What varietal innovations? *Cah. Agric.* **2012**, *21*, 77–88. [CrossRef]
7. Labouisse, J.P.; Adolphe, C. Preservation and management of the genetic resources of Arabica coffee (*Coffea arabica* L): A challenge for Ethiopia. *Cah. Agric.* **2012**, *21*, 98–105. [CrossRef]
8. Vagneron, I.; Daviron, B. Coffee in the jungle of environmental and social sustainability standards. *Cah. Agric.* **2012**, *21*, 154–161. [CrossRef]
9. Sibelet, N.; Montzieux, M. Resilience factors in the coffee sector of Kenya: From food security to product removal. *Cah. Agric.* **2012**, *21*, 179–191. [CrossRef]
10. Fournier, S. Geographical Indications: A way to perpetuate collective action processes within Localized Agrifood Systems? *Cah. Agric.* **2008**, *17*, 547–551. [CrossRef]
11. Ellis, E.A.; Baerenklau, K.A.; Marcos-Martínez, R.; Chávez, E. Land use/land cover change dynamics and drivers in a low-grade marginal coffee growing region of Veracruz, Mexico. *Agrofor. Syst.* **2010**, *80*, 61–84. [CrossRef]
12. Ávalos-Sartorio, B.; Blackman, A. Agroforestry price supports as a conservation tool: Mexican shade coffee. *Agrofor. Syst.* **2009**, *78*, 169–183. [CrossRef]
13. Leme, P.; Pinto, C. Sistemas de certificação do café sob a ótica dos Pilares da Qualidade. *Rev. Agrogeoambient.* **2019**, *10*, 9–26. [CrossRef]
14. Faure, G.; Le Coq, J.F.; Vagneron, I.; Hocde, H.; Munoz, G.S.; Kessari, M. Strategies of coffee producers' organizations in Costa Rica toward environmental and social certification processes. *Cah. Agric.* **2012**, *21*, 162–168. [CrossRef]
15. Aguilar, P.; Ribeyre, F.; Escarraman, A.; Bastide, P.; Berthiot, L. Sensory profiles of coffee in the Dominican Republic are linked to the terroirs. *Cah. Agric.* **2012**, *21*, 169–178. [CrossRef]
16. Galtier, F.; Pedregal, V.D. Can the development of Fair Trade improve justice? Some insights from the coffee case. *Cah. Agric.* **2010**, *19*, 50–57. [CrossRef]
17. Negash, M.; Starr, M.; Kanninen, M.; Berhe, L. Allometric equations for estimating aboveground biomass of *Coffea arabica* L. grown in the Rift Valley escarpment of Ethiopia. *Agrofor. Syst.* **2013**, *87*, 953–966. [CrossRef]
18. Coltri, P.P.; Zullo, J.J.; Dubreuil, V.; Ramirez, G.M.; Pinto, H.S.; Coral, G.; Lazarim, C.G. Empirical models to predict LAI and aboveground biomass of Coffea arabicaunder full sun and shaded plantation: A case study of South of Minas Gerais, Brazil. *Agrofor. Syst.* **2015**, *89*, 621–636. [CrossRef]
19. Jose, S.; Bardhan, S. Agroforestry for biomass production and carbon sequestration: An overview. *Agrofor. Syst.* **2012**, *86*, 105–111. [CrossRef]
20. Soto-Pinto, L.; Anzueto, M.; Mendoza, J.; Ferrer, G.J.; De Jong, B. Carbon sequestration through agroforestry in indigenous communities of Chiapas, Mexico. *Agrofor. Syst.* **2009**, *78*, 39–51. [CrossRef]
21. Schmitt-Harsh, M.; Evans, T.P.; Castellanos, E.; Randolph, J.C. Carbon stocks in coffee agroforests and mixed dry tropical forests in the western highlands of Guatemala. *Agrofor. Syst.* **2012**, *86*, 141–157. [CrossRef]
22. Häger, A. The effects of management and plant diversity on carbon storage in coffee agroforestry systems in Costa Rica. *Agrofor. Syst.* **2012**, *86*, 159–174. [CrossRef]
23. Pezzopane, J.R.M.; Souza, P.S.; Rolim, G.D.S.; Gallo, P.B. Microclimate in coffee plantation grown under grevillea trees shading. *Acta Sci. Agron.* **2011**, *33*. [CrossRef]
24. Alvarado-Huaman, L.; Borjas-Ventura, R.R.; Castro-Cepero, V.; Garcia-Nieves, L.; Jimenez-Davalos, J.; Julca-Otiniano, A.; Gomez-Pando, L. Dynamics of severity of coffee leaf rust (*Hemileia vastatrix*) on Coffee, in Chanchamayo (Junin-Peru). *Agron. Mesoam.* **2020**, *31*, 517–529. [CrossRef]
25. Lin, B.B. Agroforestry management as an adaptive strategy against potential microclimate extremes in coffee agriculture. *Agric. For. Meteorol.* **2007**, *144*, 85–94. [CrossRef]
26. Molin, R.; Andreotti, M.; Reis, A.; Furlani Junior, E.; Braga, G.; Scholz, M.B. Physical and sensory characterization of coffee produced in the topoclimatic conditions at Jesuítas, Paraná State (Brazil). *Acta Sci. Agron.* **2008**, *30*, 353–358. [CrossRef]
27. Peters, V.E.; Carroll, C.R. Temporal variation in coffee flowering may influence the effects of bee species richness and abundance on coffee production. *Agrofor. Syst.* **2012**, *85*, 95–103. [CrossRef]
28. Dauzat, J.; Rapidel, B.; Berger, A. Simulation of leaf transpiration and sap flow in virtual plants: Model description and application to a coffee plantation in Costa Rica. *Agric. For. Meteorol.* **2001**, *109*, 143–160. [CrossRef]

29. Dossa, E.L.; Fernandes, E.C.M.; Reid, W.S.; Ezui, K. Above- and belowground biomass, nutrient and carbon stocks contrasting an open-grown and a shaded coffee plantation. *Agrofor. Syst.* **2007**, *72*, 103–115. [CrossRef]
30. Piato, K.; Lefort, F.; Subía, C.; Calderon, D.; Pico, J.; Norgrove, L.; Caicedo, C. Effects of shade trees on robusta coffee growth, yield and quality. A meta-analysis. *Agron. Sustain. Dev.* **2020**. [CrossRef]
31. Lin, B.B. The role of agroforestry in reducing water loss through soil evaporation and crop transpiration in coffee agroecosystems. *Agric. For. Meteorol.* **2010**, *150*, 510–518. [CrossRef]
32. Flumignan, D.L.; De Faria, R.T.; Prete, C.E. Evapotranspiration components and dual crop coefficients of coffee trees during crop production. *Agric. Water Manag.* **2011**, *98*, 791–800. [CrossRef]
33. Lin, B.B.; Richards, P.L. Soil random roughness and depression storage on coffee farms of varying shade levels. *Agric. Water Manag.* **2007**, *92*, 194–204. [CrossRef]
34. Holwerda, F.; Bruijnzeel, L.A.; Barradas, V.L.; Cervantes, J. The water and energy exchange of a shaded coffee plantation in the lower montane cloud forest zone of central Veracruz, Mexico. *Agric. For. Meteorol.* **2013**, *173*, 1–13. [CrossRef]
35. Pereira, M.W.; Arêdes, A.F.; Santos, M.L. A irrigação do cafezal como alternativa econômica ao produtor. *Acta Sci. Agron.* **2010**, *32*. [CrossRef]
36. Arantes, K.R.; de Faria, M.A.; Rezende, F.C. Recuperação do cafeeiro (*Coffea arábica* L.) após recepa, submetido a diferentes lâminas de água e parcelamentos da adubação. *Acta Sci. Agron.* **2009**, *31*. [CrossRef]
37. Herpin, U.; Gloaguen, T.V.; Da Fonseca, A.F.; Montes, C.R.; Mendonça, F.C.; Piveli, R.P.; Melfi, A.J. Chemical effects on the soil–plant system in a secondary treated wastewater irrigated coffee plantation—A pilot field study in Brazil. *Agric. Water Manag.* **2007**, *89*, 105–115. [CrossRef]
38. D'haeze, D.; Raes, D.; Deckers, J.; Phong, T.A.; Loi, H.V. Groundwater extraction for irrigation of *Coffea canephora* in Ea Tul watershed, Vietnam—a risk evaluation. *Agric. Water Manag.* **2005**, *73*, 1–19. [CrossRef]
39. D'haeze, D.; Deckers, J.; Raes, D.; Phong, T.A.; Minh Chanh, N.D. Over-irrigation of Coffea canephora in the Central Highlands of Vietnam revisited. *Agric. Water Manage.* **2003**, *63*, 185–202. [CrossRef]
40. Siles, P.; Harmand, J.M.; Vaast, P. Effects of Inga densiflora on the microclimate of coffee (*Coffea arabica* L.) and overall biomass under optimal growing conditions in Costa Rica. *Agrofor. Syst.* **2009**, *78*, 269–286. [CrossRef]
41. Liu, X.; Qi, Y.; Li, F.; Yang, Q.; Yu, L. Impacts of regulated deficit irrigation on yield, quality and water use efficiency of Arabica coffee under different shading levels in dry and hot regions of southwest China. *Agric. Water Manag.* **2018**, *204*, 292–300. [CrossRef]
42. Boreux, V.; Vaast, P.; Madappa, L.; Cheppudira, K.G.; Garcia, C.; Ghazoul, J. Agroforestry coffee production increased by native shade trees, irrigation, and liming. *Agron. Sustain. Dev.* **2016**, *36*, 1–9. [CrossRef]
43. Liu, X.; Li, F.; Zhang, Y.; Yang, Q. Effects of deficit irrigation on yield and nutritional quality of Arabica coffee (*Coffea arabica*) under different N rates in dry and hot region of southwest China. *Agric. Water Manag.* **2016**, *172*, 1–8. [CrossRef]
44. Silva, N.; Assunção, W. Constatação do "efeito sombra" e economia de recursos hídricos e de energia na irrigação do cafeeiro por meio de um pivô central convencional. *Rev. Agrogeoambiental.* **2016**, *8*, 23–32. [CrossRef]
45. Marin, F.R.; Angelocci, L.R.; Nassif, D.S.P.; Costa, L.G.; Vianna, M.S.; Carvalho, K.S. Crop coefficient changes with reference evapotranspiration for highly canopy-atmosphere coupled crops. *Agric. Water Manag.* **2016**, *163*, 139–145. [CrossRef]
46. Sakai, E.; Barbosa, E.; Silveira, J.; Pires, R. Coffee productivity and root systems in cultivation schemes with different population arrangements and with and without drip irrigation. *Agric. Water Manag.* **2015**, *148*, 16–23. [CrossRef]
47. Suhartono, D.; Aditya, W.; Lestari, M.; Yasin, M. Expert System in Detecting Coffee Plant Diseases. *Int. J. Electr. Energy* **2013**, *1*, 156–162. [CrossRef]
48. Silva, M.C.; Varzea, V.; Guerra-Guimaraes, L.; Azinheira, H.; Fernandez, D.; Petitot, A.S.; Bertrand, B.; Lashermes, P.; Nicole, M. Coffee resistance to the main diseases: Leaf rust and coffee berry disease. *Braz. J. Plant Physiol.* **2006**, *18*, 119–147. [CrossRef]
49. Talhinhas, P.; Batista, D.; Diniz, I.; Vieira, A.; Silva, D.N.; Loureiro, A.; Tavares, S.; Pereira, A.P.; Azinheira, H.G.; Guerra-Guimarães, L.; et al. The Coffee Leaf Rust pathogen *Hemileia vastatrix*; One and a half centuries around the tropics. *Mol. Plant Pathol.* **2016**, *18*, 1039–1051. [CrossRef]
50. Zambolim, L. Current status and management of coffee leaf rust in Brazil. *Trop. Plant Pathol.* **2016**, *41*, 1–8. [CrossRef]
51. Capucho, A.S.; Zambolim, L.; Lopes, U.N.; Milagres, N.S. Chemical control of coffee rust in *Coffea canephora cv conilon*. *Australas. Plant Pathol.* **2013**, *42*, 667–673. [CrossRef]
52. Durand, N.; Gueule, D.; Fourny, G. Contaminants in coffee. *Cah. Agric.* **2012**, *21*, 192–196. [CrossRef]
53. Chemura, A.; Mutanga, O.; Sibanda, M.; Chidoko, P. Machine learning prediction of coffee rust severity on leaves using spectroradiometer data. *Trop. Plant Pathol.* **2018**, *43*, 117–127. [CrossRef]
54. Honorato-Junior, J.; Zambolim, L.; Aucique-Pérez, C.E.; Resende, R.S.; Rodrigues, F.A. Photosynthetic and antioxidative alterations in coffee leaves caused by epoxiconazole and pyraclostrobin sprays and *Hemileia vastatrix* infection. *Pest. Biochem. Physiol.* **2015**, *123*, 31–39. [CrossRef]
55. Dias, R.A.; Ribeiro, M.R.; Carvalho, A.M.; Botelho, C.E.; Mendes, A.G.; Ferreira, A.D.; Fernandes, F.C. Selection of coffee progenies for resistance to leaf rust and favorable agronomic traits. *Coffee Sci.* **2019**, *14*, 173–182. Available online: http://www.coffeescience.ufla.br/index.php/Coffeescience/article/view/1564 (accessed on 17 July 2021). [CrossRef]
56. De Carvalho, F.P.; França, A.C.; Lemos, V.T.; Ferreira, E.A.; Santos, J.B.; Dos Silva, A.A. Photosynthetic activity of coffee after application of glyphosate subdoses. *Acta Sci. Agron.* **2013**, *35*, 109–115. [CrossRef]

57. Oliveira, G.H.H.; Corrêa, P.C.; Botelho, F.M.; Goneli, A.L.D.; Afonso Júnior, P.C.; Campos, S.C. Modeling of the shrinkage kinetics of coffee berries during drying. *Acta Sci. Agron.* **2011**, *33*, 423–428. [CrossRef]
58. Tully, K.L.; Wood, S.A.; Lawrence, D. Fertilizer type and species composition affect leachate nutrient concentrations in coffee agroecosystems. *Agrofor. Syst.* **2013**, *87*, 1083–1100. [CrossRef]
59. Silva, C.F.; Ramos, D.M.B.; Batista, L.R.; Schwan, R.F. Inibição in vitro de fungos toxigênicos por *Pichia* sp. e *Debaryomyces* sp. isoladas de frutos de café (*Coffea arabica*). *Acta Sci. Agron.* **2010**, *32*, 397–402. [CrossRef]
60. Santa-Cecília, L.V.C.; Correa, L.R.B.; Souza, B.; Prado, E.; Alcantra, E. Desenvolvimento de Planococcus citri (Risso, 1813) (Hemiptera: Pseudococcidae) em cafeeiros. *Acta Sci. Agron.* **2009**, *31*, 13–15. [CrossRef]
61. Rodarte, M.P.; Dias, D.R.; Vilela, D.M.; Schwan, R.F. Proteolytic activities of bacteria, yeasts and filamentous fungi isolated from coffee fruit (*Coffea arabica* L.). *Acta Sci. Agron.* **2011**, *33*, 457–464. [CrossRef]
62. Bedimo, J.A.; Dufour, B.P.; Cilas, C.; Avelino, J. Effects of shade trees on Coffea Arabica pests and diseases. *Cah. Agric.* **2012**, *21*, 89–97. [CrossRef]
63. Malta, M.R.; Chagas, S.J. Avaliação de compostos não-voláteis em diferentes cultivares de cafeeiro produzidas na região sul de Minas Gerais. *Acta Sci. Agron.* **2009**, *31*, 57–61. [CrossRef]
64. Pedrosa, A.W.; Prieto Martinez, H.E.; Cruz, C.D.; DaMata, F.M.; Clemente, J.M.; Paula Neto, A. Characterizing zinc use efficiency in varieties of Arabica coffee. *Acta Sci. Agron.* **2013**, *35*, 57–61. [CrossRef]
65. Pozza, A.A.; Guimarães, P.T.; Silva, E.D.; Bastos, A.R.; Nogueira, F.D. Adubação foliar de sulfato de zinco na produtividade e teores foliares de zinco e fósforo de cafeeiros arábica. *Acta Sci. Agron.* **2009**, *31*, 49–55. [CrossRef]
66. Mora, A.; Beer, J. Geostatistical modeling of the spatial variability of coffee fine roots under Erythrina shade trees and contrasting soil management. *Agrofor. Syst.* **2012**, *87*, 365–376. [CrossRef]
67. Gomes, J.; Ponzo, A.; Oliveira, A. Viability of a terrace covered with porous concrete paving blocks for coffee bean drying. *Rev. Agrogeoambient.* **2021**, *12*, 98–109. [CrossRef]
68. Santos, F.L.; Queiroz, D.M.; Pinto, F.D.; Santos, N.T. Analysis of the coffee harvesting process using an electromagnetic shaker. *Acta Sci. Agron.* **2010**, *32*, 373–378. [CrossRef]
69. Greco, M.; Campos, A.T.; Klosowski, E.S. Variação de diferentes tempos de revolvimento em secador de camada fixa para café. *Acta Sci. Agron.* **2010**, *32*, 577–583. [CrossRef]
70. Resende, O.; Arcanjo, R.V.; Siqueira, V.C.; Rodrigues, S. Modelagem matemática para a secagem de clones de café (*Coffea canephora* Pierre) em terreiro de concreto. *Acta Sci. Agron.* **2009**, *31*, 189–196. [CrossRef]
71. Greco, M.; Campos, A.T.; Klosowski, E.S. Perdas térmicas em secador de café. *Acta Sci. Agron.* **2010**, *32*, 209–212. [CrossRef]
72. Chalfoun, S.M.; Pereira, M.C.; Carvalho, G.R.; Savian, T.V. Multivariate analysis of sensory characteristics of coffee grains (*Coffea arabica* L.) in the region of upper Paranaíba. *Acta Sci. Agron.* **2010**, *32*, 635–641. [CrossRef]
73. Fujisawa, N.; Roubik, D.W. Inoue, Makoto. Farmer influence on shade tree diversity in rustic plots of *Coffea canephora* in Panama coffee-agroforestry. *Agrofor. Syst.* **2020**, *94*, 2301–2315. [CrossRef]
74. Chaiyarat, R.; Sripho, S.; Ardsungnoen, S. Small mammal diversity in agroforestry area and other plantations of Doi Tung Development Project, Thailand. *Agrofor. Syst.* **2020**, *94*, 2099–2107. [CrossRef]
75. Mahata, A.; Samal, K.T.; Sharat, K.P. Butterfly diversity in agroforestry plantations of Eastern Ghats of southern Odisha, India. *Agrofor. Syst.* **2018**, *93*, 1423–1438. [CrossRef]
76. McDermott, M.E.; Rodewald, A.D.; Matthews Stephen, N. Managing tropical agroforestry for conservation of flocking migratory birds. *Agrofor. Syst.* **2015**, *89*, 383–396. [CrossRef]
77. Caudill, S.A.; Vaast, P.; Husband, T.P. Assessment of small mammal diversity in coffee agroforestry in the Western Ghats, India. *Agrofor. Syst.* **2014**, *88*, 173–186. [CrossRef]
78. Mukashema, A.; Veldkamp, T.; Amer, S. Sixty percent of small coffee farms have suitable socio-economic and environmental locations in Rwanda. *Agron. Sustain. Dev.* **2016**, *36*, 31. [CrossRef]
79. Valencia, V.; Naeem, S.; Garcia, B. Conservation of tree species of late succession and conservation concern in coffee. *Agric. Ecosyst. Environ.* **2016**, *219*, 32–41. [CrossRef]
80. Sri, A.; Kemp, R.C. The Impact of Coffee Certification on the Economic Performance of Indonesian Actors. *Asian J. Agric. Dev.* **2015**, *12*, 1–16. [CrossRef]
81. Martins, M.; Mendes, A.N.; Alvarenga, M. Incidência de pragas e doenças em agroecossistemas de café orgânico de agricultores familiares em Poço Fundo-MG. *Cienc. Agrotec.* **2004**, *28*, 1306–1313. [CrossRef]
82. Clarivate Web of Science. Available online: http://www.webofknowledge.com/ (accessed on 22 May 2021).
83. Vega-Muñoz, A.; Arjona-Fuentes, J.M. Social networks and graph theory in the search for distant knowledge in the field of industrial engineering. In *Handbook of Research on Advanced Applications of Graph Theory in Modern Society*; Pal, M., Samanta, S., Pal, A., Eds.; IGI-Global: Hershey, PA, USA, 2020; Volume 17, pp. 397–418. [CrossRef]
84. Price, D. A general theory of bibliometric and other cumulative advantage processes. *J. Assoc. Inf. Sci.* **1976**, *27*, 292–306. [CrossRef]
85. Dobrov, G.M.; Randolph, R.H.; Rauch, W.D. New options for team research via international computer networks. *Scientometrics* **1979**, *1*, 387–404. [CrossRef]
86. Bulik, S. Book use as a Bradford-Zipf Phenomenon. *Coll. Res. Libr.* **1978**, *39*, 215–219. [CrossRef]

87. Morse, P.M.; Leimkuhler, F.F. Technical note—Exact solution for the Bradford distribution and its use in modeling informational data. *Oper. Res.* **1979**, *27*, 187–198. [CrossRef]
88. Pontigo, J.; Lancaster, F.W. Qualitative aspects of the Bradford distribution. *Scientometrics* **1986**, *9*, 59–70. [CrossRef]
89. Swokowski, E.W. *Calculus with Analytic Geometry*, 4th ed.; Grupo Editorial Planeta: Mexico City, Mexico, 1988.
90. Kumar, S. Application of Bradford's law to human-computer interaction research literature. *DESIDOC J. Libr. Inf. Technol.* **2014**, *34*, 223–231.
91. Shelton, R.D. Scientometric laws connecting publication counts to national research funding. *Scientometrics* **2020**, *123*, 181–206. [CrossRef]
92. Lotka, A.J. The frequency distribution of scientific productivity. *J. Wash. Acad. Sci.* **1926**, *16*, 317–321.
93. Hirsch, J.E. An index to quantify an individual's scientific research output. *Proc. Natl. Acad. Sci. USA* **2005**, *102*, 16569–16572. [CrossRef]
94. Crespo, N.; Simoes, N. Publication performance through the lens of the h-index: How can we solve the problem of the ties? *Soc. Sci. Q.* **2019**, *100*, 2495–2506. [CrossRef]
95. Zipf, G.K. *Selected Studies of the Principle of Relative Frequency in Language*; Harvard University Press: Cambridge, MA, USA, 1932.
96. Moravcsik, M.J. Applied Scientometrics: An Assessment Methodology for Developing Countries. *Scientometrics* **1985**, *7*, 165–176. [CrossRef]
97. Frenken, K.; Hardeman, S.; Hoekman, J. Spatial scientometrics: Towards a cumulative research program. *J. Informetr.* **2009**, *3*, 222–232. [CrossRef]
98. Albort-Morant, G.; Henseler, J.; Leal-Millán, A.; Cepeda-Carrión, G. Mapping the field: A bibliometric analysis of green innovation. *Sustainability* **2017**, *9*, 1011. [CrossRef]
99. Mikhaylov, A.; Mikhaylova, A.; Hvaley, D. Knowledge Hubs of Russia: Bibliometric Mapping of Research Activity. *J. Scientometr. Res.* **2020**, *9*, 1–10. [CrossRef]
100. Acevedo-Duque, Á.; Vega-Muñoz, A.; Salazar-Sepúlveda, G. Analysis of Hospitality, Leisure, and Tourism Studies in Chile. *Sustainability* **2020**, *12*, 7238. [CrossRef]
101. Uribe-Toril, J.; Ruiz-Real, J.L.; Haba-Osca, J.; de Pablo Valenciano, J. Forests' First Decade: A Bibliometric Analysis Overview. *Forests* **2019**, *10*, 72. [CrossRef]
102. Bondanini, G.; Giorgi, G.; Ariza-Montes, A.; Vega-Muñoz, A.; Andreucci-Annunziata, P. Technostress Dark Side of Technology in the Workplace: A Scientometric Analysis. *Int. J. Environ. Res. Public Health* **2020**, *17*, 8013. [CrossRef]
103. Köseoglu, M.A.; Okumus, F.; Putra, E.D.; Yildiz, M.; Dogan, I.C. Authorship Trends, Collaboration Patterns, and Co-Authorship Networks in Lodging Studies (1990–2016). *J. Hosp. Mark. Manag.* **2018**, *27*, 561–582. [CrossRef]
104. Lojo, A.; Li, M.; Cànoves, G. Co-authorship Networks and Thematic Development in Chinese Outbound Tourism Research. *J. Chin. Tour. Res.* **2019**, *15*, 295–319. [CrossRef]
105. Vega-Muñoz, A.; Arjona-Fuentes, J.M.; Ariza-Montes, A.; Han, H.; Law, R. In search of 'a research front' in cruise tourism studies. *Int. J. Hosp. Manag.* **2020**, *85*, 102353. [CrossRef]
106. Gureev, V.N.; Mazov, N.A. Themes of the publications of an organization as a basis for forming an objective and optimal repertoire of scientific periodicals. *Sci. Tech. Inf. Proc.* **2013**, *40*, 195–204. [CrossRef]
107. Vega-Muñoz, A.; Gónzalez-Gómez-del-Miño, P.; Espinosa-Cristia, J.F. Recognizing New Trends in Brain Drain Studies in the Framework of Global Sustainability. *Sustainability* **2021**, *13*, 3195. [CrossRef]
108. Karakose, T.; Demirkol, M. Exploring the emerging COVID-19 research trends and current status in the field of education: A bibliometric analysis and knowledge mapping. *Educ. Process Int. J.* **2021**, *10*, 7–27. [CrossRef]
109. Zhu, X.A.; Liu, W.J.; Chen, J.; Bruijnzeel, L.A.; Mao, Z.; Yang, X.D.; Cardinael, R.; Meng, F.R.; Sidle, R.C.; Seitz, S.; et al. Reductions in water, soil and nutrient losses and pesticide pollution in agroforestry practices: A review of evidence and processes. *Plant Soil* **2020**, *453*, 45–86. [CrossRef]
110. Kullenberg, C.; Kasperowski, D. What is citizen science?—A scientometric meta-analysis. *PLoS ONE* **2016**, *11*, e0147152. [CrossRef]
111. Arun, K.B.; Madhavan, A.; Sindhu, R.; Binod, P.; Pandey, A.; Reshmy, R.; Sirohi, R. Remodeling agro-industrial and food wastes into value-added bioactives and biopolymers. *Ind. Crop. Prod.* **2020**, *154*, 112621. [CrossRef]

Article

Advances in Precision Coffee Growing Research: A Bibliometric Review

Lucas Santos Santana [1], Gabriel Araújo e Silva Ferraz [1], Alberdan José da Silva Teodoro [2], Mozarte Santos Santana [3], Giuseppe Rossi [4] and Enrico Palchetti [4,*]

1 Agricultural Engineering Department, Federal University of Lavras, P.O. Box 3037, Lavras 37200-000, Brazil; lucas.unemat@hotmail.com (L.S.S.); gabriel.ferraz@ufla.br (G.A.e.S.F.)
2 Department of Administration and Economics, Federal University of Lavras, P.O. Box 3037, Lavras 37200-000, Brazil; alberdan.teodoro@estudante.ufla.br
3 Institute of Natural Sciences, Federal University of Lavras, P.O. Box 3037, Lavras 37200-000, Brazil; ss.mozarte@gmail.com
4 Department of Agriculture, Food, Environment and Forestry (DAGRI), University of Florence, Via San Bonaventura, 13, 50145 Florence, Italy; giuseppe.rossi@unifi.it
* Correspondence: enrico.palchetti@unifi.it

Abstract: Precision coffee-growing technologies contribute to increased yield, operational efficiency, and final product quality. In addition, they strengthen coffee growing in the global agricultural scenario, which makes this activity increasingly competitive. Scientific research is essential for technological development and offering security regarding its application. For relevant research identification, bibliometric revision methods expose the best studies and their relationships with countries and authors, providing a complete map of research directions. This study identified the main contributions and contributors to academic research generation about precision coffee growing from 2000 to 2021. Bibliometric analysis was performed in VOSViewer software from the referential bases Scopus and Web of Science that identified 150 articles. Based on the number of citations, publications about precision coffee-growing showed Brazilian institutions at the top of the list, and Brazil's close relationships with North American and South African institutions. Geostatistical analysis, remote sensing and spatial variability mapping of cultivation areas were used in most experimental research. A trend in research exploring machine learning technologies and autonomous systems was evident. The identification of the main agents of scientific development in precision coffee growing contributes to objective advances in the development and application of new management systems. Overall, this analysis represents wide precision coffee growing research providing valuable information for farmers, policymakers, and researchers.

Keywords: precision agriculture; analysis; bibliometry; coffee farm; systematic review

1. Introduction

Coffee growing is among the primary agricultural activities in the world [1,2]. It represents an essential source of income for many countries [3,4]. Coffee is produced in about 60 countries, where tropical regions favor its development. Countries like Brazil, Vietnam and Colombia are the main world producers [5].

High rates of coffee yield result from application of technological practices during production and processing stages. Modern agriculture is characterized by the rapid expansion of information technologies arising from monitoring and control of storage, organization and agricultural activities [6]. Using techniques and technologies aimed at high levels of productivity combined with sustainability is known as precision agriculture [7]. Precision agricultural practical can maximize the potential of each region, making the crop more productive and favoring cost reduction [8].

Technological advances in precision agriculture contribute to obtaining accurate and reliable measurements in a crop. This can facilitate monitoring edaphoclimatic variables on a more accurate scale. Thus, designing fertilization plans, seedling selection and agricultural activities make agricultural production more effective [9]. Smart agriculture is crucial to maximizing crop yields and revenues and preserving natural resources [10].

Technologies drive the creation and segmentation of specific classes of precision agriculture. In coffee crops, such technological approaches are known as precision coffee growing. Alves et al. [11] described precision coffee growing as a set of techniques aimed at optimizing agricultural input (fertilizers, correctives, seeds and pesticides) in a function of spatial and temporal variability of factors associated with the ecosystem (water, soil, plant). Recently Kouadio et al. [12] described precision coffee growing as optimization of agricultural inputs (fertilizers, corrective and defensive) related to spatial and temporal variability of factors associated with the water-soil-plant and atmospheric system.

Crop coffee is cultivated mainly by small farmers, contributing to the low implementation of technology in the field, due to the absence of technical and financial inputs and pilot projects. The practical application in precision agriculture techniques was variable rate distribution, initially used in annual crops and adapted to other crops. Generally, cultures that depend on specific equipment for handling use solutions designed for other cultures, and these adaptations can take years.

The insertion of efficient precision coffee techniques in coffee crops can be found in many studies. When evaluating the transversal application of variable rate fertilizers, Andrade et al. [13] defined optimal lateral fertilizer distribution, and created an efficient and practical method for this type of analysis. Mapping plant attributes in a coffee crop, Ferraz et al. [14], demonstrated the importance of this mapping category for coffee crop management. Using aerial image obtained by remotely piloted aircraft, Santos et al. [15] proposed methods for estimating coffee biophysical parameters. Barros et al. [16] evaluated the operational performance of a fertilizer distribution system. These are some of the practical contributions in the literature.

Evaluating publications about precision coffee growing allows the analysis of studies carried out from planting to a producing the final product. Analyzing trends in research, perspectives and contributions of different actors is essential for assessing scientific literature concerning the development of precision coffee growing. Using techniques applied to literature reviews can create an overview of the subject. Applications of systematic reviews in agriculture are recent but have been shown to be effective in synthesizing knowledge about agricultural literature and indicating priorities for future research [17].

Making systematic reviews allows the selection of studies about a specific topic or interest area, highlighting what is already known and exposing future opportunities [18,19]. These studies establish explicit and rigorously applied criteria, facilitating their later reproduction [20]. Systematic reviews aim to answer a specific research question with a particular search strategy and a literature synthesis presentation [21]. It is essential to emphasize the criteria adopted during a systematic review to minimize bias or personal influences of the researcher in the results [22].

During the research process, scholars are interested in finding publications most relevant in a study area. Thus, researchers use citation tracking to identify the most relevant articles or journals for a particular area [23]. The bibliometric analysis technique contributes to searches by considering the differences between articles by levels of relevance [24,25].

Citation number, publication volume and relevant journals, among other categories, facilitate the scientific diagnosis of a specific area study [26]. Bibliometric analysis makes it possible to identify dynamics and possible trends in scientific production [27]. This method organizes the existing literature, showing its publications trajectory as well as traditional and emerging fields of research [28,29].

There are some bibliometric studies on agriculture in the scientific literature. Among them, Pallottino et al. [30] reported the importance of studies involving precision agriculture over a twenty year period, while Velasco-Muñoz et al. [31] portrayed global research about rainwater use concerning applications in irrigation systems for conservation and sustainability strategies. In another study that used bibliometrics with modeling topic,

Kane et al. [32] mapped research about perennial cultures by four scientific research bases. However, no bibliometric studies exist concerning issues related to precision coffee growing.

Mapping research about precision coffee growing has become important, given the significant technological advances reported in several studies carried out at different coffee cultivation stages. Identifying the most important literature about precision coffee growing can facilitate referential search processes and the identification of theoretical premises for future studies.

Given this importance, the objective of this study was to identify the main contributions of studies, researchers, entities and countries, most relevant in academic research about precision coffee growing over the last 20 years by exploring the referential bases Scopus and Web of Science. The results of this study may provide insights into research trends and contribute to research and scientific production practices.

2. Research Methodology

The evolution of precision coffee growing in scientific publications was evaluated by bibliometric analysis according to the procedures described in Figure 1. Bibliometric studies allow identification of possible theoretical trends, intellectual structures of a discipline or study area [33,34]. The work sequence in a bibliometric analysis is divided into data recovery, preprocessing, network extraction, normalization, mapping and visualization analysis [35,36].

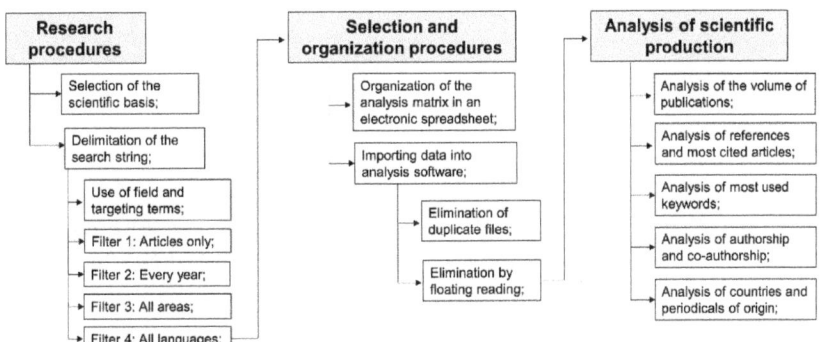

Figure 1. Processes systematization for bibliometric analysis.

2.1. Research Procedure

Scopus and Web of Science were selected for conducting the searches, aiming at a representative metadata content. The use of Scopus and Web of Science bases, due to their relevance in bibliometric studies, was a necessary prerequisite [37]. Searches in different scientific bases are essential for correct interpretation and bibliometric indicator use in scientific research evaluation [38,39]. Scientific approaches that adopted bibliometric analysis on other themes used at least one of these bases [40].

When starting a bibliometric analysis, it is necessary to define the search terms to eliminate the generalization of the results. For this, the series of key terms should be not too restrictive but sufficient to include only the topics of related studies [41]. Precision farming practices aimed at growing coffee are called "precision coffee farming" [11]. This definition contributes to string delimitation, selecting key terms and filtering only those files that depict precision agriculture in coffee culture. The key terms used were "spatial variability", "precision agriculture", "remote sensing", "soil mapping", "RPA", "UAV", "UAS" and "variable rate". Only publications that contained the key terms in the title, abstract or keywords were used.

In SCOPUS, the string TITLE-ABS-KEY (coffee) AND TITLE-ABS-KEY ("spatial variability" OR "precision agriculture" OR "remote sensing" OR "soil mapping" OR "RPA" OR "UAV" OR "UAS" OR "Variable rate") was used. In the WEB OF SCIENCE (WOS) database, the string was TS = (Coffee) AND TS = ("precision agriculture" OR "spatial variability" OR "remote sensing" OR "soil mapping" OR "RPA" OR "UAV" OR "UAS" OR "Variable rate"). Searches were not restricted in terms of academic area or languages. However, the selection of the document was restricted to articles published between 2000 and 2021/1st semester.

2.2. Selection and Organization Procedures

Selection and organization process consisted of reviewing the bibliometric data obtained. The searches resulted in 449 documents, 253 papers in Scopus and 196 papers in Web of Science. The next step was to remove duplicate articles because searches with similar parameters can find the same article. Then, documents were submitted to reading the abstracts and verifying similarity with the research theme. After these selections, 299 articles were excluded and 150 articles were chosen for use in this study.

Data were organized in an electronic spreadsheet and imported into VOSviewer bibliographic analysis software for identification and bibliometric networks analysis. VOSviewer is software for constructing and visualizing bibliometric networks. These networks can include journals, researchers and individual publications built on citation, bibliographic coupling, cocitation or coauthorship relationships [42]. In addition, they offer text mining functionality used in the construction and visualization of networks and co-occurrences of terms extracted from scientific literature [43].

2.3. Bibliometric Mapping and Clustering

Based on a multidimensional mapping technique VOSviewer locates the words in a dimensional space, portraying the distance between items according to their similarity. Results are presented in circle form, representing items found in the survey. These items are clustered and represented by color, forming a bibliometric map [44].

Quality criteria for research and journals are citations and scientific impact, as reported by Merton [45]. This rule was used for bibliometric mappings, which took account of annual evolution of publications and citations, leading researchers, most influential countries in publications related to this field, most notable journals, most relevant authors, main keywords used by authors, main keywords found in the most important publications, universities, entities related to these topics, the main areas of knowledge involved, and the trends and terms that indicate future lines of research.

3. Results and Discussion

3.1. Evolution of Publications

Bibliometric analyses found 150 articles about the precision management of coffee growing from 2000 to 2021/1st sem. The evolution of these publications is shown in Figure 2, illustrating the publications for each year.

Precision coffee research is relatively recent, as the first research found in a journal database was from 2004. This initial step in coffee research was performed by Herwitz et al. [46]. Although it was published in 2004, the experiments were carried out in 2002.

Four publications were found in the first years (2000 to 2006). Two articles were published in 2004 by the Herwitz and Johnson research groups, who used the same equipment and experimental field. In an analysis based on unmanned aerial vehicle (UAV) application to monitoring coffee trees, the authors advanced an essential step towards monitoring coffee fields by UAVs. Despite the pioneering nature of this technology in coffee growing, this type of analysis was not adopted by research groups in the coming years. The first hypothesis was related to the impossibility of carrying out similar experiments, because of high costs and few image capture and processing resources.

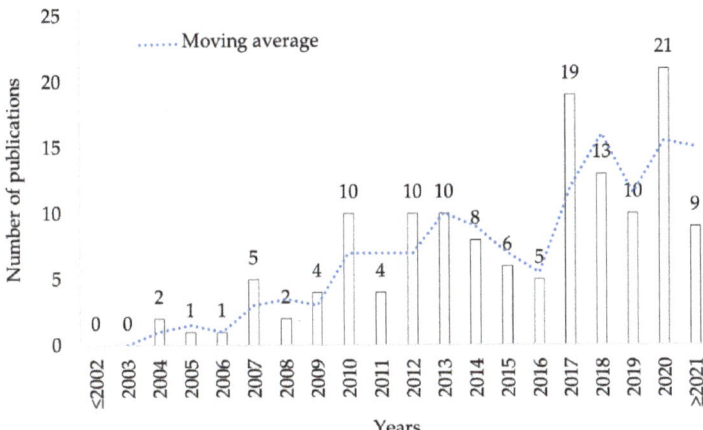

Figure 2. Evolution in precision coffee growing research publications from 2000 to 2021/1st sem.

In the following years, there was a significant increase in publications. From 2007 to 2013, most articles identified referred to spatial variability. During the period, essential discoveries were identified about variability, primary nutrient behavior and new ways of collecting soil for analysis.

A significant increase in research about precision coffee-growing demonstrated the coffee industry's interest in technological advances. Studies on the application of precise techniques in coffee management over the years have changed the technicians' and farmers' perceptions about coffee-growing. The development of such research is closely related to technological advances in agriculture. An important trend of publications on precision agriculture was presented in the research by Pallottino et al. [30], where a linear growth of publications about precision agriculture between 2000 and 2016 was demonstrated. When analyzing the academic progress of the precision coffee growing theme, a reduction in the number of publications between the years 2013 and 2016 stands out. These different publications concern the themes of "precision agriculture" and "precision coffee-growing" and how they may be related to a crop's characteristics, since in perennial crops, like coffee, vegetative development is reduced, making it time-consuming to obtain data compared to annual crops.

Another important aspect is the amount of research on the same topic. In some cases, the apparent research possibilities are exhausted in a few years. This may have happened in research related to the mapping of soil spatial variability in coffee crops, which led to a volume reduction in publications after 2013 and returning to a high level in 2017.

From 2017 onwards, the publication of articles on precision coffee growing showed a significant increase due to the application of new technologies in agriculture. The main finding after 2017 was the use of remote sensing for monitoring coffee production. In this period, the use of images obtained by Remotely Piloted Aircraft (RPA) was systematically explored.

3.2. Relevant Publications and Characteristics of Papers

Among the 150 files analyzed, ten papers were selected that stood out for having more than 20 citations from 2000 to 2021/1st sem (Table 1). The most cited author in the 20 years of analysis was Herwitz et al. [46]. This is due to the high level of technology used in the experiment available at the time. Furthermore, the authors' findings were applied again with the advent of UAVs in agriculture. The great impact of the research also is related to the journal in which it was published. Computers and Electronics in Agriculture journal is an important journal in agriculture.

Table 1. Top 20 publications scientific on precision coffee growing from 2000 to 2021/1st sem, ranked by citation number.

R	Title	Authors	PY	Journal	NC
1°	Imaging From An Unmanned Aerial Vehicle: Agricultural Surveillance And Decision Support	Herwitz, et al. [46]	2004	Computers and Electronics in Agriculture	277
2°	Separability Of Coffee Leaf Rust Infection Levels With Machine Learning Methods At Sentinel-2 Msi Spectral Resolutions	Chemura, et al. [47]	2017	Precision Agriculture	45
3°	Spatial Variability Of Leaf Wetness Duration In Different Crop Canopies	Sentelhas, et al. [48]	2005	International Journal of Biometeorology	45
4°	Spatial Variability Of Chemical Attributes And Productivity In The Coffee Cultivation	Silva [2], et al. [49]	2007	Ciencia Rural	40
5°	Spatial Variability Of Chemical Attributes And Coffee Productivity In Two Harvests	Silva [2], et al. [50]	2008	Ciencia e Agrotecnologia	41
6°	Spectral Analysis And Classification Accuracy Of Coffee Crops Using Landsat And A Topographic-Environmental Model	Cordero-Sancho and Sader [51]	2007	International Journal of Remote Sensing	38
7°	Spatial Variability Of Chemical Attributes Of An Oxisol Under Coffee Cultivation	Silva [1], et al. [52]	2010	Revista Brasileira de Ciencia do Solo	36
8°	Geostatistical Analysis Of Fruit Yield And Detachment Force In Coffee	Ferraz, et al. [53]	2012a	Precision Agriculture	33
9°	Feasibility Of Monitoring Coffee Field Ripeness With Airborne Multispectral Imagery	Johnson, et al. [46]	2004	Applied Engineering in Agriculture	32
10°	Spatial And Temporal Variability Of Phosphorus, Potassium And Of The Yield Of A Coffee Field	Ferraz, et al. [54]	2012b	Engenharia Agricola	31

R: Ranking; Silva [2]: Silva F.M.; Silva [1]: Silva S.D.A; PY: Publication Year and NC: Number of citations.

The most cited study, Herwitz et al. [46], demonstrated the positive aspects of agricultural areas monitored by unmanned aerial vehicles (UAV). The study described field data from combinations of red and infrared image aerial images, resulting in the definition of higher productivity zones, attesting to the efficiency of aerial remote sensing for agricultural monitoring with orbital imaging applications. Despite being published 16 years ago, this research is still used as a basis for various agricultural applications due to the nature of the techniques used.

Advances in remote sensing have been observed in coffee management. Relevant analyses about this technology are described in research by Chemura et al [47]. The authors evaluated applications of a Sentinel 2 sensor combined with Random Forest (RF) algorithms in the evaluation of coffee leaf rust (CLR) fungus, and demonstrated through vegetation indices the potential of remote sensing applications in identifying and discriminating levels of this fungus.

Among the most cited publications, the research developed by Sentelhas et al. [48] presented reliable methods for monitoring the duration of leaf wetness. Their results were based on installing sensors at different heights and evaluation by geometric mean regression. These results made important contributions to accurate precision irrigation practices and microclimate monitoring and evidenced spatial variability in the duration of wetness by rain, dew, and irrigation.

Pioneering various applications in coffee growing, Silva [2] et al. [49] characterized the spatial variability of chemical attributes of soil by georeferenced sampling and geostatistical techniques. Using the same experimental field, Silva [2] et al. [50] evaluated productivity of the 2002/2003 and 2003/2004 coffee harvests in georeferenced grids of 25×25 m^2. The data obtained were sufficient for geostatistical analysis such as semivariogram adjustments and kriging interpolation. In this study, the researchers defined the spatial dependence of chemical attributes and coffee crop yield. Silva's research clarified the wide range of soil chemical attributes justifying the study of variable rate fertilizer application in coffee plantations, which in one the best discoveries about the spatial variability of soil in coffee cultivation.

Among the most cited research, an article by Cordero-Sancho, Sader [51] contributed to precision coffee growing development using remote sensing technologies. Using Landsat satellite images combined with geoprocessing techniques, the authors defined optimal regions for growing coffee, which was the first of several analyzes on remote sensing applications in spatial variability for coffee growing.

Regarding mapping studies of soil variability in coffee culture, Silva [1] et al. [52] evaluated the main chemical attributes including available P, Na, and S, exchangeable Ca, Mg and Al, pH, H + Al, SB, t, T, V, m, MO, ISNa, P-remnant and micronutrients (Zn, Fe, Mn, Cu and B). Multivariate analysis techniques associated with geostatistics facilitated the assessment of soil variability. These authors demonstrated the applicability of mapping the behavior of these nutrients in the soil.

Equipment adjustments for mechanized harvesting operations in coffee farming require extensive information about plant physiology and anatomical factors. The paper of Ferraz et al. [53] used geostatistics to evaluate the detachment strength of coffee fruits in a study carried out on 22 hectares of Arabica coffee. The authors showed the possibility of detachment strength for characterizing spatial patterns of coffee fruits, classified as green or ripe by semivariogram and kriging. They found that exponential functions adjusted in the semivariogram described the structure and magnitude of spatial variation of release strength of green fruits and coffee yield.

Johnson et al. published in 2004 a pioneering article for monitoring coffee maturation by a UAV. It proposed a method to identify the coffee fruit maturation through reflectance in the aerial image. Field collections aggregated the results. The average maturation index per field was significantly correlated with soil-based counts recorded by the producer. This work is still the basis for research using aerial scenes to monitor coffee tree.

Using precision agriculture technologies, localized data collection, and geostatistical analysis techniques, Ferraz et al. [54] monitored chemical soil attributes during three consecutive harvests to optimize application of phosphorus and potassium. The study showed that semivariograms allow estimates of the spatial variability of soil chemical attributes, such as amounts of phosphorus and potassium, and their effects on coffee crop yield. This research complemented previous results on the relationship between spatial variability and yield.

The primary research related to precision coffee growing was mainly associated with soil variability (Table 1), but the essential contribution of remote sensing for the mapping of variability in the coffee crop is evident.

3.3. Most Influential Journals

Journals are ranked in order of importance by number of citations (Table 2). When analyzing the journals in Table 2, variations in their specificities were observed, but there was a predominance of journals with technological approaches. The journals "Computers and Electronics in Agriculture" and "Precision Agriculture" significantly contributed to technological development in agriculture. Pallottino et al. [30] carried out bibliometric research to demonstrate advances in precision agriculture and showed that the journals "Computers and Electronics in Agriculture" and "Precision Agriculture" predominate among the most important journals. A journal linked to remote sensing also appeared in this classification, indicating the potential use of this technology in coffee production.

Table 2. Top 6 sources of publications in word on precision coffee growing from 2000 to 2021/1st sem.

R	Journal	SJR [1]	CiteScore [2]	JCR [3]	H-i	ISSN	ND	NC
1°	Computers and Electronics in Agriculture [46]	1.208	8.6	3.858	115	0168-1699	5	409
2°	Precision Agriculture [8–10]	1.023	8.7	4.454	63	1385-2256	9	398
3°	Revista Brasileira de Ciência do Solo [52]	0.505	2.5	1.2	51	0100-0683	8	291
4°	Engenharia Agrícola [54]	0.289	1.4	0.603	27	0100-6916	11	256
5°	IEEE Journal of Selected Topics in Applied Earth Observations and Remote Sensing [8]	1.246	7.2	3.827	88	1939-1404	4	190
6°	Ciência e Agrotecnologia [49,52]	0.437	2.3	1.144	30	1413-70	4	152

[1]: Web of Science index, [2]: Scopus index, [3]: Scopus index, H-i: H index, ND: Number of documents and NC: Number of citations.

Table 2 shows that the majority of the obtained journals are from Brazil, probably because of intensive coffee production in the country.

Even with greater inclusion in the best journals, the country does not occupy first place. This is due to the quality of the journals (H index). The journals "Computers and Electronics in Agriculture" and "Precision Agriculture" are considered emerging in studies for technological application in agriculture as reported by [55]. It was observed that despite having fewer publications, these journals had a larger number of citations. This indicates high interest in searching for publications involving specialized applications in agriculture.

3.4. Publications by Authors

The H-index, which is obtained by the ratio of the number publications and their citations, was used to determine the author's impact on the topic of precision coffee growing. From the H-index values, the Scopus and WoS bases, and the volume of publications, the main authors of publications related to "precision coffee growing" were selected. Among the 186 identified, only 28 authors met the selection criteria established in the bibliometric selection methodology

According to established premises, Professor Fábio Moreira da Silva, from the Agricultural Engineering Department of Federal University of Lavras was the author with the greatest academic impact, with an H-Index of 12 (Scopus and WoS), 20 published documents and 303 citations, followed by Professor Gabriel Araújo e Silva Ferraz also from the Agricultural Engineering Department of Federal University of Lavras, with an H-index of 10 (Scopus) and 5 (WoS), 16 documents published and 203 citations. Details of the other authors can be seen in Table 3.

Table 3. Top six relevant authors of publications on precision coffee growing from 2000 to 2021/1st sem.

R	Authors	Id.	H-i (Scopus)	H-i (WoS)	NC	ND
1°	Fábio Moreira da Silva [13–16,52]	Silva, F. M.	12	12	303	20
2°	Gabriel Araújo e Silva Ferraz [6,13–15]	Ferraz, G. A. S.	10	5	203	16
3°	Marcelo Silva de Oliveira [14,50]	Oliveira, M. S.	10	9	192	11
6°	Ivoney Gontijo [55]	Gontijo, I.	6	6	139	8
4°	Julião Soares de Souza Lima [52]	Lima, J. S. S.	11	10	129	9
5°	Samuel de Assis Silva [52]	Silva, S. A.	11	5	117	9

NC: Number of citations, ND: Number of documents, H-i: H index.

By identifying the main authors with documents indexed in the Scopus and WoS databases, the relationships among them were obtained. Only authors who had at least nine citations were selected. This criterion made it possible to classify the 44 authors shown in Figure 3.

The cocitation network is represented by circle charts, in which the size represents the author's influence, and the color of the circle represents the cluster (knowledge area) to which it was grouped. Therefore, it was possible to establish similarities, differences, relations and relevance between members that represent the intellectual base concerning the "precision coffee growing" theme.

By analyzing the cocitation network among the authors, three large clusters were determined. The first cluster, in green, is formed by the presence of three main researchers linked to Federal University of Lavras, with the largest volume of documents. Its main approaches refer to spatial variability of the coffee crop from an agricultural engineering perspective, such as collection network, variable rate application, and yield mapping. Numerical systems and models needed to support decisions about soil fertilization and agricultural management were also observed in this cluster (Figure 3).

In the second cluster, in red, the main focus funded in the research was soil attributes. These authors are linked to North American universities and their research covers topics that aim to understand the location of these nutrients in the soil and their physicochemical characteristics, aimed at better nutrient use and soil conservation. In this cluster, geostatistical techniques for mapping spatial variability stand out. The use of geostatistical techniques in precision coffee growing was also observed in the bibliometric analyzes carried out by [56].

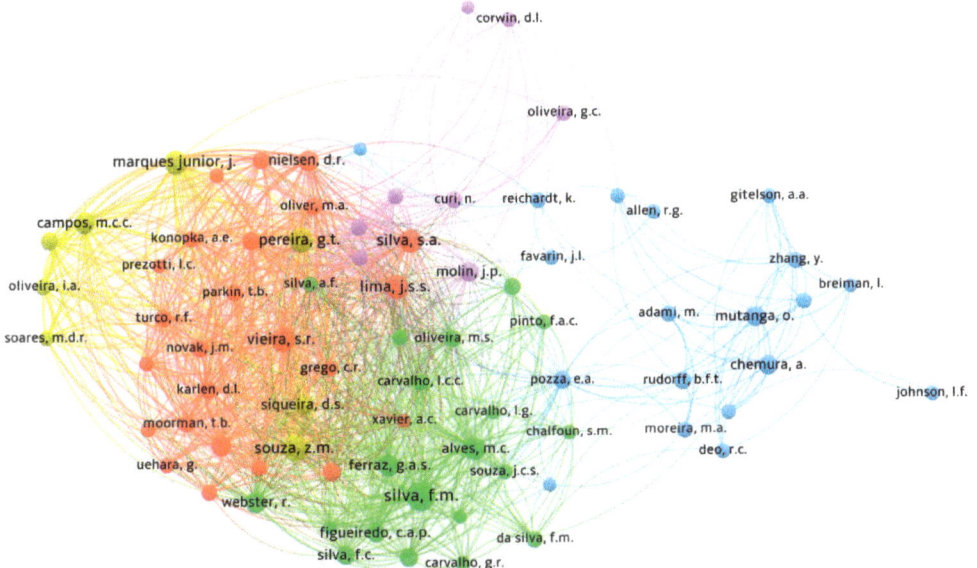

Figure 3. Scientific mapping of the cocitation of authors most relevance in precision coffee growing research. Red and yellow: Solo. Green: variable rate application and productivity mapping. Blue: remote sensing and Purple: plant nutritional status.

The researchers related to the third cluster, in blue, are characterized by research in coffee-growing by remote sensing analysis. Mapping coffee plantations by remote sensing aims to contribute to the identification of spatial variability using spectral responses [57].

3.5. Most Influential Countries

Evaluation of knowledge-producing nations on precision coffee-growing allowed them to be classified according to the number of citations over the years. Publications by country about "precision coffee growing" is shown in Figure 4. The main countries that produce the most scientific knowledge about precision coffee growing were identified. The predominance of Brazilian researchers in the top positions of publications by authors made Brazil the main country contributing to the development of precision coffee farming (Table 3). The 42 most impactful publications about precision coffee growing were carried out by Brazilian researchers.

The economic importance of coffee growing in Brazil, and the large number of research and teaching organizations related to coffee research in the country, impacts directly knowledge development about precision coffee growing. Brazil stands out as one of the countries with the highest investment in research and development in agriculture. These characteristics, associated with great territorial extension, has kept Brazil the leader in agricultural exports [58].

At the date of this study, Brazil is followed by countries such as the United States (four documents) and Colombia (three documents). The extensive presence of Brazilian researchers and journals also made Brazil the top country in producing scientific studies about precision coffee growing.

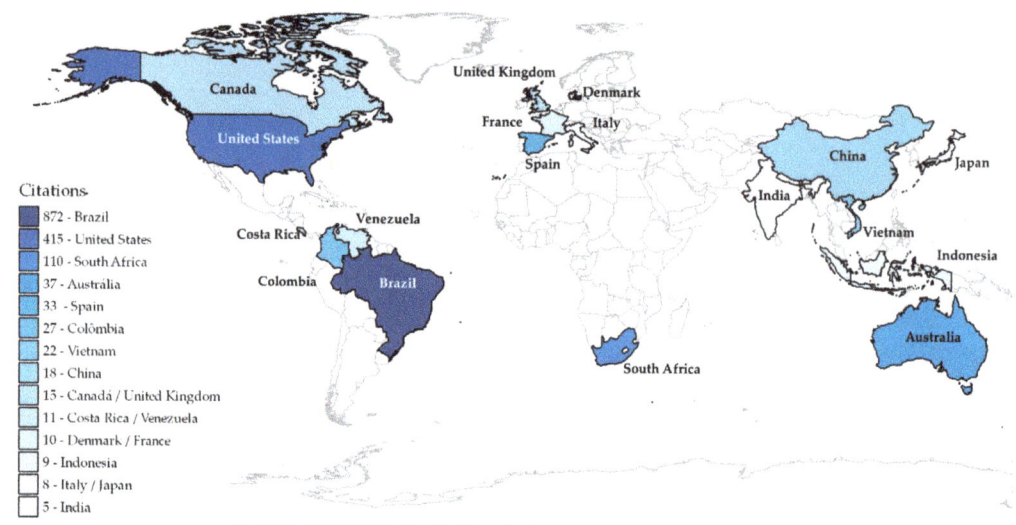

Figure 4. Number of citations by Country.

Although the cultivation of coffee in the United States is not expressive, this country is the second-largest producer of knowledge about precision coffee growing. This is due to the coffee area present in the Hawaii region, and the large number of educational and research organizations related to agricultural sciences in USA. It is important to highlight that pioneering work about precision coffee growing was carried out by Herwitz et al. [46] and Johnson, et al. [59], both in the American state of Hawaii (Figure 4).

3.6. Organizations Related to Precision Coffee Growing' Research

Identifying the organizations responsible for developing a knowledge area is of fundamental importance in biometric analysis, as it allows establishing trends and relationships between these organizations.

Research entities responsible for developing knowledge about precision coffee growing were identified. The relationships among scientific organizations that produce knowledge about this theme is presented in Figure 5. In this study, 31 organizations were highlighted with the highest volume of publications among 155 organizations identified and linked to authors (Figure 5).

Five groups were defined showing the great contribution of Brazilian universities in research development on precision coffee growing. The main institution was Federal University of Lavras, identified in the center region of the map in red. In the map, this university is linked with almost all other institutions. Directly or indirectly, this university shares research with institutions and internationals research centers, evidencing a strong relationship between Brazil and international institutions. The exchange of research within the country can be seen by the proximity between the red and blue groups, which occurs by the geolocation of these institutions. This geographic proximity facilitates the exchange of congresses and events.

Despite showing low association with each other, the grouping in green demonstrates the proximity between institutions from the United States of America and institutions from South Africa. In this grouping, a Brazilian university is seen as the "Federal University of Alfenas". This connection occurred due to the proximity of researchers to institutions in the United States of America and South Africa.

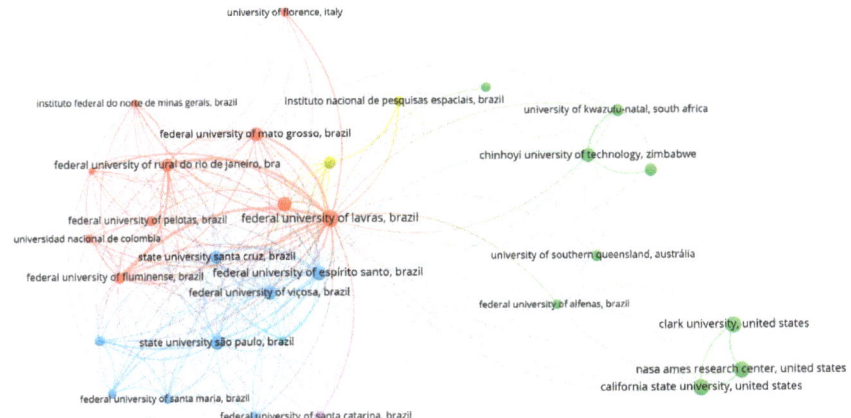

Figure 5. Scientific mapping network of educational and/or research organizations that produce knowledge about precision coffee growing. Red: the main institution was Federal University of Lavras. Yellow: "Company of technical as-sistance and rural extension of the state of Minas Gerais" and "National Institute for Space Research -I NPE". Green:proximity between institutions from the United States of America and institutions from South Africa. Blue:exchange of research within the country.

The group in yellow is represented by two institutions "Company of technical assistance and rural extension of the state of Minas Gerais" and "National Institute for Space Research -I NPE". Despite connections, this shows that these institutions follow different directions from Brazilian universities.

The analysis shows the relevance of Brazilian organizations in scientific research development about precision coffee growing, with emphasis on the Federal University of Lavras. A systematic bibliometric analysis of literature carried out by Cruz-O'Byrne et al. [60], showed the strong relationship of the Federal University of Lavras (UFLA) with coffee research. In searches performed on the Web of Science and Scopus databases, Pabon et al. [61], organized bibliometric data on coffee growing in which they also highlighted UFLA's contributions to scientific approaches to coffee crops.

The location of the Federal University of Lavras in the south of Minas Gerais state, a region with the largest coffee production in Brazil, contributed to UFLA assuming a very important role in coffee research. In the 2020 harvest, Minas Gerais produced more than 51% of national coffee production (Conab, 2020). The high productivity of this region, favored and driven by edaphoclimatic conditions, attracts researchers and installations concerning the coffee crop. Bibliometric studies about coffee growing presented by Sott et al. [56], highlighted Brazilian research dominance on coffee growing and its important role in agribusiness development.

3.7. Keywords Related to Precision Coffee Growing

Another way of investigating the study field is to analyze authors' keywords with the highest occurrence rates in all documents. In this phase, words with at least two occurrences are selected. Figure 6 presents analysis of cooccurrence of authors' keywords in analyzed documents.

Among 369 keywords identified in the studies, only 64 met adopted criteria. As a result, the "precision agriculture" term appeared most frequently, with 42 occurrences, followed by the terms "geostatistics" (40 occurrences), "remote sensing" (17 occurrences), "coffee" (14 occurrences), "Coffea arabica" (13 occurrences) and "spatial variability" (10 occurrences). In this figure it is possible to identify four distinct groups: red, representing technological applications; blue, analyses of canephore coffee; green, research related to monitoring of soil properties, and yellow, remote sensing applications. The groups have a strong connection with the areas of precision agriculture and geostatistics. This indicates that all applications for improvement in management are aimed at precise practices in coffee growing. The presentation of this map also contributes to searches for publications related to specific fields of precision coffee growing and how authors should organize their keywords for easy viewing.

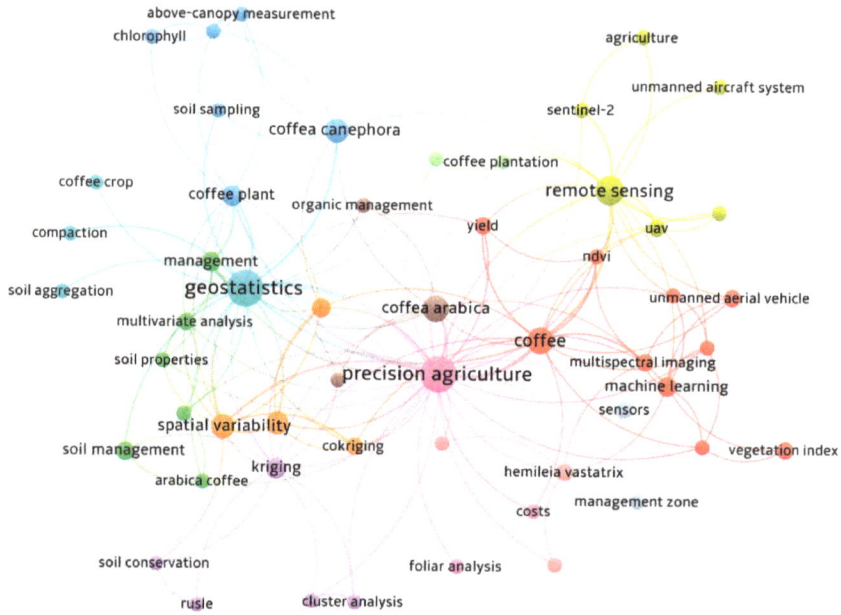

Figure 6. Map of network among author's keywords. Lines indicate co-occurrences between terms. Yellow: remote sensing. Red: remote sensing and machine learning. Green and purple: spatial variability of soil attributes. Azul: technologies applied to the cultivation of coffee *canephora*. Orange: application of techniques for mapping soil attributes.

3.8. Trends in Precision Coffee Growing Research

The surveys followed trends according to equipment availability, use of technologies and the value of theme to region. A map was created using a fractional counting method based on bibliographic data in the authors' keyword co-occurrences to understand trends (Figure 7). This map uses different colors to highlight the most commonly used author keywords over the last 20 years.

The information presented in Figure 7 demonstrates the characterization of predominant groups. Three prominent circles stood out: "precision agriculture", "remote sensing" and "geostatistic."

Precision agriculture appears as a trend in precision coffee growing. This occurs because techniques used in precision agriculture are tested in coffee growing, providing a basis for the development of several methods. From 2010 to 2020, there is a grouping in yellow colors and the relationship between "precision agriculture", "geostatistics", and "spatial variability" systematically explored at that time. The saturation of these keywords in searches began in 2018, making this technique well researched. In the following years, remote sensing techniques were again used with the advance of unmanned aerial vehicles.

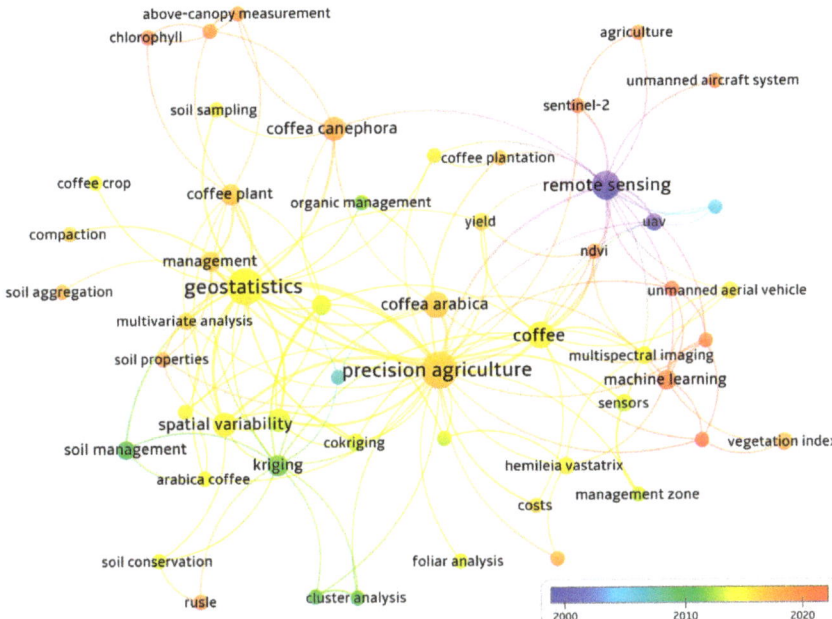

Figure 7. Map based on the co-occurrence of the authors' keywords and evolution from 2000 to 2021/1st sem. The color scale represents the year of keyword predominance.

Research related to remote sensing applications in precision coffee farming is considered pioneering. However, remote sensing technology has been exploited for the last 20 years and continues to be used. Figure 7 shows new trends in this technology, namely the words "multispectral imaging", "unmanned aerial vehicle", "ndvi", "sentinel" and "machine learning". The emergence of these trends is directly related to applications of remotely piloted aircraft (RPA) in agriculture, bringing to this field technological trends about machine learning.

New research involving precision coffee growing has explored automation profiles, aimed at improvements of crop management, such as mini sensors use to monitor coffee crops in real time [62], capacity evaluation of an Extreme Learning Machine (ELM) model when analyzing soil fertility properties, and the precise estimate of Robusta coffee yield [12]. Spatial determination of nitrogen content in coffee leaves has been made using remotely piloted aircraft, with machine learning techniques to classify aerial images [63]. Orbital sensors are used as a new methodology for obtaining maps about growth deficit (with up to 5 cm precision and 1m spatial resolution), as well as the use of Differential Interferometric Synthetic Aperture Radar—D-InSAR [64].

4. Conclusions

Intellectual base analysis by bibliometric methods allowed evaluation of scientific evolution, research, and authorial references about precision coffee growing. It was possible to infer current conditions and trends regarding the research and scientific publication theme. The main countries, journals, scientific organizations, researchers, and cocitations networks with the greatest relevance about precision coffee growing were highlighted.

There has been a significant increase in scientific publications about precision coffee growing in the last 20 years (2000 to 2021/1st sem). This research solved essential obstacles in the sector and proposed sustainable management methods. The development of precision coffee growing was mainly marked by research to solve spatial variability in soils and plants, contributing to essentials technological advancements such as agricultural input application at a variable rate.

Among the most used technologies in precision coffee growing, remote sensing stands out. This knowledge area has contributed to coffee-growing development since initial research efforts. Furthermore, an emerging area with the advent of remotely piloted aircraft (RPA) has been developed.

The advance of technologies applied in precision coffee growing was demonstrated by keyword mappings in the most important scientific journals. The main keywords used in studies in recent years were "remote sensing," "machine learning," "vegetation index," and "remotely piloted aircraft", which demonstrates strong trends in automated applications using remote sensing technologies.

The development of this research is mainly linked to coffee-producing countries. Brazil's relevance to scientific knowledge development about precision coffee growing is evident since the country was the leader in terms of publication numbers about precision coffee growing. The Brazilian institution Federal University of Lavras (UFLA) was responsible for the origin of most studies. Most of the studies developed about precision techniques and practices adopted in coffee cultivation have been carried out in the last five years, culminating in the emergence of research produced by countries in the American, European, and African continents.

Supplementary Materials: The following are available online at https://www.mdpi.com/article/10.3390/agronomy11081557/s1, supplementary file 1: scopus, WoS.csv; supplementary file 2: papers used in study.

Author Contributions: Conceptualization, L.S.S. and G.A.e.S.F.; methodology, A.J.d.S.T. and M.S.S.; software, A.J.d.S.T.; validation, L.S.S., G.A.e.S.F. and A.J.d.S.T.; formal analysis, E.P. and G.R.; investigation, M.S.S.; resources, L.S.S. and A.J.d.S.T.; data curation, M.S.S. and G.A.e.S.F.; writing—original draft preparation, L.S.S. and M.S.S.; writing—review and editing, G.A.e.S.F. and L.S.S.; visualization, G.R. and E.P.; supervision, G.A.e.S.F.; project administration, G.A.e.S.F.; funding acquisition, G.A.e.S.F., G.R. and E.P. All authors have read and agreed to the published version of the manuscript.

Funding: This research was funded by the Embrapa Café—Consórcio Pesquisa Café, project approved no 234/2019, the National Council for Scientific and Technological Development (CNPq), the Research Support Foundation of the State of Minas Gerais (FAPEMIG), the Coor-dination for the Improvement of Higher Education Personnel (CAPES), the Federal University of Lavras (UFLA) and University of Firenze (UniFI).

Conflicts of Interest: The authors declare no conflict of interest.

References

1. Chain-Guadarrama, A.; Martínez-Salinas, A.; Aristizábal, N.; Ricketts, T.H. Ecosystem services by birds and bees to coffee in a changing climate: A review of coffee berry borer control and pollination. *Agric. Ecosyst. Environ.* **2019**, *280*, 53–67. [CrossRef]
2. Marin, D.B.; Alves, M.D.C.; Pozza, E.A.; Belan, L.L.; Freitas, M.L.D.O. Multispectral radiometric monitoring of bacterial blight of coffee. *Precis. Agric.* **2018**, *20*, 959–982. [CrossRef]
3. Belan, L.L.; Junior, W.C.D.J.; De Souza, A.F.; Zambolim, L.; Filho, J.C.; Barbosa, D.H.S.G.; Moraes, W.B. Management of coffee leaf rust in Coffea canephora based on disease monitoring reduces fungicide use and management cost. *Eur. J. Plant Pathol.* **2020**, *156*, 683–694. [CrossRef]
4. Júnior, P.P.; Moreira, B.C.; Silva, M.D.C.S.D.; Veloso, T.G.R.; Stürmer, S.L.; Fernandes, R.B.A.; Mendonca, E.; Kasuya, M.C.M. Agroecological coffee management increases arbuscular mycorrhizal fungi diversity. *PLoS ONE* **2019**, *14*, e0209093. [CrossRef]
5. U.S. Department of Agriculture. *Coffee: World Markets and Trade*; U.S. Department of Agriculture, U.S. Government Printing Office: Washington, DC, USA, 2021.
6. Santana, L.; Ferraz, G.; Cunha, J.; Santana, M.; Faria, R.; Marin, D.; Rossi, G.; Conti, L.; Vieri, M.; Sarri, D. Monitoring Errors of Semi-Mechanized Coffee Planting by Remotely Piloted Aircraft. *Agronomy* **2021**, *11*, 1224. [CrossRef]
7. Cadenas, J.; Garrido, M.; Martínez-España, R.; Guillén-Navarro, M. Making decisions for frost prediction in agricultural crops in a soft computing framework. *Comput. Electron. Agric.* **2020**, *175*, 105587. [CrossRef]
8. Murugan, D.; Garg, A.; Singh, D. Development of an Adaptive Approach for Precision Agriculture Monitoring with Drone and Satellite Data. *IEEE J. Sel. Top. Appl. Earth Obs. Remote Sens.* **2017**, *10*, 5322–5328. [CrossRef]
9. Paccioretti, P.; Córdoba, M.; Balzarini, M. FastMapping: Software to create field maps and identify management zones in precision agriculture. *Comput. Electron. Agric.* **2020**, *175*, 105556. [CrossRef]

10. Yost, M.A.; Kitchen, N.R.; Sudduth, K.A.; Massey, R.E.; Sadler, E.J.; Drummond, S.T.; Volkmann, M.R. A long-term precision agriculture system sustains grain profitability. *Precis. Agric.* **2019**, *20*, 1177–1198. [CrossRef]
11. Alves, E.A.; Queiroz, D.M.; Pinto, F.A.C. Cafeicultura de precisão. In *Boas Práticas Agrícolas na Produção de Café*; Zambolim, L., Ed.; UFV: Viçosa, Brazil, 2007; p. 234. ISBN 8560027157.
12. Kouadio, L.; Deo, R.C.; Byrareddy, V.; Adamowski, J.F.; Mushtaq, S.; Nguyen, V.P. Artificial intelligence approach for the prediction of Robusta coffee yield using soil fertility properties. *Comput. Electron. Agric.* **2018**, *155*, 324–338. [CrossRef]
13. Andrade, A.D.; Ferraz, G.A.E.S.; De Barros, M.M.; Faria, R.D.O.; Da Silva, F.M.; Sarri, D.; Vieri, M. Characterization of the Transverse Distribution of Fertilizer in Coffee Plantations. *Agronomy* **2020**, *10*, 601. [CrossRef]
14. Ferraz, G.A.E.S.; Da Silva, F.M.; De Oliveira, M.S.; Custódio, A.A.P.; Ferraz, P. Variabilidade espacial dos atributos da planta de uma lavoura cafeeira. *Rev. Cienc. Agron.* **2017**, *48*, 81–91. [CrossRef]
15. Dos Santos, L.M.; Ferraz, G.A.E.S.; Barbosa, B.D.D.S.; Diotto, A.V.; Maciel, D.T.; Xavier, L.A.G. Biophysical parameters of coffee crop estimated by UAV RGB images. *Precis. Agric.* **2020**, *21*, 1227–1241. [CrossRef]
16. Barros, M.M.; Volpato, C.E.S.; Silva, F.C.; Palma, M.A.Z.; Spagnolo, R.T. Avaliação de um sistema de aplicação de fertili-zantes a taxa variável adaptado à cultura cafeeira. *Coffee Sci.* **2015**, *10*, 223–232.
17. Koutsos, T.M.; Menexes, G.C.; Dordas, C.A. An efficient framework for conducting systematic literature reviews in agricultural sciences. *Sci. Total Environ.* **2019**, *682*, 106–117. [CrossRef]
18. Pollock, M.; Fernandes, R.M.; Becker, L.A.; Featherstone, R.; Hartling, L. What guidance is available for researchers conducting overviews of reviews of healthcare interventions? A scoping review and qualitative metasummary. *Syst. Rev.* **2016**, *5*, 1–15. [CrossRef]
19. Sharma, G.; Bansal, P. Partnering Up: Including Managers as Research Partners in Systematic Reviews. *Organ. Res. Methods* **2020**, 1–30. [CrossRef]
20. Souza, V.H.S.; Dias, G.L.; Santos, A.A.R.; Costa, A.L.G.; Santos, F.L.; Magalhães, R.R. Evaluation of the interaction between a harvester rod and a coffee branch based on finite element analysis. *Comput. Electron. Agric.* **2018**, *150*, 476–483. [CrossRef]
21. Centobelli, P.; Cerchione, R.; Chiaroni, D.; Del Vecchio, P.; Urbinati, A. Designing business models in circular economy: A systematic literature review and research agenda. *Bus. Strat. Environ.* **2020**, *29*, 1734–1749. [CrossRef]
22. Coman, M.A.; Marcu, A.; Chereches, R.M.; Leppälä, J.; Broucke, S.V.D. Educational Interventions to Improve Safety and Health Literacy Among Agricultural Workers: A Systematic Review. *Int. J. Environ. Res. Public Health* **2020**, *17*, 1114. [CrossRef]
23. Seuring, S.; Gold, S. Conducting content-analysis based literature reviews in supply chain management. *Supply Chain Manag. Int. J.* **2012**, *17*, 544–555. [CrossRef]
24. Daim, T.U.; Rueda, G.; Martin, H.; Gerdsri, P. Forecasting emerging technologies: Use of bibliometrics and patent analysis. *Technol. Forecast. Soc. Chang.* **2006**, *73*, 981–1012. [CrossRef]
25. Liu, W.; Gu, M.; Hu, G.; Li, C.; Liao, H.; Tang, L.; Shapira, P. Profile of developments in biomass-based bioenergy research: A 20-year perspective. *Scientometrics* **2013**, *99*, 507–521. [CrossRef]
26. Andrade-Valbuena, N.A.; Merigo-Lindahl, J.M.; Olavarrieta, S. Bibliometric analysis of entrepreneurial orientation. *World J. Entrep. Manag. Sustain. Dev.* **2019**, *15*, 45–69. [CrossRef]
27. Sharifi, A.; Simangan, D.; Kaneko, S. Three decades of research on climate change and peace: A bibliometrics analysis. *Sustain. Sci.* **2020**, *16*, 1079–1095. [CrossRef]
28. Mallett, R.; Hagen-Zanker, J.; Slater, R.; Duvendack, M. The benefits and challenges of using systematic reviews in international development research. *J. Dev. Eff.* **2012**, *4*, 445–455. [CrossRef]
29. Chain, C.P.; Dos Santos, A.C.; De Castro, L.G.; Prado, J.W.D. Bibliometric analysis of the quantitative methods applied to the measurement of industrial clusters. *J. Econ. Surv.* **2018**, *33*, 60–84. [CrossRef]
30. Pallottino, F.; Biocca, M.; Nardi, P.; Figorilli, S.; Menesatti, P.; Costa, C. Science mapping approach to analyze the research evolution on precision agriculture: World, EU and Italian situation. *Precis. Agric.* **2018**, *19*, 1011–1026. [CrossRef]
31. Velasco-Muñoz, J.F.; Aznar-Sánchez, J.A.; Batlles-Delafuente, A.; Fidelibus, M.D. Rainwater Harvesting for Agricultural Irrigation: An Analysis of Global Research. *Water* **2019**, *11*, 1320. [CrossRef]
32. Kane, D.A.; Rogé, P.; Snapp, S.S. A Systematic Review of Perennial Staple Crops Literature Using Topic Modeling and Bibliometric Analysis. *PLoS ONE* **2016**, *11*, e0155788. [CrossRef]
33. Madani, F.; Weber, C. The evolution of patent mining: Applying bibliometrics analysis and keyword network analysis. *World Pat. Inf.* **2016**, *46*, 32–48. [CrossRef]
34. Muhuri, P.K.; Shukla, A.K.; Abraham, A. Industry 4.0: A bibliometric analysis and detailed overview. *Eng. Appl. Artif. Intell.* **2019**, *78*, 218–235. [CrossRef]
35. Noyons, E.C.M.; Moed, H.F.; Luwel, M. Combining mapping and citation analysis for evaluative bibliometric purposes: A bibliometric study. *J. Am. Soc. Inf. Sci.* **1999**, *50*, 115–131. [CrossRef]
36. Börner, J.; Marinho, E.; Wunder, S. Mixing Carrots and Sticks to Conserve Forests in the Brazilian Amazon: A Spatial Probabilistic Modeling Approach. *PLoS ONE* **2015**, *10*, e0116846. [CrossRef]
37. Garfield, P.E. Citation indexes for science. A new dimension in documentation through association of ideas†. *Int. J. Epidemiol.* **2006**, *35*, 1123–1127. [CrossRef]
38. Moed, H.F.; Markusova, V.; Akoev, M. Trends in Russian research output indexed in Scopus and Web of Science. *Scientometrics* **2018**, *116*, 1153–1180. [CrossRef]

39. Bakkalbasi, N.; Bauer, K.; Glover, J.; Wang, L. Three options for citation tracking: Google Scholar, Scopus and Web of Science. *Biomed. Digit. Libr.* **2006**, *3*, 1–8. [CrossRef]
40. Pizzi, S.; Caputo, A.; Corvino, A.; Venturelli, A. Management research and the UN sustainable development goals (SDGs): A bibliometric investigation and systematic review. *J. Clean. Prod.* **2020**, *276*, 124033. [CrossRef]
41. Barbara, K.; Charters, S.; Budgen, D.; Brereton, P.; Mark, T.; Linkman, S.; Jørgensen, M.; Mendes, E.; Visaggio, G. Guidelines for performing Systematic Literature Reviews in Software Engineering. Version 2.3. Durham UK. 2007. Available online: https://citeseerx.ist.psu.edu/viewdoc/summary?doi=10.1.1.117.471 (accessed on 22 March 2021).
42. Nardi, P.; Di Matteo, G.; Palahi, M.; Mugnozza, G.S. Structure and Evolution of Mediterranean Forest Research: A Science Mapping Approach. *PLoS ONE* **2016**, *11*, e0155016. [CrossRef]
43. Van Eck, N.J.; Waltman, L. Software survey: VOSviewer, a computer program for bibliometric mapping. *Scientometrics* **2009**, *84*, 523–538. [CrossRef]
44. Van Eck, N.J.; Waltman, L. A Comparison of TwoTechniques for Bibliometric Mapping: Multidimensional Scaling and VOS Nees. *J. Am. Soc. Inf. Sci. Technol.* **2010**, *61*, 2405–2416. [CrossRef]
45. Merton, R. The sociology of science: An episodic memoir. In *The Sociology of Science in Europe*; Southern Illinois University Press: Carbondale, IL, USA, 1977; pp. 3–141.
46. Herwitz, S.; Johnson, L.; Dunagan, S.; Higgins, R.; Sullivan, D.; Zheng, J.; Lobitz, B.; Leung, J.; Gallmeyer, B.; Aoyagi, M.; et al. Imaging from an unmanned aerial vehicle: Agricultural surveillance and decision support. *Comput. Electron. Agric.* **2004**, *44*, 49–61. [CrossRef]
47. Chemura, A.; Mutanga, O.; Dube, T. Separability of coffee leaf rust infection levels with machine learning methods at Sentinel-2 MSI spectral resolutions. *Precis. Agric.* **2016**, *18*, 859–881. [CrossRef]
48. Sentelhas, P.C.; Gillespie, T.J.; Batzer, J.C.; Gleason, M.L.; Monteiro, J.E.B.A.; Pezzopane, J.; Pedro, M.J. Spatial variability of leaf wetness duration in different crop canopies. *Int. J. Biometeorol.* **2005**, *49*, 363–370. [CrossRef]
49. Silva, F.M.; De Souza, Z.M.; De Figueiredo, C.A.P.; Júnior, J.M.; Machado, R.V. Spatial Variability Of Chemical Attributes And Productivity In The Coffee Cultivation. *Ciência Rural* **2007**, *37*, 401–407. [CrossRef]
50. Silva, F.M.; Souza, Z.M.; Figueiredo, C.A.P.; Vieira, L.H.D.S.; Oliveira, E. Spatial variability of chemical attributes and coffee productivity in two harvests. *Cienc. Agrotecnol.* **2008**, *32*, 231–241. [CrossRef]
51. Cordero-Sancho, S.; Sader, S.A. Spectral analysis and classification accuracy of coffee crops using Landsat and a topographic-environmental model. *Int. J. Remote Sens.* **2007**, *28*, 1577–1593. [CrossRef]
52. Silva, S.D.A.; Lima, J.S.D.S.; da Silva, J.M.; Teixeira, M.M. Spatial variability of chemical attributes of an Oxisol under coffee cultivation. *Rev. Bras. Ciência Solo.* **2010**, *34*, 16–23. [CrossRef]
53. Ferraz, G.A.E.S.; da Silva, F.M.; Alves, M.D.C.; Bueno, R.D.L.; da Costa, P.A.N. Geostatistical analysis of fruit yield and detachment force in coffee. *Precis. Agric.* **2011**, *13*, 76–89. [CrossRef]
54. Ferraz, G.A.E.S.; Da Silva, F.M.; Carvalho, L.C.C.; Alves, M.D.C.; Franco, B.C. Spatial And Temporal Variability Of Phosphorus, Potassium And Of The Yield Of A Coffee Field. *Eng. Agric.* **2012**, *32*, 140–150. [CrossRef]
55. Armenta-Medina, D.; Ramirez-Delreal, T.A.; Villanueva-Vásquez, D.; Mejia-Aguirre, C. Trends on Advanced Information and Communication Technologies for Improving Agricultural Productivities: A Bibliometric Analysis. *Agronomy* **2020**, *10*, 1989. [CrossRef]
56. Sott, M.K.; Furstenau, L.B.; Kipper, L.M.; Giraldo, F.D.; Lopez-Robles, J.R.; Cobo, M.J.; Zahid, A.; Abbasi, Q.H.; Imran, M.A. Precision Techniques and Agriculture 4.0 Technologies to Promote Sustainability in the Coffee Sector: State of the Art, Challenges and Future Trends. *IEEE Access* **2020**, *8*, 149854–149867. [CrossRef]
57. Santana, L.S.; Ferraz, G.A.E.S.; Santos, L.M.; Maciel, D.A.; Barata, R.A.P.; Reynaldo, É.F.; Rossi, G. Vegetative vigor of maize crop obtained through vegetation indexes in orbital and aerial sensors images. *Braz. J. Biosyst. Eng.* **2019**, *13*, 195–206. [CrossRef]
58. Pivoto, D.; Waquil, P.D.; Talamini, E.; Finocchio, C.P.S.; Corte, V.; Mores, G.D.V. Scientific development of smart farming technologies and their application in Brazil. *Inf. Process. Agric.* **2018**, *5*, 21–32. [CrossRef]
59. Johnson, L.F.; Herwitz, S.R.; Lobitz, B.M.; Dunagan, S.E. Feasibility of monitoring coffee field ripeness with airborne multi-spectral imagery. *Appl. Eng. Agric.* **2004**, *20*, 845–849. [CrossRef]
60. Cruz-O'Byrne, R.; Piraneque-Gambasica, N.; Aguirre-Forero, S.; Ramirez-Vergara, J. Microorganisms in coffee fermentation: A bibliometric and systematic literature network analysis related to agriculture and beverage quality (1965-2019). *Coffee Sci.* **2020**, *15*, 1–14. [CrossRef]
61. Pabon, C.D.R.; Sánchez-Benitez, J.; Rosero, J.R.; Ramirez-Gonzalez, G.A. Coffee crop science metric: A review. *Coffee Sci.* **2020**, *15*, 1–11. [CrossRef]
62. Sales, F.O.; Marante, Y.; Vieira, A.B.; Silva, E.F. Energy Consumption Evaluation of a Routing Protocol for Low-Power and Lossy Networks in Mesh Scenarios for Precision Agriculture. *Sensors* **2020**, *20*, 3814. [CrossRef] [PubMed]
63. Marin, D.; Ferraz, G.; Guimarães, P.; Schwerz, F.; Santana, L.; Barbosa, B.; Barata, R.; Faria, R.; Dias, J.; Conti, L.; et al. Remotely Piloted Aircraft and Random Forest in the Evaluation of the Spatial Variability of Foliar Nitrogen in Coffee Crop. *Remote Sens.* **2021**, *13*, 1471. [CrossRef]
64. Oré, G.; Alcântara, M.S.; Góes, J.A.; Oliveira, L.P.; Yepes, J.; Teruel, B.; de Castro, V.L.B.; Bins, L.S.; Castro, F.; Luebeck, D.; et al. Crop Growth Monitoring with Drone-Borne DInSAR. *Remote Sens.* **2020**, *12*, 615. [CrossRef]

Article

Worldwide Research on the Ozone Influence in Plants

Lucia Jimenez-Montenegro, Matilde Lopez-Fernandez and Estela Gimenez *

Department of Biotechnology-Plant Biology, School of Agricultural, Food and Biosystems Engineering, Universidad Politécnica de Madrid, 28040 Madrid, Spain; lucia.jmontenegro@alumnos.upm.es (L.J.-M.); matilde.lopez@upm.es (M.L.-F.)
* Correspondence: mariaestela.gimenez@upm.es; Tel.: +34-910670865

Abstract: Tropospheric ozone (O_3) is a secondary air pollutant and a greenhouse gas, whose concentration has been increasing since the industrial era and is expected to increase further in the near future. O_3 molecules can be inhaled by humans and animals, causing significant health problems; they can also diffuse through the leaf stomata of plants, triggering significant phytotoxic damage that entails a weakening of the plant, reducing its ability to cope with other abiotic and biotic stresses. This eventually leads to a reduction in the yield and quality of crops, which is a serious problem as it puts global food security at risk. Due to the importance of this issue, a bibliometric analysis on O_3 in the plant research field is carried out through the Web of Science (WoS) database. Different aspects of the publications are analysed, such as the number of documents published per year, the corresponding scientific areas, distribution of documents by countries, institutions and languages, publication type and affiliations, and, finally, special attention is paid to O_3 study in plants by means of studies about the word occurrence frequency in titles and abstracts, and the articles most frequently cited. The bibliometric study shows the great effort made by the scientific community in order to understand the damages caused by O_3 in plants, which will help reduce the big losses that O_3 causes in agriculture.

Keywords: ozone; plant; environment; health

Citation: Jimenez-Montenegro, L.; Lopez-Fernandez, M.; Gimenez, E. Worldwide Research on the Ozone Influence in Plants. *Agronomy* 2021, *11*, 1504. https://doi.org/10.3390/agronomy11081504

Academic Editor: Channa B. Rajashekar

Received: 14 June 2021
Accepted: 24 July 2021
Published: 28 July 2021

Publisher's Note: MDPI stays neutral with regard to jurisdictional claims in published maps and institutional affiliations.

Copyright: © 2021 by the authors. Licensee MDPI, Basel, Switzerland. This article is an open access article distributed under the terms and conditions of the Creative Commons Attribution (CC BY) license (https://creativecommons.org/licenses/by/4.0/).

1. Introduction

Ozone (O_3) is a gas present in the Earth's atmosphere and 90% of O_3 is found in the stratosphere. The remaining 10% is present in the troposphere, but in a lower concentration, which is known as natural background. However, since the beginning of the 21st century, the concentration of background O_3 is increasing by 0.5–2% per year in the Northern Hemisphere [1], and this upward trend is predicted to continue with mean O_3 concentrations increasing by 20–25% by 2050 and by 40–60% by 2100 [2].

The increase in tropospheric O_3 is mainly due to the rapid industrialization process of the last century and the use of fossil fuels which have caused higher emissions of O_3 precursors, such as nitrogen oxide (NOx) or volatile organic compounds (VOCs) [1,3,4]. O_3 is thus an important second air pollutant and a greenhouse gas. Its highly oxidant nature causes damage to tree species in terms of forest areas affected and their extent [5], as well as in losses in biodiversity in natural grasslands [6–8]. O_3 oxidative stress also affects most commercial crops, leading to a reduction in their yield and quality.

Wheat (*Triticum aestivum* L.), rice (*Oryza sativa* L.), and maize (*Zea mays* L.) are the three most important cereals in the world [9–12]. These crops stand out for their high sensitivity to O_3. O_3 damages wheat grain quality [13] and causes yield losses of approximately 7.1%, which can reach up to 15% in regions such as China and India [14]. For rice, yield losses due to O_3 are approximately 4.4%, and in regions such as China, India, Indonesia, and Bangladesh, losses can reach up to 7.5–12.5% [14]; while in maize, it is estimated that yield losses are approximately 6.1%, reaching up to 15% in countries such as China and the US [14].

The legume family is widely accepted as containing some of the most ozone-sensitive species that have been tested [15,16]; peas and beans are the most sensitive crops to O_3 [17]. Ten years ago, losses caused by O_3 were indicated to be 19.0% for beans, and 20% higher in the concentration range predicted for 2030 in Europe [17]. In addition, O_3-induced effects on fruit quality, changing the nutritional value, have been observed in beans [18], resulting in an important decrease in their market value in countries such as the UK [19]. Soybean (*Glycine max* L.) is also a species of legume that stands out for its high nutritional value. It is, together with wheat and bean, the crop that presents a greater sensitivity to O_3. Production losses due to O_3 are predicted to be approximately 12.4%, and can reach up to 20% in the US, which is the region with the greatest producer [14].

On the other hand, horticultural crop species, although covering relatively small cultivation areas, have great economic relevance. Tomato (*Solanum lycopersicum* L.) is a horticultural crop of great importance worldwide [9], whose yield and fruit composition are affected by O_3 [20]. Lettuce (*Lactuca sativa* L.), the most important green leaf vegetable at an economic and commercial level [9], was one of the first species recognized to have high sensitivity to O_3-induced oxidative stress [21], with mean reductions of approximately 15% of the market yield [22]. Potato (*Solanum tuberosum* L.) is the fourth staple food in the world [14] and an important source of energy, due to its high content of carbohydrates (mainly starch) but long-term exposure to high concentrations of O_3 can cause a drop in starch content, leading to significant yield losses [23]. The genus Brassicaceae includes numerous plants that are used as fresh food or fodder but can also be destined for industrial use or as medicinal and ornamental plants. It includes the rapeseed (*Brassica napus* L.), an extensively cultivated species [9], whose yield losses by O_3 are due to a reduction of pollination by insects [24], which consequently reduces 20% of its market value. It should be noted that this family also includes the species *Arabidopsis thaliana* (L.), which constitutes a model plant for research, as well as for the evaluation of O_3 injury.

The knowledge and understanding of changes caused by O_3 in plant species are essential to fix the damage generated, with the aim of avoiding or reducing yield losses. Therefore, it is important to know all the alterations at the genetic, cellular, and physiological levels that O_3 provokes in plants. To study the metabolic pathways and physiological effects caused by O_3 in plants, model organisms, such as *Arabidopsis thaliana* [25,26] or *Nicotiana tabacum* [27] have mainly been used.

First, O_3 enters the plant tissues through leaf stomata, which are the "first line of defence" [25]. The stomatal pore size and closure are mainly regulated by the activity of ion channels of guard cells [28,29]. One of the earliest responses to O_3 is an increase of cytosolic-free calcium (Ca^{2+}) in guard cells, as a consequence of the activation of Ca^{2+} channels [30]. This Ca^{2+} is a crucial second messenger in stress signalling and in the activation of O_3 response genes [31]. Similarly, potassium (K^+) channels of guard cells are required for fast stomatal closure induced by the reactive oxygen species (ROS) [32].

Once inside leaf tissues, O_3 triggers the formation of ROS that can lead to lipid peroxidation [33] and ROS accumulation in the apoplast. To cope with the damage of ROS, plants activate antioxidant systems, in which several enzymes play a crucial role. Among them, the activity of the enzymes ascorbate peroxidase, dehydroascorbate reductase, and glutathione reductase, all of which play a critical role in ROS-scavenging, can be highlighted [34,35]. It is also worth noting that both superoxide dismutase (SOD) and catalase play an essential role in ROS detoxification [36].

However, at a certain point, the antioxidant capacity of the apoplast is exceeded, and ROS is spread within the cell through NADPH-oxidases that generate superoxide ion (O^2) and type II cell wall peroxidases that generate H_2O_2 [37,38]. Once inside the cell, ROS triggers multiple signalling pathways that are integrated to achieve a proper response. The role of mitogen-activated protein kinases (MAPK) and calcium in the regulation of ethylene production and signalling is remarkable [39–41], causing a rapid accumulation of ethylene in leaf tissues [25]. Ethylene production is one of the earliest responses of plants to O_3 and it promotes lesion formation and cell death [40,42]. In addition to ethylene biosynthesis, O_3 exposure also induces endogenous production of nitric oxide (NO) in guard cells of some plant species, such as *Arabidopsis*. Salicylic acid (SA) induced protein kinase (SIPK, a tobacco MPK3 orthologue) and calcium are also involved in the regulation of NO signalling pathways [25]. NO is involved together with ROS in the activation of stress responses such as hypersensitive response (HR)-like lesions [26]. This molecule is a signal inducer that enhances O_3-induced cell death, possibly by altering the ROS–NO balance. The main impact of NO is the attenuation of SA biosynthesis and other SA-related genes [26]. However, in wheat, it has been proven that NO increases the activity of both antioxidant enzymes SOD and peroxidase in leaves, allowing them to increase photosynthetic rates and to alleviate yield reduction caused by O_3 [43].

O_3 exposure also damages the electron transport chain of the thylakoid membrane of chloroplasts and alters the non-cyclical photophosphorylation process of photosynthesis, causing a decrease in photosynthetic rates and yield [44,45]. In addition, O_3 causes the inactivation of some enzymes of the Calvin Cycle, such as the small subunit of Rubisco (ribulose-1,5-bisphosphate carboxylase oxygenase), decreasing, again, photosynthesis and yields [45,46]. Rubisco plays a crucial role in carbon (C) fixation and is also the main storage protein of foliar nitrogen (N), constituting 50–70% of the total soluble protein of leaves. Therefore, the decrease of this enzyme also causes accelerated foliar senescence [45,46]. Even though the impact of O_3 on plant growth and biomass is variable, it has been proven that the damage to the photosynthetic apparatus and the low assimilation of C previously mentioned causes a lower accumulation of biomass [45] and less fertilizer efficiency [47], which results in the significant yield loss of commercial crops [47].

It has also been described that O_3 exposure can modify biogenic volatile organic compound (BVOCs) emissions from plant leaves, which could alter flower recognition [48] by insect pollinators. This could lead to a decrease in insect pollination [24], and therefore to a reduction of the market value of some crops [24,49].

All alterations caused by O_3 lead to high losses in both quality and yield of commercial crops, which already face the challenge of producing 60% more food by 2050 [50]. With the aim of understanding how the scientific community worldwide attempts to solve the losses caused by O_3 in crops, we have performed a deep bibliometric study.

2. Materials and Methods

There are diverse bibliographic databases used for bibliometric analysis such as Web of Science (WoS), Scopus, Google Scholar, and Microsoft Academic. For this project, we selected WoS since it makes downloading data easy for bibliometric purposes, fits with the scientific coverage of our research area [51], and offers robust tools for measuring science [52].

The search was conducted in January 2021 to collect academic publications from all of the databases available in the WoS from Thomson Reuters, containing the terms "Ozone" or "Ozone" and "Plant" in the title, abstract and/or keywords. The search was limited from the first publication year to 2020. The publications obtained were assessed and classified based on the following aspects: number of publications per year, subject area, countries, institutions, languages, type of document, journals, and type of publication. The results obtained were processed to allow an easier display of the results through graphs obtained with Microsoft Excel. VOSviewer was the tool used to analyse the core content and research object of the academic literature. VOSviewer is a free computer program used for the construction and presentation of bibliometric maps [53]. The frequency of word occurrence was shown by the size of the circle under the word. Additionally, the free software WordArt [54] was used to elaborate a specific word cloud using only vegetal species named in the title of the publications related to O_3 in the plant field. In these maps, the word size is directly proportional to its frequency of occurrence in the literature reports.

The number of citations allows for the assessment of the relative impact that a single publication has on the scientific community. To measure the professional quality of journals, the tool used was the 5-years Impact Factor generated by Journal Citation Reports (JCR), based on the number of citations that their scientific articles have received [55,56]. A higher value of the H-index generally indicates greater scientific attainment. These methodologies were used successfully in other bibliometric studies [57,58].

3. Results

3.1. Evolution of Scientific Output and Distribution in Subject Categories

A total of 145,538 documents with the "ozone" term in titles, abstracts, or keywords were recovered. Remarkably, no article about O_3 appeared until the year 1855, with barely any documents until 1961. Since then, there has been a progressive increase in O_3-related publications. During this period, the year 2010 stands out due to a sudden decrease in the number of publications. Today, the number of O_3-related publications has not reached the amount of O_3 documents that were published previously to the decrease (Figure 1). The drop in the publication number could reflect the severe worldwide financial and food crises of 2008. The ways that the economic and food crises interfaced with the environment and agriculture could affect the research in said issues [59], altering the number of publications for this particular sector. A similar diminution was observed in the year 2013 in a bibliometric study about O_3 in the period 2000–2015 [60]. However, the results shown in this publication and our results are not comparable, since [60] performed a bibliometric study about O_3 using the Directory of Open Access Journals database in a limited period of fifteen years, which only retrieved 1831 articles versus 145,538 documents recovered from WoS in this project.

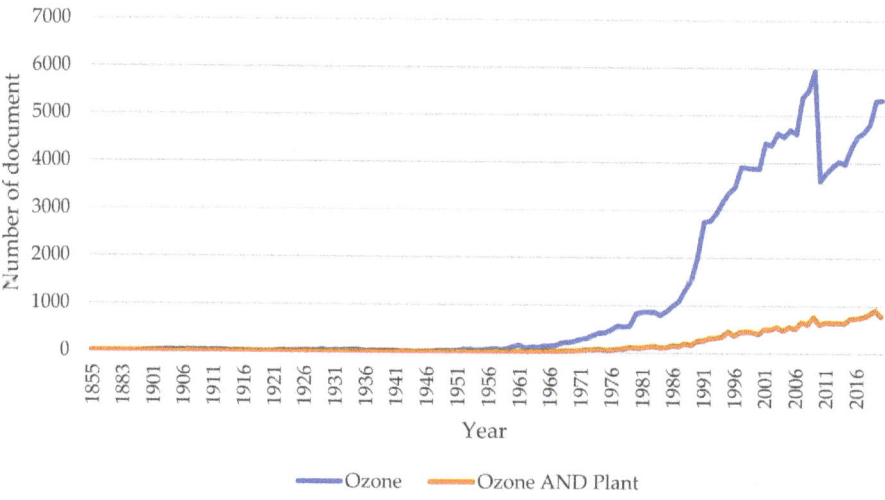

Figure 1. Trends in publications on ozone research in absolute terms (blue line) and in the plant field (orange line) in the period 1855–2020.

If the term "plant" is included in the research, only 19,202 documents are retrieved, representing 13% of O_3-related studies. This low percentage is likely due to the fact that the first report on O_3 in the plant field was published in 1902 (nearly fifty years later than the first publication of O_3), and no more than ten articles appeared until the year 1962. In this period, no manuscript was published in the years 1855–1901, 1903–1906, 1908–1909, 1912–1923, 1925–1926, 1928–1930, 1936, 1938–1939, 1941, and 1943–1946. Since 1962, the number of publications has risen slowly and continuously, until reaching a maximum of 777 documents published in 2020 (Figure 1). The results suggest that interest in O_3 should be initially focused in research areas other than the plant research field. However, an increase in publications on O_3 research in the plant studies field has been experienced from the second half of the twentieth century to the present (Figure 1). It suggests that the interest of the scientific community on the effects of O_3 in plants is relatively recent and that a constant increment in the number of O_3 publications in the plant research field is expected in the coming years.

Based on the WoS classification, the distribution of publications in the O_3 research field covered a total of 154 subject areas. However, only 17 areas included more than 10,000 articles on O_3. The largest number of documents corresponded to Environmental Sciences Ecology (54,020 records), while the second-largest area in terms of the number of publications was Chemistry (51,924 records). The third area was Engineering (47,601 records), followed by the Meteorology Atmospheric Sciences area (40,467 records), and the fifth area was Public Environmental Occupational Health (31,982 records). The first fifteen areas were mainly related to Environmental, Technology, Engineering, Health, Physics, and Chemistry, which can be explained by the O_3 involvement in climate change and human health. No area related to Plant Science or Agriculture could be found before the sixteenth position (Figure 2). Dissimilar results were shown by [60], which showed as the first discipline Engineering and Technology followed by Chemistry, Physics, Earth and Environmental Sciences, Medicine, Biosciences, and Agriculture. Again, the different results obtained can be due to the different databases and time periods used in both projects.

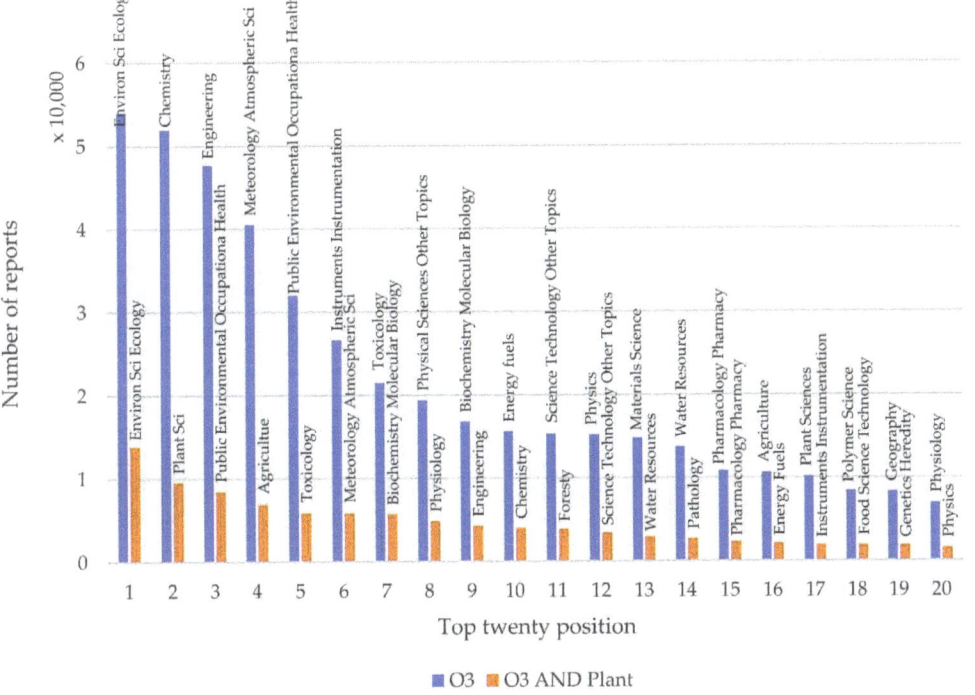

Figure 2. Distribution (number of reports) of worldwide research on Ozone (O_3) and O_3 and plant by subject area, as classified by WoS.

Distribution by areas changed when publications on O_3 were studied in the plant research field. The first area was the same as those shown by the global O_3 research field i.e., Environmental Sciences Ecology (13,796 records), while the Plant Sciences area (9523 records) and Agriculture (6832 records) moved up to second and fourth positions, respectively. In addition, the areas Public Environmental Occupational Health (8381 records) and Toxicology (5807 records) reached third and fifth positions, respectively (Figure 2). The increase in both areas implied a reduction in areas related to Environmental, Technology, Engineering, Physics, and Chemistry. However, the search query "Ozone" and "Plant" continues considering numerous articles related to Environmental research. Therefore, the bibliometric study from this point forward will be performed considering only the areas of Plant Sciences and Agriculture. The limitation to the plant sciences and agriculture areas implies a reduction of 7658 reports, however, it should be noted that a document can be assigned to more than one area at the same time.

3.2. Publication Distribution by Countries, Institutions, and Languages

O_3 in the plant research field has been studied by 111 countries, highlighting the great impact that O_3 effects in plants have worldwide. Countries such as the United States (US), the United Kingdom (UK), China, Germany, Italy, Japan, Spain, Canada, Finland, and India (in order according to the number of O_3 publications in the plant research field) have disseminated more than 400 publications in the period analysed and together accounted for 70% of total publications about O_3 in the plant field (Figure 3). The great interest of these ten countries could be due to the high O_3 levels existent in them because of the industrialization process and the use of fossil fuel [1]. Countries in the Northern Hemisphere have experienced a continuous increase of O_3 concentration from the beginning of the 21st century (Intergovernmental Panel on Climate Change, 2013) and accumulated high O_3 concentrations, mainly during spring and summer seasons, according to the model of surface O_3 from the present-day simulation (see Figure 1 in [61]). Consequently, countries with more research about O_3 in plants are placed in the Northern Hemisphere (Figure 3). It has been described that China can reach crop yield losses due to high O_3 levels near to 15% for wheat and maize and 12.5% for rice, the US reaches losses of 15% and 20% for maize and soya, respectively [14], and the UK suffers important economic losses due to O_3 effects in crops [19]. This is consistent with the fact that the US, the UK, and China show the highest number of publications related to O_3 in the plant research field. In a similar way, high O_3 levels in the Mediterranean basin cause significant losses in the quality and yield of horticultural crops such as tomato [19], lettuce [21], potato [23], etc., which coincide with an increase in the research of O_3 in plants in the Mediterranean countries, where these kind of horticultural crops are economically and nutritionally very important.

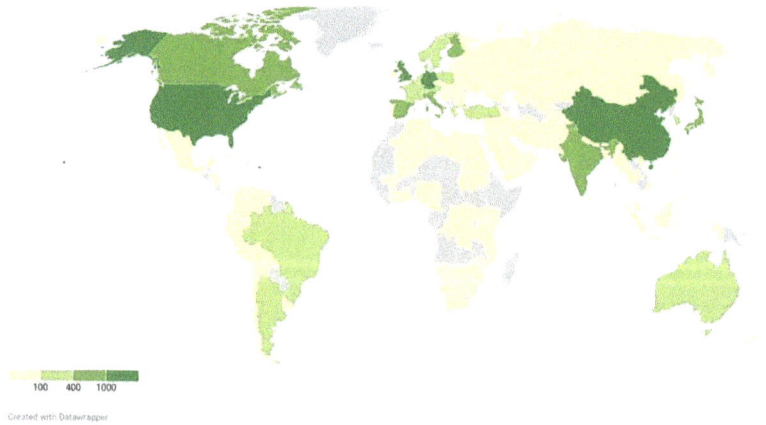

Figure 3. Contribution by country to research documents on ozone in the plant field in the period 1855–2020.

As expected, the most productive institutions were from the US and China. Table 1 shows the institutions which have more than 150 publications of O_3 in the plant field.

Eleven of these twenty institutions were from the US, accounting for 62% of total publications. It is also remarkable that all of the institutions from the top twenty belong to countries placed in the Northern Hemisphere.

Table 1. Ranking of the 20 most productive institutions in the research field of ozone in plants.

Institution	Records	Country
United States Department of Agriculture (USDA) *	866	US
University of California System	404	US
Chinese Academy of Sciences (CAS)	354	China
United States Forest Service	332	US
University of North Carolina	321	US
North Carolina State University	290	US
University of Eastern Finland	263	Finland
Consiglio Nazionale delle Ricerche (CNR)	250	Italy
Helmholtz Association	249	Germany
University of California Riverside	215	US
Pennsylvania Commonwealth System of Higher Education (PCSHE)	193	US
Pennsylvania State University	188	US
Technical University of Munich	169	Germany
l'Institut national de recherche pour l'agriculture, l'alimentation et l'environnement (INRAE) *#	168	France
Helmholtz Center Munich German Research Center for Environmental Health #	157	Germany
Centre for Ecology Hydrology (CEH) #	157	UK
University of Pisa	154	Italy
United States Environmental Protection Agency #	152	US
United States Department of Energy (DOE) #	151	US
Cornell University	147	US

* Research Institutions focused on Agronomy. # Research Institutions related to environmental research.

Furthermore, two out of the twenty most productive centres in the O_3 research field were specific to the agricultural research field, five were related to environmental research and the rest were multidisciplinary institutes.

Research studies on O_3 in the plant field have been published in 26 different languages. The most common language on this topic is English because this is the international language for science and technology. Furthermore, if the number of publications around the world is considered, the US and the UK are the countries where more scientific documents on O_3 in plants are published, and this is in accordance with the fact that English is the most used language. Despite the low number of O_3 documents published by France, the second most used language is French. This extended use of French is related to the high number of publications on this research in Canada, where French is, together with English, an official language of the country since the "Official Languages Act of 1969". It is also remarkable that German and Chinese are the third and fourth most used languages in O_3 and plant field documents, respectively, which is also consistent with the fact that China and Germany are the countries that published the third and fourth highest number of O_3 reports, respectively, only behind two English-speaking countries (the US and the UK). The remaining documents are classified as "other". This category includes languages such as Portuguese, Spanish, Russian, Korean, Japanese, or Polish (Figure 4). Again, these results prove the high interest and importance of this trend around the world, and particularly in Northern Hemisphere countries.

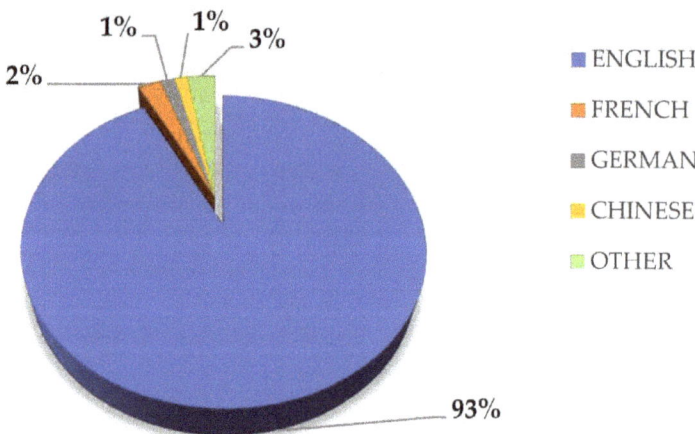

Figure 4. Languages used in the research documents on ozone in the plant field.

3.3. Distribution of Output in Journals and Types of Publications

In this research, nearly 12,000 articles were found in over 2000 different journals, although the vast majority published less than ten. This high diversity of publications, along with the heterogeneity of the journals, suggests, again, wide interest among the scientific community regarding this topic. Only 20 journals displayed more than 100 articles, comprehending 37% of the published articles overall. Of these 20 journals, 45% are exclusively related to environmental matters, another 45% on plant science, and 10% relate to plants and environment interaction. Remarkably, almost half of the journals are exclusively focused on environmental topics. This could be explained by the progressive increase in research on environmental matters in the last 30 years, due to the increasing importance of climate change and the challenges it poses to the future of agriculture and human feeding. In this sense, with regards to the publication number by source, Table 2 lists the top ten journals in which results from these topics have been published. The journal Environmental Pollution leads this list with 859 publications, comprising more than twice as many articles as the following journals. This could be due to the fact that this journal covers "all aspects of environmental pollution and its effects on ecosystems and human health", so it includes very diverse fields of study and therefore accepts a great variety of scientific topics. By the number of publications, it is followed by New Phytologist and Phytopathology, two journals about much more specific fields, however, with a deep connection to the impact of O_3 on plants. It is worth pointing out that in this top ten list of journals, none has a multidisciplinary character, possibly as the result of the particularity of the topic researched.

In reference to their scientific impact, the generic plant-related journals generally present a higher impact factor than the environmental ones, as can be seen with New Phytologist (8.795) and Plant Physiology (7.52) (Table 2). The elevated impact factor of these two journals suggests that the impact of the papers about O_3 in the plant field published in these sources is higher than those papers published in environmental-related journals with a minor impact factor such as Environmental pollution (6939) and Science of the total environment (6.419). These results point out that a journal, for instance, Environmental pollution, can publish a great number of articles about this topic (859), but have, in return, a comparatively low impact on the scientific community.

Table 2. Distribution of publications by source.

Position	Source Titles	Records	%	5 Years JCR	Country
1	Environmental pollution	859	7.45	6.939	UK
2	New phytologist	327	2.84	8.795	UK
3	Phytopathology	290	2.52	3.492	US
4	Science of the total environment	270	2.34	6.419	Netherlands
5	Atmospheric environment	267	2.32	4.633	UK
6	Plant physiology	227	1.97	7.52	US
7	Environmental and experimental botany	219	1.90	4.744	UK
8	Plant cell and environment	188	1.63	7.044	UK
9	Water air and soil pollution	178	1.54	2.041	Switzerland
10	Physiologia plantarum	172	1.50	3.947	Denmark

Documents on O_3 in the plant field recovered from the WoS database can be divided into 17 document types. The most frequently used document type was "article", which accounted for 10,848 records (65% of total publications). Another important category is "meeting", which accounted for 1373 records (8%), followed by "review", which accounted for 727 records (4%), "abstract" with 428 records (3%), "patent" with 223 records (1%), "book" with 216 records (1%), and, finally, followed by "editorial" with 113 records which accounted for less than 1% of total publications. The remaining documents are classified as "others" and correspond to minor categories such as "letter", "correction", "clinical trial", "biography", "early access", "news", and "case report". These minor categories individually contribute between 0.006–0.145%, and together accounted for 16% of total publications (Figure 5). It is important to note that a document can be included in more than one category of document type. These results indicate that most authors prefer mainly to publish their important findings in article format. Nevertheless, a lot of authors have chosen meetings and reviews as a way to disseminate the scientific research in this field, with reviews being a good way to summarize and assemble important findings and studies of specific research. It is worth pointing out the high number of patents registered in which the use of O_3 as a tool for developing equipment and procedures is included.

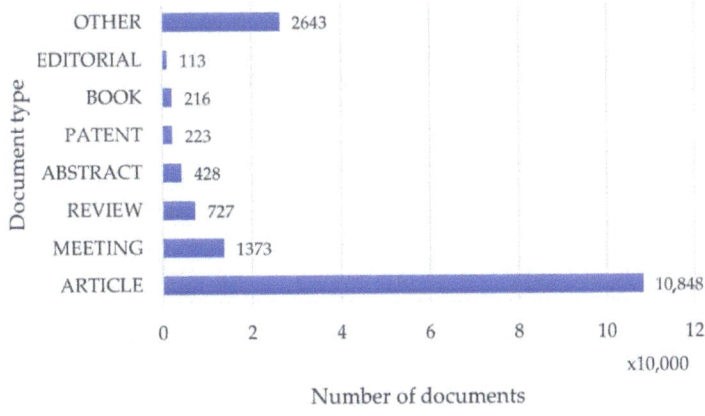

Figure 5. The most frequently used document type in the research documents on ozone in the plant field.

3.4. Analysis of Terms Used in Titles and Abstracts

To identify trends in scientific research on the topic of study, we analysed and represented the words from titles and abstracts using the VOSviewer computer program. To perform a general analysis, the research of terms mentioned more than 100 times in titles and abstracts from articles about O_3 in the plant research field was performed. The retrieved terms were divided into five clusters by the VOSviewer program, according to the relationship between the items (Figure 6).

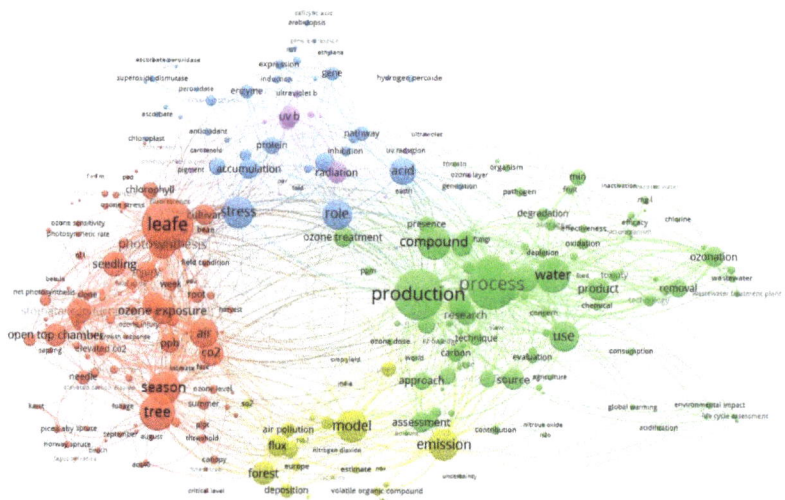

Figure 6. Word clouds based on worldwide research on ozone in the plant field. Terms named more than 100 times in titles and abstracts. The size of the circle under the word indicates the frequency of word occurrence. The five clusters, made by the VOSviewer program according to the relationship between the items, are represented in different colors.

In the first cluster, the green one, the terms "process", "production", "water", "compound" and "use" were the most frequently mentioned. These five words are habitual terms in agronomic research. However, when we analysed the term combination of the green cluster, numerous terms related to the environment and sustainability could be found: "climate change", "global warming", "degrees C", "environmental impact", "lca" (life cycle assessment), "wastewater", and so on (Figure 6). These terms are associated with O_3 research since O_3 affects climate change, because of its relationship with the ozone hole in the stratosphere and by being a greenhouse gas. In addition, terms such as "oxidation", "toxicity", "food", and "agriculture" indicate that O_3 is an important air pollutant whose oxidant capacity causes significant damage to forests and crops [5,14].

The second cluster with the most words is the red one, where the terms "leafe", "tree", "season", "air", and "seedling" are highlighted because they appear more frequently in articles about O_3 in the plant field. This group included a series of concepts related to the physiological processes affected by high O_3 levels ("photosynthesis", "rubisco", "ozone injure", "stomatal conductance", and so on), with the assays that are carried out to study such physiological processes ("open-top chamber", "otc", "aot40" (accumulated O_3 exposure over a threshold of 40 ppb), "non-filtered air", etc.), the plant tissues and organs affected ("stomata", "root", "shoot", "stem", "foliage", and so forth), as well as the altered parameters ("grow respond", "age", "dry weight", "height", "senescence", etc.) (Figure 6). As mentioned above, the first defence line of the plants against high O_3 levels is the stomata [25]. Once overcome, the photosynthesis [44] and the Calvin cycle [46] are affected, cell death [26] and the senescence are accelerated [45], and, therefore, the plants show damage in different tissues that alter the quality and yield of the species affected [45]. The assays for analysing this kind of damage are usually carried out in open-top chambers (otc) using diverse O_3 concentrations (Aot40, non-filtered air, ambient air, etc.) [62,63]. All terms included in the red cluster speak about plant physiological processes, tissues, and parameters damaged with high O_3 levels and the assays performed to analyse them. In addition, numerous vegetal species, which will be analysed later on, are included in this cluster (Figure 6).

The third broader cluster, the blue one, highlighted words like "stress", "role", "acid", "accumulation", and "tolerance". However, this group included numerous words related to molecular and cellular alterations shown when plants are exposed to high O_3 levels such as "oxidative stress", "ascorbate peroxidase", "superoxide dismutase", "glutathione", "salicylic acid", "ethylene", "ros" (reactive oxygen species), and "H_2O_2" (hydrogen peroxide), "cell death", etc. (Figure 6). As mentioned in the introduction, the plant metabolic and molecular pathways altered by high O_3 levels are mainly the ethylene and salicylic acid biosynthesis pathways [25], and antioxidant systems [34,35], all of them represented in the terms included in this cluster. This cluster includes only two vegetal species, tobacco and *Arabidopsis* (Figure 6), which are species widely used to perform genetic and molecular studies. This is due to their exceptional characteristics, such as a small genome, short life cycle, and accessible transformation methods, all of which make them both suitable as model systems [64,65].

In the fourth cluster, the yellow one, the words "model", "emission", "forest", "flux", and "air pollution" are highlighted as frequently named in titles and abstracts of articles on O_3 in the plant research field. This cluster is characteristic since it includes words related to environmental contamination and O_3 precursors such as "NOx", "volatile organic compound", "isoprene and monoterpene", "air pollution", "nitrogen dioxide", "climate", etc. (Figure 6). As described previously, the industrialization process and the use of fossil fuels have increased emissions of O_3 precursors such as NOx or VOCs [1,3,4]. In addition, isoprene and monoterpene are the most abundant BVOCs emitted by terrestrial vegetation, particularly by forests. Similar to VOCs, BVOCs can lead to changes in the production of tropospheric O_3, depending on the NOx concentration [66]. All terms in this cluster are related to O_3 precursors that cause environmental pollution and increased O_3 levels.

The last cluster, the purple one, includes only words related to stratosphere O_3 ("ozone layer", "ultraviolet b radiation", "active radiation", "par" (photosynthetic active radiation), etc.) and with plant pigment ("flavonoid" and "carotenoid") (Figure 6). This cluster is likely generated due to the fact that O_3 in the stratosphere protects life on Earth from harmful ultraviolet radiation. The appearance of pigments in this cluster could be due to different studies about damage in the plant pigments caused by UV radiation [67].

In conclusion, this word analysis indicates that the most important issues for the scientific community on O_3 in the plant research field are the ozone hole, climate change, environmental damage in general, the stratospheric O_3 formation and its precursors, as well as the genetic, metabolic, and physiological changes suffered by plants.

These interests are easy to understand, since all damages induced by O_3 in plants cause a weakening of the plant, reducing its ability to cope with other abiotic and biotic stresses. This eventually leads to a reduction in the yield and quality of cash crops, which is a serious problem as it puts global food security at risk.

To perform an analysis about the plant species studied more by the scientific community in the O_3 research field, a word cloud using plant species named more than 25 times in titles was carried out using the program WordArt (Figure 7). First, we can appreciate that the plant species mentioned more frequently in titles is "bean" ("*Phaseolus vulgaris* L."). This is because of the high importance of beans in the human diet, representing 50% of grain legumes consumed worldwide [68], and the great impact of O_3 on their growth. Almost 78% of the tested species showed detrimental effects on their total biomass relative to their growth rate due to O_3 [69]. In addition, "*Phaseolus vulgaris* L." is widely used as a bioindicator system to detect ambient O_3 effects [70,71] and is a model species for other leguminous plants such as "soybean" [68,72], which are also represented in the cloud but with less importance (Figure 7). Other model species such as "*Arabidopsis*" and "tobacco", both widely used in molecular and metabolic research studies as mentioned above, are also included among species of interest for O_3 research in plants, mainly *Arabidopsis* (Figure 7). Similar to beans, tobacco is used as a biomarker system [71]. According to those described in the introduction, two cereals ("rice"/"*Oriza sativa* L." and "*Triticum aestivum* L."/"spring wheat"/"winter wheat") stood out among plant species mentioned more frequently in the titles of O_3 studies in plants (Figure 7). Both cereals have a high economical and commercial importance and show high sensibility to high O_3 levels, which provoke big yield losses of both cereals. Alternatively, numerous trees species are mentioned in the titles from articles on O_3 in plants (Figure 7). It is worth mentioning the presence of "betula/birch", "picea abies/Norway spruce/spruce", "Fagus sylvatica/European beech/beech", "aspen", and three species from the pinaceae family, "Scots pine", "ponderosa pine", and "eastern white pine". These findings make sense because the high oxidant capacity of O_3 damages tree species [5].

Figure 7. Word clouds based on worldwide research on ozone in the plant field. Plant species named more than 25 times in titles. The size of the word indicates the frequency of word occurrence. Leguminous plants are represented in green, the trees in pink, the cereals in yellow, and the rest, including model species, in blue.

According to this, tropospheric O_3 is a likely contributing factor to tree decline in some North American and European forests [5,73–77]. Some of the effects caused are the alterations of BVOCs mentioned in the introduction, which could alter communication among themselves or insect pollination [24,78]. However, establishing a cause and effect relationship for ambient O_3 exposure and tree growth in forests is a difficult task [79,80]. This is due to different sensitivities between species from the same natural community, which induce the selection of resistant species versus sensitive species [3,81]. Among the forest masses that suffer O_3 impact, we can underline those located in southern California, where the *Pinus ponderosa* presents high sensitivity to O_3 [82], and it is estimated that by the year 2074 it will be close to disappearing, thus changing the dominance of this ecosystem in favour of the *Quercus kelloggii* [83]. All of this reveals the considerable interest of the scientific community in the relationship between trees and O_3.

3.5. The Most Cited Articles

Another way to evaluate the main interests of the scientific community on O_3 in the plant research field is by analysing the most cited articles. Papers receiving more than 1000 citations are listed in Table 3. Two papers stand out with nearly 2000 citations. The most cited paper was published in 2015 by Lelieveld [84] in Nature (n = 1951). The topic of this paper is the impact of air pollutants, including O_3, on premature mortality. The implication of the plants in this article is due to the fact that agricultural emissions make a relative contribution to outdoor air pollution in some countries. The second most cited paper (n = 1766) was published in Science in 1990 by Crutzen and Andreae [85]. The paper explains how biomass burning (to convert forests to agricultural and pastoral lands, control of pests and insects, prevention of brush, nutrient mobilization, etc.) in the tropics leads to high concentrations of air pollutants, including O_3, which implies damage to trees and vegetation, in addition to affecting climate, atmospheric chemistry, and ecology in the tropical regions. Two papers with nearly 1000 citations, published in Plant Cell and Environment (top three; n = 1182) and Plant Physiology (top five; n = 1011), respectively, discuss metabolic and genetic alterations generated by plant exposition to high O_3 levels. Finally, the top four articles most cited (n = 1063), published in Nature Protocols, describe an assay to estimate the total phenolic content and other oxidation substrates in plant tissues (Table 3). Despite the selection of Plant Sciences and Agriculture research areas to perform this project, the most cited paper discusses O_3 involvement in human health. Similarly, the second most cited paper is mainly focused on environmental changes caused by air pollutants, including O_3, although this paper also describes the damage caused by O_3 in plants. However, the other three most cited papers are focused on metabolic and genetic pathways affected in plants exposed to high O_3 levels and assays to analyse these changes. Results emphasise the high interest of the scientific community, not only in the damage caused by O_3 in plants but also to human health and to the environment.

Table 3. Ranking of the five most cited articles in the research field of ozone in plants.

Cites Number	Title	Publication Year	Journal	Reference
1951	The contribution of outdoor air pollution sources to premature mortality on a global scale	2015	Nature	[84]
1766	Biomass burning in the tropics—impact on atmospheric chemistry and biogeochemical cycles	1990	Science	[85]
1182	Oxidant and antioxidant signalling in plants: a re-evaluation of the concept of oxidative stress in a physiological context	2005	Plant cell and environment	[86]
1063	Estimation of total phenolic content and other oxidation substrates in plant tissues using Folin–Ciocalteu reagent	2007	Nature protocols	[87]
1011	Ultraviolet-B- and ozone-induced biochemical changes in antioxidant enzymes of *Arabidopsis thaliana*	1996	Plant physiology	[88]

3.6. Databases

This study has been carried out by assembling information from the WoS database. However, this database, besides having the WoS core collection (10,988 records), also includes other sources of information such as Business Cycle Indicators (BCI) (9547 records), BIOSIS (9546 records), Current Contents Connect (CCC) (7057 records), and Medical Literature Analysis and Retrieval System Online (MEDLINE) (4570 records) (Figure 8). Most documents can be included in more than one database, so the final number accounted for 42,069 records, despite the 11,529 documents that have been selected for the present study. BCI, with 9547 records, is a database that provides economic indicators and statistical information. A lot of publications of O_3 and plant research are included in this database, which could be due to the fact that O_3 damage causes important economic losses. BIOSIS Previews is part of the Clarivate Analytics WoS database and contains abstracts and citation indexing. CCC is a WoS platform, allowing current awareness and notifying alerts when new issues of a journal are released. Finally, MEDLINE (Medical Literature Analysis and Retrieval System Online) is a bibliographic database that includes life science and biomedical information. Therefore, O_3 documents are registered in MEDLINE because this oxidative gas is associated with adverse health outcome damage, especially on respiratory and cardiovascular systems [89].

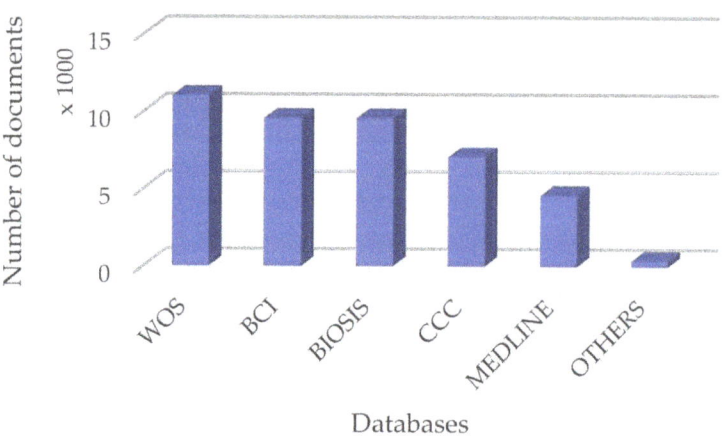

Figure 8. Number of documents on O3 in the plant research field included in the different databases of Web of Sciences.

4. Conclusions

O_3 is an important air pollutant, a greenhouse gas, and a highly reactive oxygen species that cause significant economic losses in crops, forests, and ecosystems. Even though the interest of the scientific community on O_3 in the plant research field, measured as document number per year, has been increasing since the second half of the past century, this rise is slow in relation to O_3 studies in other research areas. Research in O_3 has been studied most in the health area. Both research areas, health and environmental, are fields of high interest for humanity, and both are affected by high O_3 levels, contributing to climate change, damaging human health, and putting global food security at risk.

The countries with more studies and publications are placed in the Northern Hemisphere, where the rapid industrialization process of the last century and the use of fossil fuels have caused higher emissions of O_3 precursors such as NOx or VOCs [1,3,4]. Knowing the causes that provoke the increase of tropospheric O_3 can help avoid the continuous increase of this pollutant.

With regard to plants, alterations caused by O_3 at genetic, metabolic, and physiological levels end in important losses of quality and yield in commercial crops. Among crops studied more frequently in the O_3 research field, species with high commercial value can be underlined, such as horticultural crops (tomato, lettuce), beans, or genus *Brassicaceae* species, as well as staple food crops (soybean, maize, wheat, and rice). The latter ones are the base of the human diet, meaning that losses caused by O_3 could put at risk global food security. This important problem claims to be solved through the implementation of breeding programs, with the ultimate goal of developing commercial varieties tolerant to this contaminant. Because of this, it is paramount that the scientific community acknowledges the issue, and develops studies through which to determine the genetic and metabolic pathways that are damaged by high O_3 levels, as well as the genes that regulate tolerance to O_3. In accordance with this objective, some pre-breeding programs have been developed in rice [90,91] and soybean [92], in which some genes that confer resistance to O_3 have been identified. However, very little progress has been made with horticultural crops. Therefore, Booker et al. [93] claim the need to identify molecular markers related to O_3 tolerance in modern cultivars. With the aim of developing breeding programs, model species have previously been used, such as *Arabidopsis* or *Nicotiana tabacum*, in which numerous metabolic and physiological pathways are known, and the effects of O_3 on them are easier to study [30,35].

In addition, numerous studies have been carried out about O_3 damage on trees, forests, and ecosystems. The effects of O_3 accumulate over time and, together with other stresses (prolonged drought, excess nitrogen deposition), may exacerbate the direct effect of O_3 on ecosystems. These alterations could influence competitive interactions among species [94], and finally, cause the disappearance of sensitive species through the selection of varieties resistant to O_3.

Author Contributions: Conceptualization, E.G.; methodology, L.J.-M. and E.G.; software, L.J.-M. and E.G.; formal analysis, E.G.; investigation, L.J.-M., M.L.-F. and E.G.; writing—original draft preparation, L.J.-M., M.L.-F. and E.G.; writing—review and editing, L.J.-M., M.L.-F. and E.G. All authors have read and agreed to the published version of the manuscript.

Funding: The authors were funded by the Spanish Ministry of Science and Innovation (Grant No. PID2019-109089RB-C32), by Comunidad de Madrid (Spain) and Structural EU Funds 2014-2020 (ERDF and ESF) (Grant No. AGRISOST-CM S2018/BAA-4330) and by IMIDRA-Comunidad de Madrid (Spain) (Grant No. PDR-18 OZOCAM).

Data Availability Statement: Data sharing not applicable.

Acknowledgments: The authors at grateful to P. Giraldo and E. Benavente for their helpful comments on a former version of the manuscript.

Conflicts of Interest: The authors declare no conflict of interest.

References

1. Vingarzan, R. A review of surface ozone background levels and trends. *Atmos. Environ.* **2004**, *38*, 3431–3442. [CrossRef]
2. Wagg, S.; Mills, G.; Hayes, F.; Wilkinson, S.; Cooper, D.; Davies, W.J. Reduced soil water availability did not protect two competing grassland species from the negative effects of increasing background ozone. *Environ. Pollut.* **2012**, *165*, 91–99. [CrossRef]
3. Aschmann, S.M.; Reisseil, A.; Atkinson, R.; Arey, J. Products of the gas phase reactions of the OH radical with α-and β-pinene in the presence of NO. *J. Geophys. Res. Atmos.* **1998**, *103*, 25553–25561. [CrossRef]
4. Calogirou, A.; Larsen, B.R.; Kotzias, D. Gas-phase terpene oxidation products: A review. *Atmos. Environ.* **1999**, *33*, 1423–1439. [CrossRef]
5. Skärby, L.; Ro-Poulsen, H.; Wellburn, F.A.; Sheppard, L.J. Impacts of ozone on forests: A European perspective. *New Phytol.* **1998**, *139*, 109–122. [CrossRef]
6. Bassin, S.; Volk, M.; Fuhrer, J. Factors affecting the ozone sensitivity of temperate European grasslands: An overview. *Environ. Pollut.* **2007**, *146*, 678–691. [CrossRef] [PubMed]
7. Bassin, S.; Volk, M.; Suter, M.; Buchmann, N.; Fuhrer, J. Nitrogen deposition but not ozone affects productivity and community composition of subalpine grassland after 3 years of treatment. *New Phytol.* **2007**, *175*, 523–534. [CrossRef]
8. Bermejo, V.; Gimeno, B.S.; Sanz, J.; De la Torre, D.; Gil, J.M. Assessment of the ozone sensitivity of 22 native plant species from Mediterranean annual pastures based on visible injury. *Atmos. Environ.* **2003**, *37*, 4667–4677. [CrossRef]
9. FAOSTAT. Agriculture Organization of the United Nations Statistics Division. Production. Available online: http://www.fao.org/faostat/es/#data/QC/visualize (accessed on 15 March 2021).
10. Abbate, P.; Cardos, M.; Campaña, L. El trigo, su difusión, importancia como alimento y consumo. In *Manual del Cultivo de Trigo*; Y García FO Instituto Internacional de Nutrición de Plantas Programa Latinoamérica Cono Sur (IPNI); Divito, A., Ed.; Acassuso: Buenos Aires, Argentina, 2017; pp. 7–19.
11. Ram, P.C. Maclean, J.L.; Dawe, D.C.; Hardy, B. and Hettel, G.P. (Eds.) Rice almanac. 3rd edn. *Ann. Bot.* **2003**, *92*, 739. [CrossRef]
12. Paliwal, R.L. Introducción al maíz y su importancia. In *El maíz en los Trópicos: Mejoramiento y Producción*; Food & Agriculture Organization: Rome, Italy, 2001.
13. Broberg, M.C.; Feng, Z.; Xin, Y.; Pleijel, H. Ozone effects on wheat grain quality—A summary. *Environ. Pollut.* **2015**, *197*, 203–213. [CrossRef]
14. Mills, G.; Sharps, K.; Simpson, D.; Pleijel, H.; Frei, M.; Burkey, K.; Emberson, L.; Uddling, J.; Broberg, M.; Feng, Z. Closing the global ozone yield gap: Quantification and cobenefits for multistress tolerance. *Glob. Chang. Biol.* **2018**, *24*, 4869–4893. [CrossRef]
15. Mills, G.; Buse, A.; Gimeno, B.; Bermejo, V.; Holland, M.; Emberson, L.; Pleijel, H. A synthesis of AOT40-based response functions and critical levels of ozone for agricultural and horticultural crops. *Atmos. Environ.* **2007**, *41*, 2630–2643. [CrossRef]
16. Hayes, F.; Jones, M.; Mills, G.; Ashmore, M. Meta-analysis of the relative sensitivity of semi-natural vegetation species to ozone. *Environ. Pollut.* **2007**, *146*, 754–762. [CrossRef]

17. Mills, G.; Harmens, H. *Ozone Pollution: A Hidden Threat to Food Security*; NERC/Centre for Ecology & Hydrology: Lancaster, UK, 2011.
18. Iriti, M.; Di Maro, A.; Bernasconi, S.; Burlini, N.; Simonetti, P.; Picchi, V.; Panigada, C.; Gerosa, G.; Parente, A.; Faoro, F. Nutritional traits of bean (*Phaseolus vulgaris*) seeds from plants chronically exposed to ozone pollution. *J. Agric. Food Chem.* **2009**, *57*, 201–208. [CrossRef] [PubMed]
19. Mills, G.; Hayes, F.; Norris, D.; Hall, J.; Coyle, M.; Cambridge, H.; Cinderby, S.; Abott, J.; Cooke, S.; Murrells, T. Impacts of Ozone Pollution on Food Security in the UK: A Case Study for Two Contrasting Years, 2006 and 2008; Report for Defra Contract AQ08610. 2011. Available online: https://icpvegetation.ceh.ac.uk/impacts-ozone-pollution-food-security-uk-case-study-two-contrasting-years-2006-and-2008 (accessed on 13 July 2021).
20. González-Fernández, I.; Calvo, E.; Gerosa, G.; Bermejo, V.; Marzuoli, R.; Calatayud, V.; Alonso, R. Setting ozone critical levels for protecting horticultural Mediterranean crops: Case study of tomato. *Environ. Pollut.* **2014**, *185*, 178–187. [CrossRef]
21. Reinert, R.A.; Tingey, D.T.; Carter, H.B. Ozone induced foliar injury in lettuce and radish cultivars. *J. Am. Soc. Hortic. Sci.* **1972**, 97.
22. Marzuoli, R.; Finco, A.; Chiesa, M.; Gerosa, G. A dose-response relationship for marketable yield reduction of two lettuce (*Lactuca sativa* L.) cultivars exposed to tropospheric ozone in Southern Europe. *Environ. Sci. Pollut. Res.* **2017**, *24*, 26249–26258. [CrossRef] [PubMed]
23. Vorne, V.; Ojanperä, K.; De Temmerman, L.; Bindi, M.; Högy, P.; Jones, M.B.; Lawson, T.; Persson, K. Effects of elevated carbon dioxide and ozone on potato tuber quality in the European multiple-site experiment 'CHIP-project'. *Eur. J. Agron.* **2002**, *17*, 369–381. [CrossRef]
24. Saunier, A.; Blande, J.D. The effect of elevated ozone on floral chemistry of Brassicaceae species. *Environ. Pollut.* **2019**, *255*, 113257. [CrossRef]
25. Vainonen, J.P.; Kangasjärvi, J. Plant signalling in acute ozone exposure. *Plant Cell Environ.* **2015**, *38*, 240–252. [CrossRef] [PubMed]
26. Ahlfors, R.; Brosché, M.; Kollist, H.; Kangasjärvi, J. Nitric oxide modulates ozone-induced cell death, hormone biosynthesis and gene expression in *Arabidopsis thaliana*. *Plant J.* **2009**, *58*, 1–12. [CrossRef]
27. Ederli, L.; Morettini, R.; Borgogni, A.; Wasternack, C.; Miersch, O.; Reale, L.; Ferranti, F.; Tosti, N.; Pasqualini, S. Interaction between nitric oxide and ethylene in the induction of alternative oxidase in ozone-treated tobacco plants. *Plant Physiol.* **2006**, *142*, 595–608. [CrossRef] [PubMed]
28. Hedrich, R. Ion channels in plants. *Physiol. Rev.* **2012**, *92*, 1777–1811. [CrossRef]
29. Vahisalu, T.; Puzõrjova, I.; Brosché, M.; Valk, E.; Lepiku, M.; Moldau, H.; Pechter, P.; Wang, Y.; Lindgren, O.; Salojärvi, J. Ozone-triggered rapid stomatal response involves the production of reactive oxygen species, and is controlled by SLAC1 and OST1. *Plant J.* **2010**, *62*, 442–453. [CrossRef]
30. Evans, N.H.; McAinsh, M.R.; Hetherington, A.M.; Knight, M.R. ROS perception in *Arabidopsis thaliana*: The ozone-induced calcium response. *Plant J.* **2005**, *41*, 615–626. [CrossRef]
31. Short, E.F.; North, K.A.; Roberts, M.R.; Hetherington, A.M.; Shirras, A.D.; McAinsh, M.R. A stress-specific calcium signature regulating an ozone-responsive gene expression network in *Arabidopsis*. *Plant J.* **2012**, *71*, 948–961. [CrossRef] [PubMed]
32. Sato, A.; Sato, Y.; Fukao, Y.; Fujiwara, M.; Umezawa, T.; Shinozaki, K.; Hibi, T.; Taniguchi, M.; Miyake, H.; Goto, D.B. Threonine at position 306 of the KAT1 potassium channel is essential for channel activity and is a target site for ABA activated SnRK2/OST1/SnRK2. 6 protein kinase. *Biochem. J.* **2009**, *424*, 439–448. [CrossRef]
33. Calatayud, A.; Barreno, E. Response to ozone in two lettuce varieties on chlorophyll a fluorescence, photosynthetic pigments and lipid peroxidation. *Plant Physiol. Biochem.* **2004**, *42*, 549–555. [CrossRef]
34. Yoshida, S.; Tamaoki, M.; Shikano, T.; Nakajima, N.; Ogawa, D.; Ioki, M.; Aono, M.; Kubo, A.; Kamada, H.; Inoue, Y. Cytosolic dehydroascorbate reductase is important for ozone tolerance in *Arabidopsis thaliana*. *Plant Cell Physiol.* **2006**, *47*, 304–308. [CrossRef]
35. Dghim, A.A.; Mhamdi, A.; Vaultier, M.; Hasenfratz-Sauder, M.; Le Thiec, D.; Dizengremel, P.; Noctor, G.; Jolivet, Y. Analysis of cytosolic isocitrate dehydrogenase and glutathione reductase 1 in photoperiod-influenced responses to ozone using *Arabidopsis* knockout mutants. *Plant Cell Environ.* **2013**, *36*, 1981–1991. [CrossRef]
36. Li, C.; Wang, T.; Li, Y.; Zheng, Y.; Jiang, G. Flixweed is more competitive than winter wheat under ozone pollution: Evidences from membrane lipid peroxidation, antioxidant enzymes and biomass. *PLoS ONE* **2013**, *8*, e60109. [CrossRef]
37. Torres, M.A.; Dangl, J.L.; Jones, J.D. *Arabidopsis* gp91phox homologues AtrbohD and AtrbohF are required for accumulation of reactive oxygen intermediates in the plant defense response. *Proc. Natl. Acad. Sci. USA* **2002**, *99*, 517–522. [CrossRef]
38. Daudi, A.; Cheng, Z.; O'Brien, J.A.; Mammarella, N.; Khan, S.; Ausubel, F.M.; Bolwell, G.P. The apoplastic oxidative burst peroxidase in *Arabidopsis* is a major component of pattern-triggered immunity. *Plant Cell* **2012**, *24*, 275–287. [CrossRef]
39. Samuel, M.A.; Walia, A.; Mansfield, S.D.; Ellis, B.E. Overexpression of SIPK in tobacco enhances ozone-induced ethylene formation and blocks ozone-induced SA accumulation. *J. Exp. Bot.* **2005**, *56*, 2195–2201. [CrossRef] [PubMed]
40. Tosti, N.; Pasqualini, S.; Borgogni, A.; Ederli, L.; Falistocco, E.; Crispi, S.; Paolocci, F. Gene expression profiles of O_3-treated *Arabidopsis* plants. *Plant Cell Environ.* **2006**, *29*, 1686–1702. [CrossRef] [PubMed]
41. Wang, P.; Du, Y.; Zhao, X.; Miao, Y.; Song, C. The MPK6-ERF6-ROS-responsive cis-acting Element7/GCC box complex modulates oxidative gene transcription and the oxidative response in *Arabidopsis*. *Plant Physiol.* **2013**, *161*, 1392–1408. [CrossRef] [PubMed]
42. Overmyer, K.; Kollist, H.; Tuominen, H.; Betz, C.; Langebartels, C.; Wingsle, G.; Kangasjärvi, S.; Brader, G.; Mullineaux, P.; Kangasjärvi, J. Complex phenotypic profiles leading to ozone sensitivity in *Arabidopsis thaliana* mutants. *Plant Cell Environ.* **2008**, *31*, 1237–1249. [CrossRef]

43. Li, C.; Song, Y.; Guo, L.; Gu, X.; Muminov, M.A.; Wang, T. Nitric oxide alleviates wheat yield reduction by protecting photosynthetic system from oxidation of ozone pollution. *Environ. Pollut.* **2018**, *236*, 296–303. [CrossRef]
44. Calatayud, A.; Ramirez, J.W.; Iglesias, D.J.; Barreno, E. Effects of ozone on photosynthetic CO_2 exchange, chlorophyll a fluorescence and antioxidant systems in lettuce leaves. *Physiol. Plant.* **2002**, *116*, 308–316. [CrossRef]
45. Lehnherr, B.; Grandjean, A.; Mächler, F.; Fuhrer, J. The effect of ozone in ambient air on ribulosebisphosphate carboxylase/oxygenase activity decreases photosynthesis and grain yield in wheat. *J. Plant Physiol.* **1987**, *130*, 189–200. [CrossRef]
46. Glick, R.E.; Schlagnhaufer, C.D.; Arteca, R.N.; Pell, E.J. Ozone-induced ethylene emission accelerates the loss of ribulose-1, 5-bisphosphate carboxylase/oxygenase and nuclear-encoded mRNAs in senescing potato leaves. *Plant Physiol.* **1995**, *109*, 891–898. [CrossRef] [PubMed]
47. Broberg, M.C.; Uddling, J.; Mills, G.; Pleijel, H. Fertilizer efficiency in wheat is reduced by ozone pollution. *Sci. Total Environ.* **2017**, *607*, 876–880. [CrossRef] [PubMed]
48. Farré-Armengol, G.; Peñuelas, J.; Li, T.; Yli-Pirilä, P.; Filella, I.; Llusia, J.; Blande, J.D. Ozone degrades floral scent and reduces pollinator attraction to flowers. *New Phytol.* **2016**, *209*, 152–160. [CrossRef] [PubMed]
49. Bommarco, R.; Marini, L.; Vaissière, B.E. Insect pollination enhances seed yield, quality, and market value in oilseed rape. *Oecologia* **2012**, *169*, 1025–1032. [CrossRef] [PubMed]
50. Shiferaw, B.; Smale, M.; Braun, H.; Duveiller, E.; Reynolds, M.; Muricho, G. Crops that feed the world 10. Past successes and future challenges to the role played by wheat in global food security. *Food Secur.* **2013**, *5*, 291–317. [CrossRef]
51. Mongeon, P.; Paul-Hus, A. The journal coverage of bibliometric databases: A comparison of Scopus and Web of Science. *J. Cover. Web Sci. Scopus Comp. Anal.* **2014**, *10*. [CrossRef]
52. Archambault, É.; Campbell, D.; Gingras, Y.; Larivière, V. Comparing bibliometric statistics obtained from the Web of Science and Scopus. *J. Am. Soc. Inf. Sci. Technol.* **2009**, *60*, 1320–1326. [CrossRef]
53. Van Eck, N.J.; Waltman, L. Software survey: VOSviewer, a computer program for bibliometric mapping. *Scientometrics* **2010**, *84*, 523–538. [CrossRef]
54. Word Cloud Art Creator. Available online: https://wordart.com (accessed on 13 July 2021).
55. Hirsch, J.E.; Buela-Casal, G. The meaning of the h-index. *Int. J. Clin. Health Psychol.* **2014**, *14*, 161–164. [CrossRef]
56. Bornmann, L.; Mutz, R.; Hug, S.E.; Daniel, H. A multilevel meta-analysis of studies reporting correlations between the h index and 37 different h index variants. *J. Informetr.* **2011**, *5*, 346–359. [CrossRef]
57. Gimenez, E.; Salinas, M.; Manzano-Agugliaro, F. Worldwide research on plant defense against biotic stresses as improvement for sustainable agriculture. *Sustainability* **2018**, *10*, 391. [CrossRef]
58. Gimenez, E.; Manzano-Agugliaro, F. DNA damage repair system in plants: A worldwide research update. *Genes* **2017**, *8*, 299. [CrossRef] [PubMed]
59. Braun, J. *Food and Financial Crises. Implications for Agriculture and the Poor*; International Food Policy Research Institute: Washington, DC, USA, 2008. [CrossRef]
60. Maity, D. Literature on ozone (2000–2015): A bibliometric analysis. *Int. J. Libr. Inf. Sci.* **2018**, *7*, 2. Available online: http://www.iaeme.com/IJLIS/issues.asp?JType=IJLIS&VType=7&IType=2 (accessed on 17 January 2021). [CrossRef]
61. Zeng, G.; Pyle, J.A.; Young, P.J. Impact of climate change on tropospheric ozone and its global budgets. *Atmos. Chem. Phys.* **2008**, *8*, 369–387. [CrossRef]
62. Heagle, A.S. Ozone and crop yield. *Ann. Rev. Phytopathol.* **1989**, *27*, 397–423. [CrossRef]
63. Schenone, G.; Botteschi, G.; Fumagalli, I.; Montinaro, F. Effects of ambient air pollution in open-top chambers on bean (*Phaseolus vulgaris* L.) I. Effects on growth and yield. *New Phytol.* **1992**, *122*, 689–697. [CrossRef]
64. Rédei, G.P. *Arabidopsis* as a genetic tool. *Ann. Rev. Genet.* **1975**, *9*, 111–127. [CrossRef]
65. Goodin, M.M.; Zaitlin, D.; Naidu, R.A.; Lommel, S.A. *Nicotiana benthamiana*: Its history and future as a model for plant–pathogen interactions. *Mol. Plant Microbe Interact.* **2008**, *21*, 1015–1026. [CrossRef] [PubMed]
66. Atkinson, R. Atmospheric chemistry of VOCs and NOx. *Atmos. Environ.* **2000**, *34*, 2063–2101. [CrossRef]
67. Searles, P.S.; Flint, S.D.; Caldwell, M.M. A meta-analysis of plant field studies simulating stratospheric ozone depletion. *Oecologia* **2001**, *127*, 1–10. [CrossRef]
68. McClean, P.E.; Lavin, M.; Gepts, P.; Jackson, S.A. Phaseolus vulgaris: A diploid model for soybean. In *Genetics and Genomics of Soybean*; Springer: Berlin/Heidelberg, Germany, 2008; pp. 55–76. [CrossRef]
69. Gimeno, B.S.; Bermejo, V.; Sanz, J.; De La Torre, D.; Elvira, S. Growth response to ozone of annual species from Mediterranean pastures. *Environ. Pollut.* **2004**, *132*, 297–306. [CrossRef] [PubMed]
70. Burkey, K.O.; Miller, J.E.; Fiscus, E.L. Assessment of ambient ozone effects on vegetation using snap bean as a bioindicator species. *J. Environ. Qual.* **2005**, *34*, 1081–1086. [CrossRef]
71. Krupa, S.; McGrath, M.T.; Andersen, C.P.; Booker, F.L.; Burkey, K.O.; Chappelka, A.H.; Chevone, B.I.; Pell, E.J.; Zilinskas, B.A. Ambient ozone and plant health. *Plant Dis.* **2001**, *85*, 4–12. [CrossRef] [PubMed]
72. Broughton, W.J.; Hernandez, G.; Blair, M.; Beebe, S.; Gepts, P.; Vanderleyden, J. Beans (*Phaseolus* spp.)—Model food legumes. *Plant Soil* **2003**, *252*, 55–128. [CrossRef]
73. Taylor, G.E.; Johnson, D.W.; Andersen, C.P. Air pollution and forest ecosystems: A regional to global perspective. *Ecol. Appl.* **1994**, *4*, 662–689. [CrossRef]

74. Sandermann, H.; Wellburn, A.R.; Heath, R.L. *Forest Decline and Ozone: A Comparison of Controlled Chamber and Field Experiments*; Springer Science & Business Media: Berlin/Heidelberg, Germany, 2012; Volume 127.
75. Chappelka, A.H.; Samuelson, L.J. Ambient ozone effects on forest trees of the eastern United States: A review. *New Phytol.* **1998**, *139*, 91–108. [CrossRef]
76. Matyssek, R.; Innes, J.L. Ozone—A risk factor for trees and forests in Europe? *Water Air Soil Pollut.* **1999**, *116*, 199–226. [CrossRef]
77. McLaughlin, S.; Percy, K. Forest health in North America: Some perspectives on actual and potential roles of climate and air pollution. *Water Air Soil Pollut.* **1999**, *116*, 151–197. [CrossRef]
78. Emberson, L. Effects of ozone on agriculture, forests and grasslands. *Philos. Trans. R. Soc.* **2020**, *378*, 20190327. [CrossRef]
79. Manning, W.J. Establishing a cause and effect relationship for ambient ozone exposure and tree growth in the forest: Progress and an experimental approach. *Environ. Pollut.* **2005**, *137*, 443–454. [CrossRef]
80. Kolb, T.E.; Matyssek, R. Limitations and perspectives about scaling ozone impacts in trees. *Environ. Pollut.* **2001**, *115*, 373–393. [CrossRef]
81. Barbo, D.N.; Chappelka, A.H.; Somers, G.L.; Miller-Goodman, M.S.; Stolte, K. Diversity of an early successional plant community as influenced by ozone. *New Phytol.* **1998**, *138*, 653–662. [CrossRef]
82. Patterson, M.T.; Rundel, P.W. Carbon isotope discrimination and gas exchange in ozone-sensitive and-resistant populations of Jeffrey pine. In *Stable Isotopes and Plant Carbon-Water Relations*; Academic Press: Cambridge, MA, USA, 1993; pp. 213–225. [CrossRef]
83. McBride, J.R.; Laven, R.D. Impact of Oxidant Air Pollutants on Forest Succession in the Mixed Conifer Forests of the San Bernardino Mountains. In *Oxidant Air Pollution Impacts in the Montane Forests of Southern California*; Springer: New York, NY, USA, 1999; pp. 338–352. [CrossRef]
84. Lelieveld, J.; Evans, J.S.; Fnais, M.; Giannadaki, D.; Pozzer, A. The contribution of outdoor air pollution sources to premature mortality on a global scale. *Nature* **2015**, *525*, 367–371. [CrossRef]
85. Crutzen, P.J.; Andreae, M.O. Biomass burning in the tropics: Impact on atmospheric chemistry and biogeochemical cycles. *Science* **1990**, *250*, 1669–1678. [CrossRef] [PubMed]
86. Foyer, C.H.; Noctor, G. Oxidant and antioxidant signalling in plants: A re-evaluation of the concept of oxidative stress in a physiological context. *Plant Cell Environ.* **2005**, *28*, 1056–1071. [CrossRef]
87. Ainsworth, E.A.; Gillespie, K.M. Estimation of total phenolic content and other oxidation substrates in plant tissues using Folin–Ciocalteu reagent. *Nat. Protoc.* **2007**, *2*, 875–877. [CrossRef]
88. Rao, M.V.; Paliyath, G.; Ormrod, D.P. Ultraviolet-B-and ozone-induced biochemical changes in antioxidant enzymes of *Arabidopsis thaliana*. *Plant Physiol.* **1996**, *110*, 125–136. [CrossRef] [PubMed]
89. Nuvolone, D.; Petri, D.; Voller, F. The effects of ozone on human health. *Environ. Sci. Pollut. Res.* **2018**, *25*, 8074–8088. [CrossRef]
90. Frei, M.; Tanaka, J.P.; Chen, C.P.; Wissuwa, M. Mechanisms of ozone tolerance in rice: Characterization of two QTLs affecting leaf bronzing by gene expression profiling and biochemical analyses. *J. Exp. Bot.* **2010**, *61*, 1405–1417. [CrossRef]
91. Frei, M. Breeding of ozone resistant rice: Relevance, approaches and challenges. *Environ. Pollut.* **2015**, *197*, 144–155. [CrossRef] [PubMed]
92. Burton, A.L.; Burkey, K.O.; Carter, T.E., Orf, J.; Cregan, P.B. Phenotypic variation and identification of quantitative trait loci for ozone tolerance in a Fiskeby III× Mandarin (Ottawa) soybean population. *Theor. Appl. Genet.* **2016**, *129*, 1113–1125. [CrossRef] [PubMed]
93. Booker, F.; Muntifering, R.; McGrath, M.; Burkey, K.; Decoteau, D.; Fiscus, E.; Manning, W.; Krupa, S.; Chappelka, A.; Grantz, D. The ozone component of global change: Potential effects on agricultural and horticultural plant yield, product quality and interactions with invasive species. *J. Integr. Plant Biol.* **2009**, *51*, 337–351. [CrossRef] [PubMed]
94. Grulke, N.E.; Heath, R.L. Ozone effects on plants in natural ecosystems. *Plant Biol.* **2020**, *22*, 12–37. [CrossRef] [PubMed]

Review

An Overview of Cooperative Robotics in Agriculture

Chris Lytridis, Vassilis G. Kaburlasos *, Theodore Pachidis, Michalis Manios, Eleni Vrochidou, Theofanis Kalampokas and Stamatis Chatzistamatis

Human-Machines Interaction (HUMAIN) Lab, Department of Computer Science, International Hellenic University (IHU), 65404 Kavala, Greece; lytridic@cs.ihu.gr (C.L.); pated@cs.ihu.gr (T.P.); m.manios@cs.ihu.gr (M.M.); evrochid@teiemt.gr (E.V.); theokala@cs.ihu.gr (T.K.); stami@emt.ihu.gr (S.C.)
* Correspondence: vgkabs@teiemt.gr; Tel.: +30-2510-462-320

Abstract: Agricultural robotics has been a popular subject in recent years from an academic as well as a commercial point of view. This is because agricultural robotics addresses critical issues such as seasonal shortages in manual labor, e.g., during harvest, as well as the increasing concern regarding environmentally friendly practices. On one hand, several individual agricultural robots have already been developed for specific tasks (e.g., for monitoring, spraying, harvesting, transport, etc.) with varying degrees of effectiveness. On the other hand, the use of cooperative teams of agricultural robots in farming tasks is not as widespread; yet, it is an emerging trend. This paper presents a comprehensive overview of the work carried out so far in the area of cooperative agricultural robotics and identifies the state-of-the-art. This paper also outlines challenges to be addressed in fully automating agricultural production; the latter is promising for sustaining an increasingly vast human population, especially in cases of pandemics such as the recent COVID-19 pandemic.

Keywords: agricultural robots; agriculture 4.0/5.0; cooperative robots; farming automation

1. Introduction

The popular term Precision Agriculture, or PA for short, has been defined as "a management strategy that uses electronic information and other technologies to gather, process, and analyze spatial and temporal data for the purpose of guiding targeted actions that improve efficiency, productivity, and sustainability of agricultural operations" [1]. Based on this definition, the introduction of robots in agricultural tasks can serve the purpose of PA by taking advantage of sophisticated equipment for accurate measurements, management, and operations. Hence, an analysis of the effects of the introduction of agricultural robots in the workforce was presented in [2]. Systematic reviews that widely cover research in the field of agricultural robots can be found in [3–5]. These studies show a wide range of agricultural applications that can be achieved by replacing humans with autonomous machines. The objective of introducing robots in agriculture are mainly (a) to improve efficiency and productivity, (b) to counter labor shortages of seasonal workers, and (c) to perform laborious and possibly dangerous tasks. Those developments in agriculture can be interpreted in a more general, industrial context as follows.

Industry 1.0 or, equivalently, the (classic) industrial revolution, has been called the transition from manual production to mechanical (steam) production from the late 18th century to the early 19th century. The second industrial revolution (Industry 2.0), from the late 19th century to the early 20th century, was shaped by the widespread use of electricity. The third industrial revolution (Industry 3.0), in the second half of the 20th century, was shaped by the widespread use of digital computers. Currently, the fourth industrial revolution (Industry 4.0) is driven by advanced artificial intelligence as well as by the Internet. Corresponding developments can be observed in agricultural technology whose most recent developments are outlined next.

The term "Agriculture 3.0" has been proposed as an alternative to "Precision Agriculture" [6]. Agriculture 3.0 can be interpreted as a domain-specific extension of Industry

3.0 in agriculture. Note that PA, which includes the application of personalized practices (i.e., inputs) based on local measurements, may not be suitable for all agricultural tasks. In particular, the production of certain high-quality agricultural products may require manual skills based on empirical knowledge. For example, vinicultural tasks such as harvesting, pruning, spraying, tying, etc. require the aforementioned skills.

Lately, the term "Agriculture 4.0" has been proposed as a domain-specific extension of "Industry 4.0" to agriculture [6,7]. More specifically, among other things, "Agriculture 4.0" refers to a massive automation of skillful manual agricultural tasks. The work here has been motivated by an ongoing project regarding the development of a team of cooperative robots, including ground robot vehicles as well as unmanned aerial vehicles, for vinicultural applications [6], where emphasis is given to the engagement of mechanical hands with many (>20) degrees of freedom toward reproducing the skillfulness of the human hand in selected vinicultural tasks.

The cooperative robotics reviewed in this work could be regarded as a precursor, in agriculture, of a more general industrial trend toward a cooperative integration of humans with robots/machines, namely "Industry 5.0" [8]. More specifically, cooperative robotics can be a future "Agriculture 5.0" technology that integrates humans with robots in agricultural applications. In the latter context, a technological challenge regards the development of effective models to support interaction between humans and/or robots. This work, in the discussion section below, proposes a novel information processing paradigm for supporting cooperative robots in agriculture.

Technological advances in sensing and actuation as well as machine learning have allowed more agricultural tasks to be feasible by autonomous machines. Such tasks range in all stages of cultivation from land preparation and sowing, to monitoring and harvesting. Commercial agricultural robots are already available, and more are expected to appear in the next years as technologies such as machine vision and dexterous grasping become more mature. However, the introduction of multiple cooperating robots in the field can have good prospects in the reduction of production costs and the improvement of operational efficiency. This paper presents an overview of research in the field of cooperative robotics in the context of agriculture. The incentive behind this work is the fact that even though there are numerous studies regarding advances in agricultural robots and their underlying technologies, there are no studies that comprehensively survey the utilization of cooperation in agricultural robotic applications.

The paper is structured as follows. Section 2 describes the methodology followed for compiling relevant research works. Section 3 details our results regarding cooperative agricultural robots, including relevant statistics. Finally, Section 4 summarizes the contribution of this work, including discussions for potential future work.

2. Materials and Methods

The sources used for the compilation of the present review were the databases of Google Scholar, ScienceDirect, Scopus, IEEE Xplorer, and Wiley. The criteria for selecting the research to be included in this paper were the following: (a) work from the last 15 years was reviewed, (b) aerial or ground robots had to demonstrate autonomy as well as cooperative and coordination skills, and (c) the application area had to be in agriculture exclusively, i.e., no robots executing general-purpose cooperation algorithms were included. Furthermore, among articles by the same author(s) that report research results incrementally, only the most recent ones were considered in the present review.

Note that, sometimes, a team of robots is employed "in parallel" such that each robot is operating alone on a different land parcel without interacting with another robot. Such teams of robots have not been considered here. Instead, the interest of this work is in teams of robots interacting with one another.

This paper presents statistical results regarding papers based on the date of research as well as the country where the research took place. For the former, the publication year was used; for the latter, a country was defined by the affiliation of the first author, even though several articles are the result of international collaborations.

In total, 77 articles were compiled and (a) reviewed and (b) statistically analyzed regarding their (b.1) type of publication, (b.2) research topic, (b.3) geographical region of the first author, (b.4) country of origin of the first author, and (b.5) year of publication. Details regarding our review analysis are presented next.

3. Cooperative Agricultural Robotics

The reviewed papers were categorized into five main research topics where cooperation is found in agricultural robotics: (a) human–robot cooperation, or "human–robot" for short, (b) cooperative Unmanned Aerial Vehicles (UAVs), or "multi-UAV" for short, (c) cooperative Unmanned Ground Vehicles (UGVs), or "multi-UGV" for short, (d) Hybrid teams of UAVs and UGVs, or "UAV/UGV" for short, and (e) cooperative manipulation by multi-arm systems, or "manipulators" for short.

3.1. Human–Robot Cooperation (Human–Robot)

The majority of agricultural work is currently being performed by humans either manually or using human-operated machines or equipment. In recent years, there have been many attempts to automate tasks and produce fully autonomous robots. However, some tasks cannot yet be carried out by a single robot in a reliable and efficient manner. For this reason, collaboration between humans and robots has been considered [9–12].

When a human and a robot must work together towards a common objective, there are several ways in which control can be realized, such as through remote control, supervisory control, or cooperative control [13]. In this section, the focus will be on examples of cooperative human–robot control in agricultural applications.

To realize cooperative control, some interface between the robot and the human must be established, such that information is shared and some level of control from the part of the human is achieved. For example, in [14], the usability of different types of user interfaces was studied for the control of a semi-autonomous vineyard sprayer robotic system. In this case, the robot can perform some tasks autonomously, but the human operator can intervene through a user interface. In [15], tractor steering was achieved using signals from an electromyographic (EMG) human–machine interface put on a human tractor driver. Apart from interface types, other works explore the idea that, depending on the conditions, different levels of cooperation may be needed. For instance, in [16], a semi-autonomous system was presented. The system used a three-layer architecture that includes a servo control, autonomous control, and manual operation. The operator can either manually operate the vehicle or can act as a supervisor of the autonomous vehicle, able to intervene at appropriate times through an interface. In [17], automatic melon recognition was investigated. The objective was to evaluate the effectiveness of collaboration between a human and a robot in a target recognition task. Four collaboration levels were investigated, ranging from target selection, performed by the human operator, to automatic target selection, performed by the system. The study showed higher detection rates in the collaborative detection case compared to either manual or autonomous detection cases. In a later study [18], the authors used an objective function for the performance of the collaborative task (as defined in [19]) to dynamically alter the collaboration levels during the melon recognition task.

A mathematical programming framework for optimizing human–robot collaboration was proposed in [20]. The framework considers the question of when interaction of the human operator with the robotic system is most economically beneficial. To validate the framework, simulations of citrus robotic harvesting were implemented, and showed how the robotic system required human collaboration in order to compensate for inefficient components of the system.

In [21], human action recognition by robots in an agricultural task was investigated. More specifically, human participants equipped with wearable sensors for data acquisition were asked to perform the common agricultural task of lifting and carrying a crate. The objective was to determine whether the robots, using appropriate machine learning models and classification algorithms, could identify the actions of the human participants throughout the task. The authors reported an average accuracy in action recognition of 85.6%. In an earlier work [22], an omni-directional stereo vision camera mounted on a robot tractor was employed for human detection. The system was validated using field tests, which showed that the human could be detected successfully in the range of 4 to 11 m.

In the precision spraying task described in [23], the authors reported a reduction of up to 50% in terms of spraying material. The proposed human–robot collaboration framework aimed at minimizing the false positives in spraying targets, based on images collected by an on-board camera. Depending on the selected cooperation level, target detection can be fully automatic, completely manual by the remote operator, or the operator can adjust the automatically marked targets.

In [24,25], an emulated cooperative strawberry recognition task was presented. In this work, a robot navigated the environment and relayed the images with the automatically recognized targets (together with the degrees of recognition confidence) to human test operators. The user could then accept the recognized targets or not. Based on questionnaires completed by the test users, they reported that they preferred a robot behavior where automatic recognition yields more false positives as opposed to a behavior which results in more false negatives.

A model which enables coordination between humans, robots, sensors, and software agents (i.e., a cyber-physical organization) for gathering unspecified crops and fruit was introduced in [26]. The proposed model consisted of five connected layers, namely network, communication, interaction, organization, and collective intelligence. Through this layered approach, the objective was to achieve indistinguishability, i.e., to enable the system to achieve the desired goal regardless of the actor, either human or machine, that performs the task.

A human–robot skill transfer interface aimed at improving UAV pesticide delivery was proposed in [27]. In this scheme, the UAV was first instructed a trajectory by a human operator via the interface. Then, the accuracy of the trajectory derived in the demonstration phase was improved using an adaptive cubature Kalman filter. Finally, the UAV could follow the resulting trajectory using the stored position and velocity data. The methodology was tested in both simulation through SIMULINK and field experiments using an actual UAV in a commercial canola field.

The cooperative tea harvesting system proposed in [28] used a robot with a camera to detect a marker-carrying human and move by his side by estimating position differences. This coordinated motion then made it easy for the human operator to guide the robot, which had the harvesting device mounted on it, through the field, compared to the standard tea plucking machine which requires two workers.

The presence of a human in an agricultural task requires additional considerations to ensure the health and safety of the workers and to increase the level of trust in human–robot interaction among agricultural workers [29,30]. The study presented in [31] identified the main risk factors in human–robot collaboration in agricultural tasks and proposed methods for safe collaboration by minimizing potential hazards. Moreover, in the pilot study presented in [32], the authors conducted field experiments both in open and indoors environments, where field workers harvesting strawberries evaluated their work carried out alongside the Thorvald robot. The data collected from this study can be used for the design of collaborative systems in terms of the safety aspects.

Figure 1 shows examples of systems where human–robot cooperation is exploited.

(a) Agribot interface [14]   (b) Collaborative target recognition [17]   (c) Tea plucking [28]

Figure 1. Examples of human–robot cooperative systems.

Table 1 summarizes the basic features of the reviewed studies.

Table 1. Summary of the reviewed human–robot cooperation studies in agriculture.

Ref.	Task	Objective	Type of Study	Cooperation Strategy
[14]	Spraying	Vineyard	Field trial	User confirmation of machine vision
[15]	Driving	N/A	Field trial	EMG interface
[16]	Driving	N/A	Field trial	Teleoperation platform
[17–19]	Target recognition	Melon	Lab experiments	User confirmation of machine vision
[20]	Harvesting	Citrus	Simulations	Risk-averse collaboration
[21]	Transportation	N/A	Field trial	Activity recognition
[22]	Human detection	N/A	Field trial	Stereo vision
[23]	Spraying	Vineyard	Field trial	User confirmation of machine vision
[24,25]	Harvesting	Strawberry	Simulation	User confirmation of machine vision
[26]	Harvesting	N/A	Lab experiments	Layered task selection
[27]	Spraying	Canola	Simulation and field trial	Skills transfer interface
[28]	Harvesting	Tea	Field trial	Motion coordination
[29]	N/A	N/A	Correlational study	Acceptance issues
[30]	N/A	N/A	Design principles	Safety issues
[31]	N/A	N/A	Design principles	Safety and ergonomics issues
[32]	Transportation	Strawberry	Field trial	Safety issues

In conclusion, from the perspective of human–robot cooperation, it can be seen that the research has focused on two main areas: firstly, for improving the sensory limitations of current vision-based systems. In this context, the human operator complements the automatic detection capabilities of the autonomously navigating robot, by performing an additional verification and corrections of the robot perceptions; secondly, robotic support of manual labor. Here, the robot acts as an assistant to alleviate the burden of arduous and possibly hazardous tasks. In both of these areas, the level of autonomy of the robot and the division of labor between the human and the robot is an open question. This balance is dependent upon the nature of the task at hand, since the effectiveness, efficiency, and accuracy of the robot varies for each function. Additionally, issues that are currently being considered are the design of appropriate human–robot interfaces and ensuring the safety of the human when he shares a common workspace with the robot.

3.2. UAV Robot Teams (Multi-UAV)

While UAVs have been used in various tasks in agriculture such as remote sensing (e.g., [33,34]), mapping (e.g., [35]), monitoring (e.g., [36]), and pest control (e.g., [37]), using a single UAV for these tasks presents certain limitations, most prominently time efficiency and battery limitations [38]. First, the time required for a single vehicle to cover a large area can be long. Second, because of the increased task duration and the workload, the energy requirements can force the UAV to cover only a small area between frequent recharges. A solution is to use a team of cooperating UAVs so that the task is divided between different UAVs, which can coordinate their movements by partitioning the area and therefore the workload. Such an approach certainly reduces the task duration but it can also reduce the energy consumption at the individual UAV level. Because of the nature of the agricultural work that can be assigned to UAVs, the research in this area is focused on formation control and area coverage algorithms. Even though numerous such algorithms have been developed in the past for a variety of applications, in this section, the focus is specifically on agricultural applications. A recent detailed historical survey of the research in unmanned aerial vehicles in agriculture can be found in [39]. Furthermore, a formal description of managing a group of heterogeneous UAVs was proposed in [40], where parameters such as the field, various facilities, available resources, and constraints were considered.

In 2008, a multi-UAV system for water management and irrigation control was presented [41]. The system is viewed as a camera array with image reconstruction (stitching), and the bands of the images that are collected can be reconfigured depending on the mission. To ensure that the maximum number of images is acquired simultaneously, the system employs formation control where the UAVs are aligned horizontally with a certain distance in between. The paths are precomputed based on mission parameters.

The Swarm Robotics for Agricultural Applications (SAGA) project aims at employing cooperating UAVs for precision farming. In [42], a simulation of the collective behavior of a UAV team for weed monitoring and mapping was presented. The system implements a stochastic coverage and mapping that includes collision avoidance among the aerial vehicles and onboard vision. Further simulation studies on using UAV robot swarms for weed control and mapping were presented in [43]. The monitoring strategy adopted was first to divide the field in cells and assign to each agent a random-walk-based path. The individual agent then decides to move to neighboring cells according to the probability governed by a Gaussian distribution. On the other hand, the Robot Fleets for Highly Effective Agriculture and Forestry Management (RHEA) project aimed at coordinating aerial and ground vehicles in precision agriculture tasks. Specifically, in [44,45], the control structure of the aerial team, consisting of two hex-rotors and tasked with taking high resolution pictures for pest control, was described.

Recall that in [38], the design of a system to perform inspections for precision agriculture by controlling a single UAV or by coordinating multiple UAVs was presented. The system is based on the idea of a control station for on-the-fly mission planning. A heterogeneous embedded framework for small UAVs was also proposed. The work described in [46] involved simulation studies and experiments using four quadrotor aerial vehicles to evaluate a control algorithm for swarm control of agricultural UAV in pest and disease detection. The approach followed in that paper was to implement control in two layers: the first layer was teleoperation where a human operator set the velocity control and the second layer dealt with velocity and formation control as well as collision avoidance. The work in [47] dealt with a surveying task where the UAV team was controlled by a system responsible for connecting the UAVs to act as a swarm, produce flight plans, and respond to disruptive circumstances. Initially, the system divides the survey area in squares, whose size varies according to the UAV's on-board camera characteristics. Each UAV tries to find unvisited and unplanned squares and plans routes depending on both how long a square has remained without supervision and the distance of the UAV to that square. The sub-tasks selected by the UAVs can be exchanged dynamically depending on the predicted sub-task completion times communicated between the agents. A remote sensing task with a self-organizing multi-UAV team capturing georeferenced pictures was presented in [48]. A central controller divided the global task (i.e., the farm area) into sub-tasks and assigned the sub-tasks to the UAVs, based on an extension of the alternate-offers protocol. The UAVs then computed their paths. The proposed approach was validated through field experiments. Similarly, [49] proposed a task allocation and coordination strategy based on a space-based middleware. First, an area decomposition algorithm partitioned the search space so that tasks were dynamically allocated to the UAVs aiming at minimal spatial interference between the UAVs. Second, the task selection was improved by using a model of robot capabilities to extend the space-based middleware. The approach was tested on a weed control task. Another study on a remote sensing task in [50] compared four configurations of agricultural UAVs, namely autonomous versus teleoperated single and multiple UAV teams. This was essentially an area coverage task. To evaluate the performance of the system, total time, setup time, flight time, battery consumption, inaccuracy of land, haptic control effort, and coverage ratio were used as metrics. Experimental results showed that using the autonomous swarm control algorithm [51] improved the efficiency of the agricultural task.

Another path planning technique for UAV teams was proposed in [52]. In particular, the authors proposed a technique where coordinate transformations between virtual and actual workspaces were performed in order to focus on regions of interest, with conventional path planning algorithms applied to each region. The methods were demonstrated in real-world experiments using 3 UAVs in a surveying task. Table 2 summarizes the basic features of the reviewed studies.

Table 2. Summary of the reviewed multi-UAV cooperation studies in agriculture.

Ref.	Task	Objective	Type of Study	Robot Team	Cooperation Strategy
[40]	N/A	N/A	Formal description	Variable number of UAVs	N/A
[41]	Irrigation control	N/A	Field tests	Variable number of UAVs	Coverage control
[42]	Weed mapping	N/A	Simulation	Variable number of UAVs	Coverage control
[43]	Field monitoring	N/A	Simulation	Variable number of UAVs	Individual random walk
[44,45]	Pest control	Maize	Architecture design	N/A	Central robot management system
[46]	Disease detection	N/A	Simulation	Four UAVs	Formation control
[47]	Surveying	N/A	Simulation and field tests	Up to 10 UAVs	Distributed mission planning
[48]	Mapping	Vineyard	Field tests	Three UAVs	Centralized path planning
[49]	Weed control	N/A	Simulation	Three UAVs	Centralized area decomposition
[50]	N/A	N/A	Field tests	Three UAVs	Formation control
[51]	Remote sensing	N/A	Simulation	Four UAVs	Formation control
[52]	Crop health surveying	N/A	Simulation and field tests	Three UAVs	Centralized path planning

In conclusion, the type of cooperation between artificial systems depends on the characteristics of the cooperating systems. Unmanned aerial vehicles are not normally tasked with physically acting on a field, such as performing seeding or harvesting. In most cases, UAVs are typically equipped with a variety of cameras and sensors and are used for monitoring, pest detection, and mapping. The use of multiple UAVs aims at achieving more rapid field coverage compared to employing a single UAV, and cooperative algorithms aim at improving the efficiency of coverage. In addition, by reducing the task duration, battery limitations of aerial vehicles can be remedied. Consequently, research in this field has mainly focused on path planning and coordination and collision avoidance algorithms which take into account the spatial arrangement and the battery limitations of UAVs. As more technological tools become available, future research will focus on extending the utility of UAVs by improving their perception capabilities and battery autonomy as well as enhancing manipulation skills. Figure 2 shows examples of multi-UAV robot teams.

  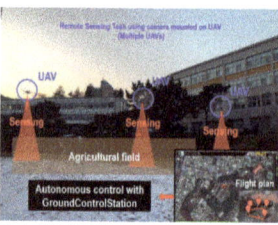

(**a**) RHEA drone [44]   (**b**) Remote sensing drone [48]   (**c**) Autonomous multi-UAVs [50]

Figure 2. Examples of multi-UAV cooperative systems.

3.3. UGV Robot Teams (Multi-UGV)

An important consideration in using multiple robots in farming operations is to maximize the efficiency and therefore the operational costs. This is achieved by effective task planning and path optimization. In [53], the effect of employing a controlled traffic strategy for a pair of agricultural robots compared to uncontrolled traffic was investigated. The simulation study assumed that one robot was an application unit while the other was a refilling unit. In various scenarios and field traffic patterns, the results showed an efficiency improvement in terms of total distance travelled when traffic control was adopted.

One of the early works in multi-robot control in agricultural robotics was presented in [54]. In that work, two algorithms for the control of two master-slave agricultural robots, namely the GOTO algorithm and a FOLLOW algorithm, were introduced. In the proposed architecture, the master is making decisions and sends commands to the slave vehicle, while the slave vehicle follows the master vehicle and broadcasts its status. Another leader–follower system for agricultural applications was demonstrated in simulation and field tests in [55,56]. The robot tractors in this system can work independently, and work cooperatively in the sense that they must keep a certain spatial arrangement during operation. Efficiency improvements were shown when the robots coordinated in either a formation maintenance strategy or a skipping path turn method. Each of the proposed algorithms is suitable for different field operations. In another study [57], the authors discussed a complete farming system which comprised multiple tractors coordinated by a robot management system and central monitoring.

A large part of the work carried out regarding cooperative UGVs in agriculture has been carried out using simulations in order to establish appropriate path planning algorithms. For example, in [58], a simulation of cooperative citrus harvesting was presented. The focus of that study was the demonstration of both a hierarchical task assignment and a trajectory planning algorithm. In the proposed planning framework, there were two main optimization iterations: the cooperative level for formation planning and the individual level for trajectory planning. A leader–follower structure was adopted for the group of agricultural robots. When a configuration trigger event was detected, a wavefront path planning algorithm is used to find an obstacle-free corridor by the leader. The leader then determines the optimal trajectory information and sends it to the followers who, in turn, determine their own local optimal trajectory. The simulation results indicated that the proposed approach is not computationally intensive and can produce optimal paths fast, even though the complex dynamics of the robots are included in the trajectory calculations. The framework is partly decentralized in the sense that some of the computational tasks are decentralized while others are not. The route planner for herbicide applications proposed in [59] considers various criteria in determining the robot teams' routes, including the distance to be travelled, herbicide tank capacity, dynamic characteristics of the heterogeneous robots, etc. A simulation study aimed at studying the planner under different optimization targets (e.g., time to completion) as well as under different conditions (e.g., number of vehicles).

The "Mobile Agricultural Robot Swarms" (MARS) project [60] aimed at employing cooperating small-sized UGVs for precision farming. The work presented in [60] demonstrated a system architecture where a centralized controller (OptiVisor) coordinated and supervised the motion of a team of low-level intelligence robot team in a field seeding task. A decentralized swarm control system for various farming operations such as ploughing, seeding, watering, etc. was proposed in [61]. The experiments were conducted in a land area replicating a farm using miniature prototype robots equipped with tools to perform farming operations. The work presented in [62] dealt with a spraying task by a team of robots using local information only. Simulations were utilized to demonstrate the proposed strategy's capacity to perform task allocation. The authors also explored the multi-robot ploughing task in [63]. Also focusing on spraying tasks, specifically for vineyards, the work described in [64] aimed at utilizing at least two robots working on either side of a vine row in order to improve accuracy. For this, the authors employed Ultra-Wide Band (UWB) sensors to achieve relative localization and synchronize the trajectories of the two robots in a leader–follower scheme.

A route planning algorithm for efficient field coverage was proposed in [65]. The objective of this work was to replace multiple large agricultural machinery with smaller autonomous robots in order to minimize soil compaction. The route planning algorithm was designed to produce efficient field coverage for finding routes with minimal costs. The framework included a mission control center to allocate sub-tasks to robot teams, to coordinate their movements, and to allow them to communicate with one another. In a simulation study [66], the authors considered teams of heterogeneous robots (harvesting and transport robots) deployed in grapevines and modeled their behavior in order to investigate the effect of team size in both harvesting and processing times.

The simulation environment named "Simulation Environment for Precision Agriculture Tasks using Robot Fleets" (SEARFS) presented in [67] allows for the investigation of multi-robot teams in precision agriculture and more specifically in a weed management task. It is a general-purpose computational tool that can model a 3D virtual agricultural environment and simulate the behavior of fleets of autonomous agricultural robots. The user is allowed to select the robots, their sensory and actuation characteristics, the type of field, and determine the specific mission. The behavior of the robot fleet can then be studied.

The cooperative two-robot system for rice harvesting proposed in [68] employs two head-feeding combines. To initialize the harvest, a human operator drives the combines a few laps of crops first, in a spiral toward the center of the field. The combines then begin harvesting autonomously according to target paths planned from the locus of the combine while the operator was driving. The robots harvest in a spiral where the second robot is located 1.2 m inward. Collision avoidance is achieved by inter-robot communication of location.

A simulation of a precision agriculture scenario was presented in [69]. The scenario explores the use of three types of robot for collecting information, sowing, and harvesting. The work focuses on (a) modeling of the robots, which is based on the open-source packages Gazebo and ROS, and (b) interaction between the robots, which is based on the smart space combined with the blockchain platform for information (represented by fuzzy sets) exchange between the robots.

In [70], a monitoring application for precision agriculture using heterogeneous ground robots was presented. The approach followed was to use a weighted directed graph to represent the robot team. The partitioning of the workspace took into account the possible heterogeneous characteristics of the robots such as speed and processing power. According to these characteristics, the robots were distributed on the virtual graph and tasked to monitor a specific region. The potential of the method was demonstrated both by simulations and by experiments on the field.

A collaborative fleet management system for coordinating the flow of operations in a field was demonstrated in simulation and field experiments in [71]. The system supports all the operating stages of a field crop and is based on a novel algorithm which assigns strips of field to each robot, then dynamically updates the state of each strip.

Figure 3 shows examples of multi-UGV robot teams.

(**a**) Tractor in leader–follower team [55]   (**b**) Combine robots [68]   (**c**) Monitoring robot [70]

Figure 3. Examples of multi-UGV cooperative systems.

Workspace partitioning for a multi-robot system operating in an orchard in a spraying task was the subject of [72]. In this work, given a map induced from a UAV-acquired image, a number of nodes for a Voronoi diagram were produced, where an orchard tree was considered to be a node. The nodes were then clustered so that partitions were computed through the Voronoi diagram. In this case, the robots were not cooperating directly; instead, indirect cooperation arose by coordination of their independent actions. Table 3 summarizes the basic features of the reviewed studies.

Table 3. Summary of the reviewed multi-UGV cooperation studies in agriculture.

Ref.	Task	Objective	Type of Study	Robot Team	Cooperation Strategy
[53]	N/A	N/A	Simulation	An application unit and a refilling unit	Leader–follower
[54]	N/A	N/A	Simulation	A master and a slave vehicle	Master–slave
[55,56]	N/A	N/A	Field trials	A master and a slave tractor	Master–slave
[57]	Planting, seeding, transplanting, and harvesting	Rice	Architecture design	A robot for data acquisition and two robot tractors for farming operations	Central robot management system
[58]	Harvesting	Citrus	Simulation	A virtual leader robot and three follower robots	Formation selection or individual trajectory selection
[59]	Herbicide application	N/A	Simulation	Multiple heterogeneous robots	Route planning
[60]	Seeding	N/A	Simulation and field tests	Variable number of robots	Central robot management system
[61]	Ploughing, irrigation, seeding, and harvesting	N/A	Lab experiments	Multiple heterogeneous robots	Central robot management system
[62,63]	Spraying, ploughing	N/A	Simulation	Variable number of robots	Use of information stored at checkpoints
[64]	Spraying	N/A	Lab experiments	A leader robot and a follower robot	Formation control
[65]	N/A	N/A	Simulation	Variable number of robots	Central robot management system
[66]	Harvesting, transport	Grapes	Simulation	One harvesting robot and two transport robots	Central robot management system
[67]	Weed management	N/A	Simulation	Variable number of robots	Central robot management system
[68]	Harvesting	Rice	Field trials	Two combine robots	Leader–follower
[69]	Harvesting	N/A	Simulation	Variable number of heterogeneous robots	Central robot management system
[70]	Monitoring	N/A	Simulation and field trials	Two robots	Route planning
[71]	Coordination	N/A	Simulation and field trials	Three robot tractors	Central robot management system
[72]	Spraying	N/A	Simulation with real data	2 to 10 robots	Central robot management system

In conclusion, in contrast to UAV systems, UGVs are more suitable for agricultural tasks traditionally demanding human intervention, such as sowing, weeding, spraying, harvesting, etc. Research in this area has focused on improving the efficiency of the aforementioned tasks by introducing multi-robot teams in the field. The introduction of multiple robots in a given area, however, demands appropriate management of the spatial allocation of robots, with several approaches proposed, some stemming from the sub-field of swarm robotics. In the special case of heterogeneous robots assigned with different operations, temporal task allocation is also required. A popular method in the literature is the leader–follower approach, with inter-robot communications coordinating motions based on progress status. More research is required to improve collaborative tasks by ground robots, e.g., transfer of a load between a harvesting and a carrier robot or individual robots harvesting a tree concurrently.

3.4. UGV and UAV Teams (UAV/UGV)

While the use of multiple UAVs has advantages such as large area coverage, speed etc., they also have limitations including uncertainty in ground measurements and power limitations. Typically, UAVs are equipped with long-range measuring equipment, such as cameras, and are used in field monitoring tasks. In contrast, UGVs can be deployed in the field to locate targets and either take short-range measurements or perform a physical action. The combined use of UAV and UGV teams has also been proposed in order for the robots to complement each other in agricultural tasks [73].

In [74], the team consisted of a UAV and a UGV for disease detection in a strawberry field. The role of the UAV was to inspect the entire crop and to mark suspect regions. The UGV then approached the marked regions to perform spectral analysis and collect samples.

The robot team presented in [75] consisted of a single UAV and a single UGV. The UGV measured nitrogen in a field, and depending on these measurements, deployed the UAV at selected locations. The UAV landed on the UGV once its mission was complete. This approach has the advantage of limiting the operating time of the UAV, which is desirable given the UAV's limited battery life. The problem is then to minimize the time the UGV needs to obtain soil samples and a path planning algorithm was proposed to this end. The methodology was validated through simulation studies with real-world data.

A bioinspired path planning strategy for coordinating a hybrid (aerial and ground-based) multi-robot team toward a target was presented in [76]. In this strategy, investigated using simulation studies, the three=dimensional terrain was modeled as a neuron topological map and a Dragonfly Algorithm (DA) optimized the movements of the robots. Although this algorithm was not developed specifically for agriculture, the scenario can have applications in agricultural robot teams consisting of UAVs and UGVs. Other examples of UAV/UGV coordination approaches can be found in [77–79].

As mentioned earlier, the RHEA project dealt with coordinating aerial and ground robots in precision agriculture [80,81]. In [81], two sub-tasks of weed and pest control missions were considered: (a) inspection missions carried out by the aerial team and (b) treatment missions carried out by the ground robots. A Mission Manager was employed to manage the collected data from the various units and centrally compute the trajectories and actions of the robots. Furthermore, the ground robot plans were optimized based on factors such as costs and time.

In [82,83], a UGV and UAV independently generated point clouds that represented a map of a field using own on-board cameras. The proposed methodology aimed at effectively merging the two individual maps, thus producing a more accurate map which included the surface model as well as the vegetation index. Therefore, collaboration was implicit and arose from the aggregate result of the individual measurements.

In [84,85], dual agricultural robot teams consisting of an aerial unit and a ground unit were proposed, but no details on the implementation of the proposed cooperation strategy were given. Similarly, the hardware design of a dual UAV/UGV robot system was proposed in [86]. The objective of the system was to collect images of a crop and then process them using various vegetation indices in order to determine the crop status.

Another approach for robot team control was followed in another simulation study [87], where the agricultural robot team consisted of three unmanned aerial vehicles and one unmanned ground robot. Each robot was modeled as a finite state automaton and the entire multi-robot system as a discrete event system. It featured a supervisory controller that enabled heterogeneous agricultural robots to perform field operations, avoid obstacles, follow a defined formation, and follow a given path. Table 4 summarizes the basic characteristics of the reviewed studies. Figure 4 shows examples of UAV/UGV robot teams.

Table 4. Summary of the reviewed UAV/UGV cooperation studies in agriculture.

Ref.	Task	Objective	Type of Study	Robot Team	Cooperation Strategy
[74]	Disease detection	Strawberry	Architecture design	One UAV and one UGV	UGV visiting locations identified by the UAV
[75]	Fertilization	Not specified	Simulation	One UAV and one UGV	UGV visiting locations identified by the UAV
[80,81]	Pest control	Winter cereal	Field trials	Two six-rotor drones and three tractors	UGVs visiting locations identified by the UAVs
[82,83]	Mapping	Not specified	Simulation with real data	One UAV and one simulated UGV	Map data fusion
[84]	Crop management	Lettuce	Architecture design	One UAV and one UGV	UGV visiting locations identified by the UAV
[85]	Inspection	Not specified	Architecture design	One UAV and one UGV	Transportation of UAV by the UGV
[86]	Crop status mapping	Not specified	Architecture design	One UAV and one UGV	Crop data fusion
[87]	N/A	N/A	Simulation	Three UAVs and one UGV	Leader-follower formation control

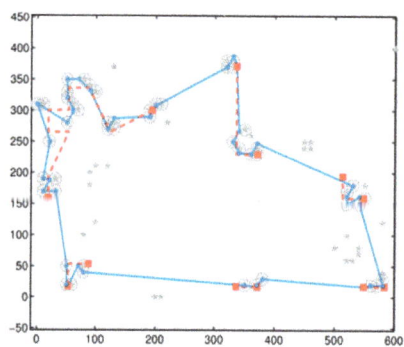
(a) Simulation of a UAV and a UGV visiting points in a field [75]

(b) RHEA UAV/UGV fleet [81]

Figure 4. Examples of UAV/UGV cooperative system.

In summary, hybrid robot teams comprised of UAVs and UGVs engaged in agricultural work are found in the literature, exploiting their relative merits. In particular, the aerial vehicle(s) first mark areas of interest and then the ground robot approaches the location and performs the necessary operations. Another cooperation strategy between the UAV and the UGV in the literature is that the UGVs can also serve as landing and charging stations for the accompanying aerial vehicle.

3.5. Cooperative Manipulation (Manipulators)

The notion of cooperative operation in agricultural robots can also be extended from cooperation between separate robotic platforms to cooperation between arms typically mounted on the same robot. The main benefit of a multi-arm robotic system performing an agricultural task is to improve efficiency and reduce the task duration. In addition, the arms can be actively collaborating toward the same goal, e.g., both working together to harvest a single fruit, therefore attempting to solve problems such as occlusion. The principles governing cooperative manipulation by robotic arms mounted on a single robot platform can also be extended to cooperative manipulation by robotic arms mounted on separate robotic platforms.

A multi-arm kiwi harvester was presented in [88]. The system was equipped with four identical 3 Degrees-of-Freedom (DoF) arms with customized grippers. Although the arms harvested the fruit independently, there was an overall task scheduler which, based on the detected fruits, created fruit clusters, determined the harvesting order, and assigned fruit clusters to each arm. Another multi-arm robot for melon harvesting was presented in [89]. The proposed system consisted of four Cartesian manipulators which reached down, picked melons, and placed them on lateral conveyors. The assignment of melons to each arm was considered to be an interval graph coloring problem, with a greedy search algorithm that calculated an optimal solution for the harvest order. The controller took into account the kinematic conditions that governed the capabilities of each arm and the hardware design was oriented toward improving the harvest ratio. A dual-arm strawberry harvesting robot was described in [90]. Two single-rail 5 DoF manipulators were controlled by a collision avoidance and harvesting order planner based on the location of detected strawberries. The authors reported reduced harvesting times with the dual-arm robot compared to harvesting with a single manipulator.

Cooperative manipulation in an apple orchard was demonstrated in [91]. The authors employed a graph-based method to guide two 6 DoF arms. Each arm was assigned a different role; the grasping arm was designated to pick the apple and the searching arm was designated to locate apples that were hidden from the point of view of the grasping arm. Both arms were equipped with depth cameras. Location information was encoded as a graph whose nodes could be used to calculate appropriate paths. The study reported that the method worked reasonably well in simulation as well as in experimental studies. Apple harvesting using dual cooperative manipulators was also proposed in [92]. In this case, simulation studies were carried out where RGB cameras, located on the manipulators' end effectors, were assumed to accurately detect and locate apples on randomly generated virtual trees. The two manipulators cooperated since one served as the searching arm which identified the other's (grasping arm's) reference points and helped determine clear paths to the detected fruit.

The dual-arm configuration proposed in [93] for aubergine harvesting assumed three modes of operation: (1) a single arm picking a single aubergine, (2) two arms picking fruits independently, and (3) arms working cooperatively to pick a single aubergine. In the second mode, a planning algorithm was developed for task scheduling and collision avoidance. The cooperative mode was employed when there was limited visibility to a fruit and so, one arm was tasked with removing any occlusions while the other arm was tasked with grasping. The performance of the system was evaluated in laboratory experiments.

In [94], a pair of collaborating manipulators were evaluated in an apple harvesting task. The first manipulator was equipped with an 8-DoF manipulator and was tasked with picking apples, while the second one was a catching manipulator with two links which could reach any drop position in the picking manipulator's workspace. Two strategies of fruit harvesting were tested in studies with a replica apple tree, in order to determine the most efficient method in terms of average picking time. However, additional tests in the field demonstrated the need for additional considerations in the picking strategies in order to limit damage to the harvested fruit.

A robot for harvesting greenhouse tomatoes was described in [95]. Two mirrored 3-DoF arms in this dual-arm system had different end-effectors: one arm was fitted with a cutter and the other one was fitted with a suction cup. A single stereo camera mounted at the top of the robot captured images that were processed by a computer responsible for tomato detection and performed a 3D world reconstruction, after which the arm with the vacuum end effector grasped the fruit while the other arm detached the fruit. A planting and watering dual-arm robot was presented in [96]. The prototype robot was equipped with two Prismatic-Revolute (PR) arms which were tasked with digging, seeding, and covering the soil. Then, there was a separate watering module which was activated depending on the readings of a soil moisture sensor. Table 5 summarizes the basic features of the reviewed studies.

Table 5. Summary of the examined cooperative manipulation studies in agriculture.

Ref.	Task	Objective	Type of Study	Mode of Operation	Manipulators
[88]	Harvesting	Kiwi	Field trials	Arm coordination	Four 3-DoF arms
[89]	Harvesting	Mellon	Simulation	Arm coordination	Variable number of 3-DoF arms
[90]	Harvesting	Strawberry	Field trials	Arm coordination	Two single-rail 5-DoF arms
[91]	Harvesting	Apple	Simulation and lab experiments	Arm collaboration	Two 6-DoF arms
[92]	Harvesting	Apple	Simulation	Arm collaboration	Two 6-DoF arms
[93]	Harvesting	Aubergine	Lab experiments	Arm collaboration and coordination	Two 6-DoF arms
[94]	Harvesting	Apple	Lab experiments	Arm collaboration	An 8-DOF arm and a 2-DoF arm
[95]	Harvesting	Tomato	Field experiments	Arm collaboration	Two mirrored 3-DoF arms
[96]	Planting and watering	N/A	Field experiments	Arm coordination	Two 2-DoF arms

In conclusion, cooperative manipulation for agricultural operations is generally applied by robotic manipulators mounted on the same vehicle in order to collaborative complete a task, mainly harvesting. Although there are several agricultural robots equipped with more than one manipulator, in most cases the robotic arms operate independently with some planning algorithm coordinating motions for task assignment and collision avoidance. Few studies examine actual cooperative manipulation tasks, such as two robotic arms cooperatively picking an individual fruit. These can be especially advantageous in cases where view of fruits is limited due to occlusion or when cutting and grasping of fruit may require two hands. For instance, recently, the European project BACCHUS [97] considered, in a cognitive mechatronics context, the development of a bi-manual robotic platform for grape harvest; detailed application results are expected in the future. Nevertheless, to the authors' knowledge, applications where cooperative manipulation must be performed by arms mounted on different mobile platforms are yet to appear in agricultural robotics, and this is another future direction for research. Such a system would render the robot team more flexible, as one of the platforms could be attending to other field work and assist the other platform only when needed. This would improve the efficiency of the work and would reduce the need for additional task-specific machines, reducing operational costs as a result. The increased complexity and sophistication of such a system could potentially be compensated to some degree by the aforementioned flexibility it would provide.

Figure 5 shows systems with cooperative manipulators.

(a) Kiwi harvester [88]

(b) Apple harvesting arms [91]

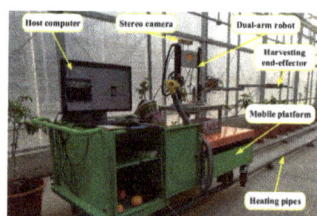
(c) Tomato harvester [95]

Figure 5. Examples of multi-arm systems.

3.6. Trends

With the 77 articles studied in this review, trends were identified in the field of cooperative agricultural robotics. More specifically, the distribution per publication type (i.e., conferences, journals, theses, and books) is shown in the pie chart in Figure 6, which shows that the 77 articles were fairly evenly distributed mainly between conference publications (33) and journal publications (42); in addition, one thesis and one book have been published.

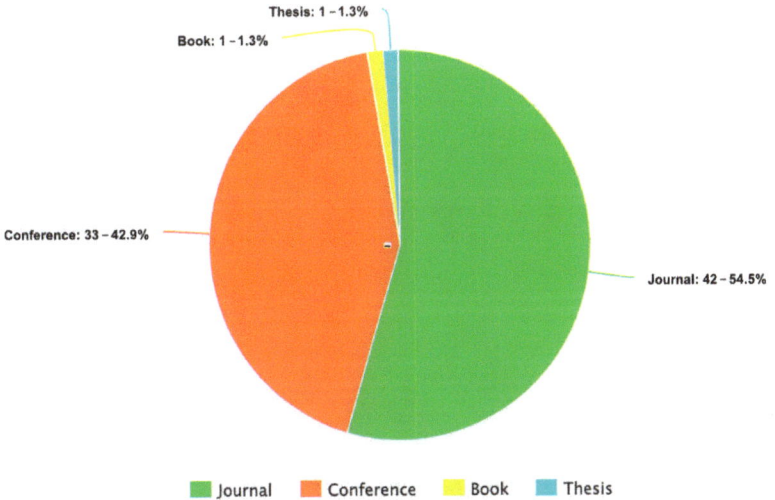

Figure 6. Number of articles per publication type (i.e., conferences, journals, theses, and books).

The pie chart in Figure 7 shows the distribution of the 77 articles in the assumed five sections, namely (a) human–robot, (b) multi-UAV, (c) multi-UGV, (d) UAV/UGV, and (e) manipulators.

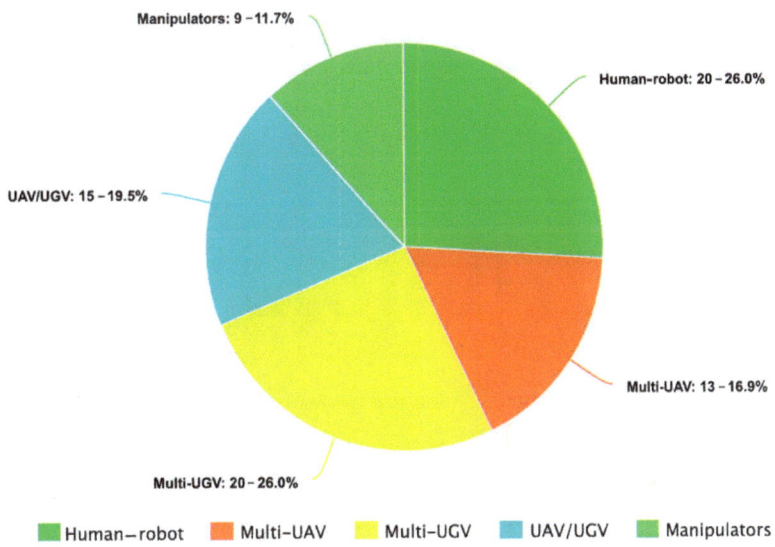

Figure 7. Distribution of articles per topic.

The bar chart in Figure 8 displays the distribution of the 77 articles in various geographical regions including North America, Asia, Europe, and elsewhere, such that, on the bar of a region, the distribution of different topics is also indicated by different colors. Furthermore, Figure 9 details the distribution of the 77 articles in different countries; again, on the bar of a country, the distribution of different topics is indicated by different colors.

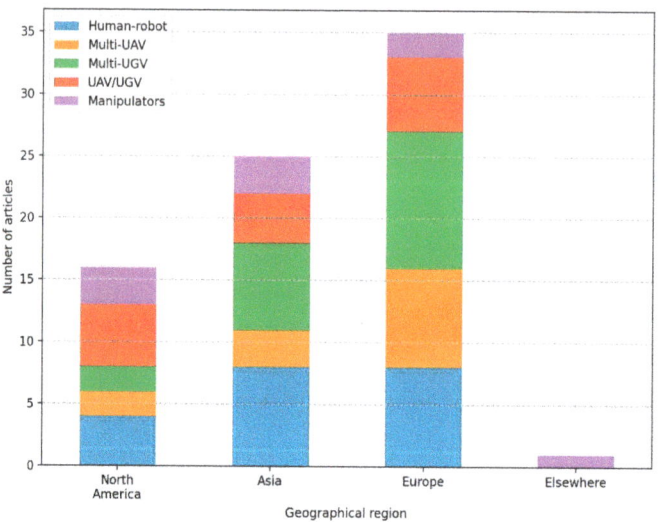

Figure 8. Geographical distribution of articles per geographical region per topic.

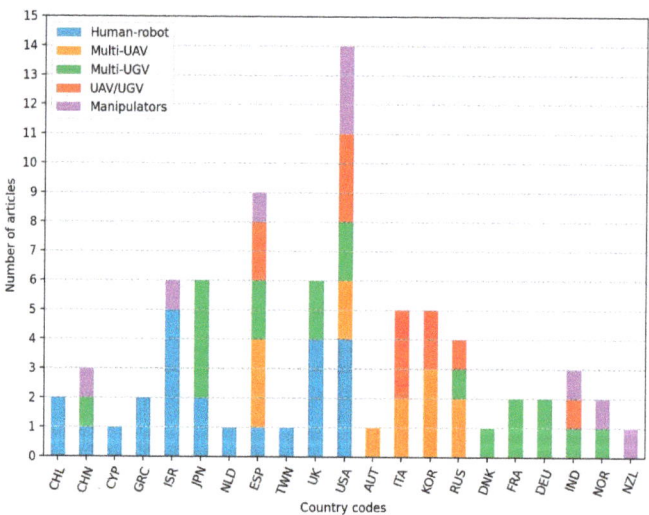

Figure 9. Geographical distribution of articles per country per topic.

Finally, the bar chart in Figure 10 displays the distribution of the 77 articles per year from the year 2003 to the year 2021, where, on the bar of a year, the distribution to different topics is indicated by a different color. An abrupt increase is obvious in Figure 10 in the year 2016. More specifically, since 2016, there has been a sustained increase (tripling or more) of the annual publications compared to the previous years, before 2016. The reason for the aforementioned abrupt increase is unknown. However, for the current year (2021), based on evidence, it is reasonable to assume that the aforementioned trend is sustainable because 6 publications have already been reported up to May. In the aforementioned sense, the recent COVID-19 pandemic does not seem to have affected interest in this technology.

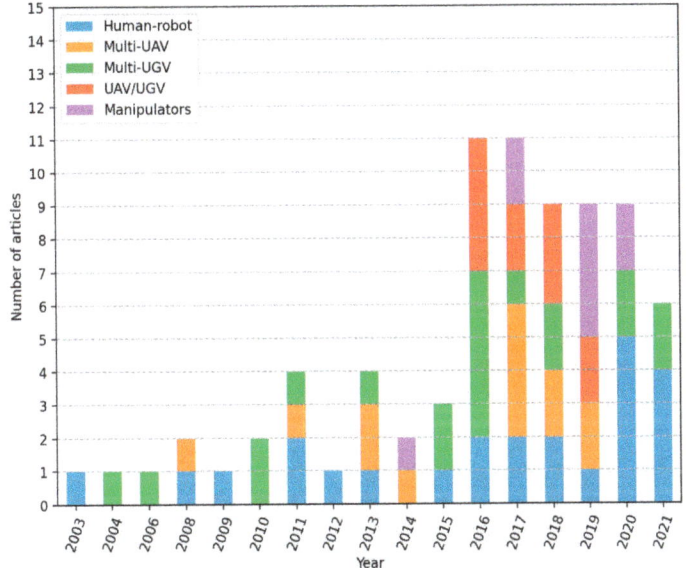

Figure 10. Number of articles in chronological order. The year 2021 was partially included (up to May).

4. Discussion and Conclusions

This paper reviewed fifteen years of work on cooperative robotics in agriculture including human–robot and robot–robot cooperation. Five different topics were identified, namely (a) human–robot cooperation (human–robot), (b) cooperative UAVs (multi–UAV), (c) cooperative UGVs (multi–UGV), (d) hybrid teams of UAVs and UGVs (UAV/UGV), and (e) cooperative manipulation by multi-arm systems (manipulators).

The compiled evidence from the literature (Figures 6–9) suggests that there is an emerging global interest in cooperative robotics in agriculture. In conclusion, Figure 10 confirms that the interest in the area has clearly increased over the last 15 years.

There are a number of important reasons for considering cooperative robotics in agriculture. One reason is food production for an increasingly vast human population, especially under seasonal human labor shortages, e.g., during harvest. Another reason is controllably minimal environmental pollution pursued by the minimization of human involvement during food production. Yet another reason is sustainable food production in the event of natural disasters, including climate changes as well as pandemics such as the recent COVID-19 pandemic.

The technology of interest here, namely "collaborative robotics in agriculture", was described as (a) an "Agriculture 4.0" technology, and (b) a precursor to "Agriculture 5.0" technology; the latter regards an integration of humans with robots in agricultural applications.

In the aforementioned context, a significant technological challenge remains, which is the development of models to effectively drive a robot during its interaction with another robot and/or a human. Note that a robot in agriculture has been described as a Cyber-Physical System (CPS) [7]. In the latter context, the "Lattice Computing (LC) information processing paradigm" has been proposed as a promising mathematical domain for rigorous modeling CPSs due to LC's capacity to accommodate rigorously both numerical data (regarding the "physical" components of a CPS) and non-numerical data (regarding the "cyber" components of a CPS) [98]. Moreover, the work in [7] considered the potential of LC in agricultural applications. Future work remains to demonstrate it further [99].

Based on the reviewed work, a number of research areas in cooperative agricultural robotics are still open to further development in order to improve the current systems, both in terms of usefulness as well as reliability, thus reaching the stage of commercial availability in the near future. In terms of human–robot collaborative teams, it is important to research appropriate interfaces so that effective collaborative control is achieved, especially by field workers with minimal technological training. In addition, human–robot coordination is still a very promising research area that will provide robots with a better perception of human actions and intentions, and therefore greatly improve coordination issues. On the other hand, monitoring tasks require the deployment of large numbers of cooperating aerial and ground vehicles. Despite the increasing availability of affordable small-sized robotic platforms, it appears that the power requirements of such robots are restricting the number of tasks they are able to perform, and, as a result, they are limited to short inspection and mapping missions. Finally, a promising research field is the cooperative manipulation of agricultural products, as applied in harvesting and transportation operations. In addition to being indispensable in some tasks, for instance occlusion removal or handling, the use of multiple arms in a cooperative manner could mitigate limitations of other underlying technologies such as vision.

Figure 11 shows the SWOT analysis of cooperative approaches in agricultural robotics.

Strengths	Weaknesses
• Reduction of task duration • Completion of tasks where heterogeneous robots are required • Reduction of soil compaction through the use of smaller machines • Redundancy in cases of hardware failure	• Limitations in power autonomy • Human-robot interfaces and action coordination • Absence of cooperation between humans and multiple robots
Opportunities	**Threats**
• Labor shortages of seasonal workers • Increased availability of low-cost robotic equipment and platforms	• Human–robot work safety

Figure 11. SWOT analysis.

Additional research, in the context of this work, regarded the engagement of collaborative robots for livestock handling. It turned out that only a few publications exist in the literature typically involving single robots. We believe that there is a promising potential in using teams of collaborative robots for livestock handling.

Author Contributions: Conceptualization, C.L. and V.G.K.; methodology, V.G.K. and T.P.; validation, M.M. and S.C.; formal analysis, C.L., T.P., M.M. and E.V.; writing—original draft preparation, C.L., V.G.K., T.P. and T.K.; writing—review and editing, C.L., T.K., E.V. and S.C.; supervision, V.G.K. All authors have read and agreed to the published version of the manuscript.

Funding: We acknowledge support of this work by the project "Technology for Skillful Viniculture (SVtech)" (MIS 5046047) which is implemented under the Action "Reinforcement of the Research and Innovation Infrastructure" funded by the Operational Program "Competitiveness, Entrepreneurship and Innovation" (NSRF 2014-2020) and co-financed by Greece and the European Union (European Regional Development Fund).

Institutional Review Board Statement: Not applicable.

Informed Consent Statement: Not applicable.

Conflicts of Interest: The authors declare no conflict of interest.

References

1. Lowenberg-DeBoer, J.; Erickson, B. Setting the Record Straight on Precision Agriculture Adoption. *Agron. J.* **2019**, *111*, 1552–1569. [CrossRef]
2. Marinoudi, V.; Sørensen, C.G.; Pearson, S.; Bochtis, D. Robotics and labour in agriculture. A context consideration. *Biosyst. Eng.* **2019**, *184*, 111–121. [CrossRef]
3. Bac, C.W.; van Henten, E.J.; Hemming, J.; Edan, Y. Harvesting Robots for High-value Crops: State-of-the-art Review and Challenges Ahead. *J. Field Robot.* **2014**, *31*, 888–911. [CrossRef]
4. Fountas, S.; Mylonas, N.; Malounas, I.; Rodias, E.; Hellmann Santos, C.; Pekkeriet, E. Agricultural Robotics for Field Operations. *Sensors* **2020**, *20*, 2672. [CrossRef]
5. Oliveira, L.F.P.; Moreira, A.P.; Silva, M.F. Advances in Agriculture Robotics: A State-of-the-Art Review and Challenges Ahead. *Robotics* **2021**, *10*, 52. [CrossRef]
6. Skillful Viniculture Technology (SVTECH), Action "Reinforcement of the Research and Innovation Infrastructure", Operational Programme "Competitiveness, Entrepreneurship and Innovation", NSRF (National Strategic Reference Framework) 2014–2020. Available online: http://evtar.eu/en/home_en/ (accessed on 10 May 2021).
7. Vrochidou, E.; Tziridis, K.; Nikolaou, A.; Kalampokas, T.; Papakostas, G.A.; Pachidis, T.P.; Mamalis, S.; Koundouras, S.; Kaburlasos, V.G. An Autonomous Grape-Harvester Robot: Integrated System Architecture. *Electronics* **2021**, *10*, 1056. [CrossRef]
8. Welfare, K.S.; Hallowell, M.R.; Shah, J.A.; Riek, L.D. Consider the Human Work Experience When Integrating Robotics in the Workplace. In Proceedings of the 2019 14th ACM/IEEE International Conference on Human-Robot Interaction (HRI), Daegu, Korea, 11–14 March 2019; pp. 75–84.
9. Vasconez, J.P.; Kantor, G.A.; Auat Cheein, F.A. Human–robot interaction in agriculture: A survey and current challenges. *Biosyst. Eng.* **2019**, *179*, 35–48. [CrossRef]
10. Cheein, F.A.; Herrera, D.; Gimenez, J.; Carelli, R.; Torres-Torriti, M.; Rosell-Polo, J.R.; Escola, A.; Arno, J. Human-robot interaction in precision agriculture: Sharing the workspace with service units. In Proceedings of the 2015 IEEE International Conference on Industrial Technology (ICIT), Seville, Spain, 17–19 March 2015; pp. 289–295. [CrossRef]
11. van Henten, E.J.; Bac, C.W.; Hemming, J.; Edan, Y. Robotics in protected cultivation. *IFAC Proc. Vol.* **2013**, *46*, 170–177. [CrossRef]
12. Bechar, A.; Vigneault, C. Agricultural robots for field operations: Concepts and components. *Biosyst. Eng.* **2016**, *149*, 94–111. [CrossRef]
13. Sheridan, T.B. Human–Robot Interaction. *Hum. Factors J. Hum. Factors Ergon. Soc.* **2016**, *58*, 525–532. [CrossRef]
14. Adamides, G.; Katsanos, C.; Constantinou, I.; Christou, G.; Xenos, M.; Hadzilacos, T.; Edan, Y. Design and development of a semi-autonomous agricultural vineyard sprayer: Human–robot interaction aspects. *J. Field Robot.* **2017**, *34*, 1407–1426. [CrossRef]
15. Gomez-Gil, J.; San-Jose-Gonzalez, I.; Nicolas-Alonso, L.F.; Alonso-Garcia, S. Steering a Tractor by Means of an EMG-Based Human-Machine Interface. *Sensors* **2011**, *11*, 7110–7126. [CrossRef] [PubMed]
16. Murakami, N.; Ito, A.; Will, J.D.; Steffen, M.; Inoue, K.; Kita, K.; Miyaura, S. Development of a teleoperation system for agricultural vehicles. *Comput. Electron. Agric.* **2008**, *63*, 81–88. [CrossRef]
17. Bechar, A.; Edan, Y. Human-robot collaboration for improved target recognition of agricultural robots. *Ind. Robot Int. J.* **2003**, *30*, 432–436. [CrossRef]
18. Tkach, I.; Bechar, A.; Edan, Y. Switching Between Collaboration Levels in a Human–Robot Target Recognition System. *IEEE Trans. Syst. Man Cybern. Part C (Appl. Rev.)* **2011**, *41*, 955–967. [CrossRef]
19. Bechar, A.; Meyer, J.; Edan, Y. An Objective Function to Evaluate Performance of Human–Robot Collaboration in Target Recognition Tasks. *IEEE Trans. Syst. Man Cybern. Part C (Appl. Rev.)* **2009**, *39*, 611–620. [CrossRef]
20. Rysz, M.W.; Mehta, S.S. A risk-averse optimization approach to human-robot collaboration in robotic fruit harvesting. *Comput. Electron. Agric.* **2021**, *182*, 106018. [CrossRef]
21. Anagnostis, A.; Benos, L.; Tsaopoulos, D.; Tagarakis, A.; Tsolakis, N.; Bochtis, D. Human Activity Recognition through Recurrent Neural Networks for Human–Robot Interaction in Agriculture. *Appl. Sci.* **2021**, *11*, 2188. [CrossRef]

22. Yang, L.; Noguchi, N. Human detection for a robot tractor using omni-directional stereo vision. *Comput. Electron. Agric.* **2012**, *89*, 116–125. [CrossRef]
23. Berenstein, R.; Edan, Y. Human-robot collaborative site-specific sprayer. *J. Field Robot.* **2017**, *34*, 1519–1530. [CrossRef]
24. Huang, Z.; Miyauchi, G.; Gomez, A.S.; Bird, R.; Kalsi, A.S.; Jansen, C.; Liu, Z.; Parsons, S.; Sklar, E. An Experiment on Human-Robot Interaction in a Simulated Agricultural Task. In *Lecture Notes in Computer Science (Including Subseries Lecture Notes in Artificial Intelligence and Lecture Notes in Bioinformatics)*; Springer: Berlin/Heidelberg, Germany, 2020; Volume 12228 LNAI, pp. 221–233. ISBN 9783030634858.
25. Huang, Z.; Gomez, A.; Bird, R.; Kalsi, A.; Jansen, C.; Liu, Z.; Miyauchi, G.; Parsons, S.; Sklar, E. Understanding human responses to errors in a collaborative human-robot selective harvesting task. In Proceedings of the UKRAS20 Conference: "Robots into the Real World" Proceedings, Lincoln, England, 17 April 2020; EPSRC UK-RAS Network, 2020. pp. 89–91.
26. Kim, M.; Koh, I.; Jeon, H.; Choi, J.; Min, B.C.; Matson, E.T.; Gallagher, J. A HARMS-based heterogeneous human-robot team for gathering and collecting. *Adv. Robot. Res.* **2018**, *3*, 201–217. [CrossRef]
27. Zhou, X.; He, J.; Chen, D.; Li, J.; Jiang, C.; Ji, M.; He, M. Human-robot skills transfer interface for UAV-based precision pesticide in dynamic environments. *Assem. Autom.* **2021**, *41*, 345–357. [CrossRef]
28. Lai, Y.-L.; Chen, P.-L.; Yen, P.-L. A Human-Robot Cooperative Vehicle for Tea Plucking. In Proceedings of the 2020 7th International Conference on Control, Decision and Information Technologies (CoDIT), Prague, Czech Republic, 29 June–2 July 2020; Volume 2, pp. 217–222.
29. Baylis, L.C. Organizational Culture and Trust within Agricultural Human-Robot Teams. Ph.D. Thesis, Grand Canyon University, Phoenix, AZ, USA, 2020.
30. Rose, D.C.; Lyon, J.; de Boon, A.; Hanheide, M.; Pearson, S. Responsible development of autonomous robotics in agriculture. *Nat. Food* **2021**, *2*, 306–309. [CrossRef]
31. Benos, L.; Bechar, A.; Bochtis, D. Safety and ergonomics in human-robot interactive agricultural operations. *Biosyst. Eng.* **2020**, *200*, 55–72. [CrossRef]
32. Baxter, P.; Cielniak, G.; Hanheide, M.; From, P. Safe Human-Robot Interaction in Agriculture. In Proceedings of the Companion of the 2018 ACM/IEEE International Conference on Human-Robot Interaction, Chicago, IL, USA, 5–8 March 2018; ACM: New York, NY, USA, 2018; pp. 59–60.
33. Alsalam, B.H.Y.; Morton, K.; Campbell, D.; Gonzalez, F. Autonomous UAV with vision based on-board decision making for remote sensing and precision agriculture. In Proceedings of the 2017 IEEE Aerospace Conference, Big Sky, MT, USA, 4–11 March 2017; pp. 1–12.
34. Santesteban, L.G.; Di Gennaro, S.F.; Herrero-Langreo, A.; Miranda, C.; Royo, J.B.; Matese, A. High-resolution UAV-based thermal imaging to estimate the instantaneous and seasonal variability of plant water status within a vineyard. *Agric. Water Manag.* **2017**, *183*, 49–59. [CrossRef]
35. Torres-Sánchez, J.; López-Granados, F.; Serrano, N.; Arquero, O.; Peña, J.M. High-throughput 3-D monitoring of agricultural-tree plantations with Unmanned Aerial Vehicle (UAV) technology. *PLoS ONE* **2015**, *10*, e0130479. [CrossRef]
36. Long, D.; McCarthy, C.; Jensen, T. Row and water front detection from UAV thermal-infrared imagery for furrow irrigation monitoring. In Proceedings of the 2016 IEEE International Conference on Advanced Intelligent Mechatronics (AIM), Banff, AB, Canada, 12–15 July 2016; pp. 300–305.
37. Faiçal, B.S.; Freitas, H.; Gomes, P.H.; Mano, L.Y.; Pessin, G.; de Carvalho, A.C.P.L.F.; Krishnamachari, B.; Ueyama, J. An adaptive approach for UAV-based pesticide spraying in dynamic environments. *Comput. Electron. Agric.* **2017**, *138*, 210–223. [CrossRef]
38. Doering, D.; Benenmann, A.; Lerm, R.; de Freitas, E.P.; Muller, I.; Winter, J.M.; Pereira, C.E. Design and Optimization of a Heterogeneous Platform for multiple UAV use in Precision Agriculture Applications. *IFAC Proc. Vol.* **2014**, *47*, 12272–12277. [CrossRef]
39. del Cerro, J.; Cruz Ulloa, C.; Barrientos, A.; de León Rivas, J. Unmanned Aerial Vehicles in Agriculture: A Survey. *Agronomy* **2021**, *11*, 203. [CrossRef]
40. Vu, Q.; Nguyen, V.; Solenaya, O.; Ronzhin, A. Group Control of Heterogeneous Robots and Unmanned Aerial Vehicles in Agriculture Tasks. In Proceedings of the International Conference on Interactive Collaborative Robotics (ICR 2017), Hatfield, UK, 12–16 September 2017; Ronzhin, A., Rigoll, G., Meshcheryakov, R., Eds.; Springer: Cham, Switzerland, 2017; Volume 10459, pp. 260–267, ISBN 978-3-319-66470-5.
41. Chao, H.; Baumann, M.; Jensen, A.; Chen, Y.; Cao, Y.; Ren, W.; McKee, M. Band-reconfigurable Multi-UAV-based Cooperative Remote Sensing for Real-time Water Management and Distributed Irrigation Control. *IFAC Proc. Vol.* **2008**, *41*, 11744–11749. [CrossRef]
42. Albani, D.; Manoni, T.; Arik, A.; Nardi, D.; Trianni, V. Field Coverage for Weed Mapping: Toward Experiments with a UAV Swarm. In *Lecture Notes of the Institute for Computer Sciences, Social-Informatics and Telecommunications Engineering, LNICST*; Springer: Berlin/Heidelberg, Germany, 2019; Volume 289, pp. 132–146, ISBN 9783030242015.
43. Albani, D.; IJsselmuiden, J.; Haken, R.; Trianni, V. Monitoring and mapping with robot swarms for agricultural applications. In Proceedings of the 2017 14th IEEE International Conference on Advanced Video and Signal Based Surveillance (AVSS), Lecce, Italy, 29 August–1 September 2017; pp. 1–6.

44. Del Cerro, J.; Barrientos, A.; Sanz, D.; Valente, J. Aerial Fleet in RHEA Project: A High Vantage Point Contributions to ROBOT 2013; Armada, M.A., Sanfeliu, A., Ferre, M., Eds.; Advances in Intelligent Systems and Computing; Springer: Cham, Switzerland, 2014; Volume 252, ISBN 978-3-319-03412-6.
45. Valente, J.; Del Cerro, J.; Barrientos, A.; Sanz, D. Aerial coverage optimization in precision agriculture management: A musical harmony inspired approach. *Comput. Electron. Agric.* **2013**, *99*, 153–159. [CrossRef]
46. Ju, C.; Son, H. Il A distributed swarm control for an agricultural multiple unmanned aerial vehicle system. *Proc. Inst. Mech. Eng. Part I J. Syst. Control Eng.* **2019**, *233*, 1298–1308. [CrossRef]
47. Skobelev, P.; Budaev, D.; Gusev, N.; Voschuk, G. Designing Multi-agent Swarm of UAV for Precise Agriculture. In *Highlights of Practical Applications of Agents, Multi-Agent Systems, and Complexity: The PAAMS Collection*; Bajo, J., Corchado, J.M., Navarro Martínez, E.M., Osaba Icedo, E., Mathieu, P., Hoffa-Dąbrowska, P., del Val, E., Giroux, S., Castro, A.J.M., Sánchez-Pi, N., et al., Eds.; Communications in Computer and Information Science; Springer: Cham, Switzerland, 2018; Volume 887, pp. 47–59, ISBN 978-3-319-94778-5.
48. Barrientos, A.; Colorado, J.; del Cerro, J.; Martinez, A.; Rossi, C.; Sanz, D.; Valente, J. Aerial remote sensing in agriculture: A practical approach to area coverage and path planning for fleets of mini aerial robots. *J. Field Robot.* **2011**, *28*, 667–689. [CrossRef]
49. Drenjanac, D.; Tomic, S.D.K.; Klausner, L.; Kühn, E. Harnessing coherence of area decomposition and semantic shared spaces for task allocation in a robotic fleet. *Inf. Process. Agric.* **2014**, *1*, 23–33. [CrossRef]
50. Ju, C.; Son, H. Multiple UAV Systems for Agricultural Applications: Control, Implementation, and Evaluation. *Electronics* **2018**, *7*, 162. [CrossRef]
51. Ju, C.; Park, S.; Park, S.; Son, H. Il A Haptic Teleoperation of Agricultural Multi-UAV. In Proceedings of the Workshop on Agricultural Robotics: Learning from Industry 4.0 and Moving into the Future at the IEEE/RSJ International Conference on Intelligent Robots and Systems (IROS), Vancouver, BC, Canada, 28 September 2017; pp. 1–6.
52. Nolan, P.; Paley, D.A.; Kroeger, K. Multi-UAS path planning for non-uniform data collection in precision agriculture. In Proceedings of the 2017 IEEE Aerospace Conference, Big Sky, MT, USA, 4–11 March 2017; pp. 1–12.
53. Bochtis, D.D.; Sørensen, C.G.; Green, O.; Moshou, D.; Olesen, J. Effect of controlled traffic on field efficiency. *Biosyst. Eng.* **2010**, *106*, 14–25. [CrossRef]
54. Noguchi, N.; Will, J.; Reid, J.; Zhang, Q. Development of a master-slave robot system for farm operations. *Comput. Electron. Agric.* **2004**, *44*, 1–19. [CrossRef]
55. Zhang, X.; Geimer, M.; Noack, P.O.; Grandl, L. Development of an intelligent master-slave system between agricultural vehicles. In Proceedings of the 2010 IEEE Intelligent Vehicles Symposium, La Jolla, CA, USA, 21–24 June 2010; pp. 250–255.
56. Zhang, C.; Noguchi, N.; Yang, L. Leader–follower system using two robot tractors to improve work efficiency. *Comput. Electron. Agric.* **2016**, *121*, 269–281. [CrossRef]
57. Noguchi, N.; Barawid, O.C. Robot Farming System Using Multiple Robot Tractors in Japan Agriculture. *IFAC Proc. Vol.* **2011**, *44*, 633–637. [CrossRef]
58. Li, N.; Remeikas, C.; Xu, Y.; Jayasuriya, S.; Ehsani, R. Task Assignment and Trajectory Planning Algorithm for a Class of Cooperative Agricultural Robots. *J. Dyn. Syst. Meas. Control* **2015**, *137*, 1–9. [CrossRef]
59. Conesa-Muñoz, J.; Bengochea-Guevara, J.M.; Andujar, D.; Ribeiro, A. Route planning for agricultural tasks: A general approach for fleets of autonomous vehicles in site-specific herbicide applications. *Comput. Electron. Agric.* **2016**, *127*, 204–220. [CrossRef]
60. Blender, T.; Buchner, T.; Fernandez, B.; Pichlmaier, B.; Schlegel, C. Managing a Mobile Agricultural Robot Swarm for a seeding task. In Proceedings of the IECON 2016-42nd Annual Conference of the IEEE Industrial Electronics Society, Florence, Italy, 23–26 October 2016; pp. 6879–6886.
61. Anil, H.; Nikhil, K.S.; Chaitra, V.; Sharan, B.S.G. Revolutionizing Farming Using Swarm Robotics. In Proceedings of the 2015 6th International Conference on Intelligent Systems, Modelling and Simulation, Kuala Lumpur, Malaysia, 9–12 February 2015; pp. 141–147.
62. Janani, A.; Alboul, L.; Penders, J. Multi Robot Cooperative Area Coverage, Case Study: Spraying. In *Lecture Notes in Computer Science*; Springer: Berlin/Heidelberg, Germany, 2016; Volume 9716, pp. 165–176, ISBN 9783319403786.
63. Janani, A.; Alboul, L.; Penders, J. Multi-agent cooperative area coverage: Case study ploughing. In Proceedings of the 2016 International Conference on Autonomous Agents & Multiagent Systems, AAMAS, Singapore, Singapore, 9–13 May 2016; pp. 1397–1398.
64. Tourrette, T.; Deremetz, M.; Naud, O.; Lenain, R.; Laneurit, J.; De Rudnicki, V. Close Coordination of Mobile Robots Using Radio Beacons: A New Concept Aimed at Smart Spraying in Agriculture. In Proceedings of the 2018 IEEE/RSJ International Conference on Intelligent Robots and Systems (IROS), Madrid, Spain, 1–5 October 2018; pp. 7727–7734.
65. Hameed, I.A. A Coverage Planner for Multi-Robot Systems in Agriculture. In Proceedings of the 2018 IEEE International Conference on Real-time Computing and Robotics (RCAR), Kandima, Maldives, 1–5 August 2018; pp. 698–704.
66. Arguenon, V.; Bergues-Lagarde, A.; Rosenberger, C.; Bro, P.; Smari, W. Multi-Agent Based Prototyping of Agriculture Robots. In Proceedings of the International Symposium on Collaborative Technologies and Systems (CTS'06), Las Vegas, NV, USA, 14–17 May 2006; Volume 2006, pp. 282–288.
67. Emmi, L.; Paredes-Madrid, L.; Ribeiro, A.; Pajares, G.; Gonzalez-de-Santos, P. Fleets of robots for precision agriculture: A simulation environment. *Ind. Robot Int. J.* **2013**, *40*, 41–58. [CrossRef]

68. Iida, M.; Harada, S.; Sasaki, R.; Zhang, Y.; Asada, R.; Suguri, M.; Masuda, R. Multi-Combine Robot System for Rice Harvesting Operation. In Proceedings of the 2017 ASABE Annual International Meeting, Spokane, WA, USA, 16–19 July 2017; American Society of Agricultural and Biological Engineers: St. Joseph, MI, USA, 2017; pp. 1–5.
69. Teslya, N.; Smirnov, A.; Ionov, A.; Kudrov, A. Multi-robot Coalition Formation for Precision Agriculture Scenario Based on Gazebo Simulator. In *Smart Innovation, Systems and Technologies*; Springer: Berlin/Heidelberg, Germany, 2021; pp. 329–341, ISBN 9789811555794.
70. Davoodi, M.; Faryadi, S.; Velni, J.M. A Graph Theoretic-Based Approach for Deploying Heterogeneous Multi-agent Systems with Application in Precision Agriculture. *J. Intell. Robot. Syst.* **2021**, *101*, 10. [CrossRef]
71. Wu, C.; Chen, Z.; Wang, D.; Song, B.; Liang, Y.; Yang, L.; Bochtis, D.D. A Cloud-Based In-Field Fleet Coordination System for Multiple Operations. *Energies* **2020**, *13*, 775. [CrossRef]
72. Kim, J.; Son, H. Il A Voronoi Diagram-Based Workspace Partition for Weak Cooperation of Multi-Robot System in Orchard. *IEEE Access* **2020**, *8*, 20676–20686. [CrossRef]
73. Vu, Q.; Raković, M.; Delic, V.; Ronzhin, A. Trends in Development of UAV-UGV Cooperation Approaches in Precision Agriculture. In *Lecture Notes in Computer Science (Including Subseries Lecture Notes in Artificial Intelligence and Lecture Notes in Bioinformatics)*; Springer: Berlin/Heidelberg, Germany, 2018; Volume 11097 LNAI, pp. 213–221, ISBN 9783319995816.
74. Menendez-Aponte, P.; Garcia, C.; Freese, D.; Defterli, S.; Xu, Y. Software and hardware architectures in cooperative aerial and ground robots for agricultural disease detection. In Proceedings of the 2016 International Conference on Collaboration Technologies and Systems (CTS), Orlando, FL, USA, 31 October–4 November 2016; pp. 354–358. [CrossRef]
75. Tokekar, P.; Hook, J.V.; Mulla, D.; Isler, V. Sensor Planning for a Symbiotic UAV and UGV System for Precision Agriculture. *IEEE Trans. Robot.* **2016**, *32*, 1498–1511. [CrossRef]
76. Ni, J.; Wang, X.; Tang, M.; Cao, W.; Shi, P.; Yang, S.X. An Improved Real-Time Path Planning Method Based on Dragonfly Algorithm for Heterogeneous Multi-Robot System. *IEEE Access* **2020**, *8*, 140558–140568. [CrossRef]
77. Li, J.; Deng, G.; Luo, C.; Lin, Q.; Yan, Q.; Ming, Z. A Hybrid Path Planning Method in Unmanned Air/Ground Vehicle (UAV/UGV) Cooperative Systems. *IEEE Trans. Veh. Technol.* **2016**, *65*, 9585–9596. [CrossRef]
78. Peterson, J.; Li, W.; Cesar-Tondreau, B.; Bird, J.; Kochersberger, K.; Czaja, W.; McLean, M. Experiments in unmanned aerial vehicle/unmanned ground vehicle radiation search. *J. Field Robot.* **2019**, *36*, 818–845. [CrossRef]
79. Wang, Z.; McDonald, S.T. Convex relaxation for optimal rendezvous of unmanned aerial and ground vehicles. *Aerosp. Sci. Technol.* **2020**, *99*, 105756. [CrossRef]
80. Conesa-Muñoz, J.; Valente, J.; del Cerro, J.; Barrientos, A.; Ribeiro, A. A Multi-Robot Sense-Act Approach to Lead to a Proper Acting in Environmental Incidents. *Sensors* **2016**, *16*, 1269. [CrossRef]
81. Gonzalez-de-Santos, P.; Ribeiro, A.; Fernandez-Quintanilla, C.; Lopez-Granados, F.; Brandstoetter, M.; Tomic, S.; Pedrazzi, S.; Peruzzi, A.; Pajares, G.; Kaplanis, G.; et al. Fleets of robots for environmentally-safe pest control in agriculture. *Precis. Agric.* **2017**, *18*, 574–614. [CrossRef]
82. Potena, C.; Khanna, R.; Nieto, J.; Nardi, D.; Pretto, A. Collaborative UAV-UGV Environment Reconstruction in Precision Agriculture. In Proceedings of the IEEE/RSJ IROS Workshop" Vision-Based Drones: What's Next, Madrid, Spain, 1–5 October 2010; pp. 1–6.
83. Potena, C.; Khanna, R.; Nieto, J.; Siegwart, R.; Nardi, D.; Pretto, A. AgriColMap: Aerial-Ground Collaborative 3D Mapping for Precision Farming. *IEEE Robot. Autom. Lett.* **2019**, *4*, 1085–1092. [CrossRef]
84. Bhandari, S.; Raheja, A.; Green, R.L.; Do, D. Towards collaboration between unmanned aerial and ground vehicles for precision agriculture. In Proceedings of the Autonomous Air and Ground Sensing Systems for Agricultural Optimization and Phenotyping II, Anaheim, CA, USA, 10–11 April 2017; Thomasson, J.A., McKee, M., Moorhead, R.J., Eds.; International Society for Optics and Photonics: Bellingham, WA, USA, 2017; Volume 10218, p. 1021806.
85. Grassi, R.; Rea, P.; Ottaviano, E.; Maggiore, P. Application of an Inspection Robot Composed by Collaborative Terrestrial and Aerial Modules for an Operation in Agriculture. In *Mechanisms and Machine Science*; Springer: Berlin/Heidelberg, Germany, 2018; Volume 49, pp. 539–546. ISBN 9783319612751.
86. Vasudevan, A.; Kumar, D.A.; Bhuvaneswari, N.S. Precision farming using unmanned aerial and ground vehicles. In Proceedings of the 2016 IEEE Technological Innovations in ICT for Agriculture and Rural Development (TIAR), Chennai, India, 15–16 July 2016; pp. 146–150.
87. Ju, C.; Son, H. Il Hybrid Systems based Modeling and Control of Heterogeneous Agricultural Robots for Field Operations. In Proceedings of the 2019 ASABE Annual International Meeting, Boston, MA, USA, 7–10 July 2019; American Society of Agricultural and Biological Engineers: St. Joseph, MI, USA, 2019; pp. 3–15.
88. Williams, H.A.M.; Jones, M.H.; Nejati, M.; Seabright, M.J.; Bell, J.; Penhall, N.D.; Barnett, J.J.; Duke, M.D.; Scarfe, A.J.; Ahn, H.S.; et al. Robotic kiwifruit harvesting using machine vision, convolutional neural networks, and robotic arms. *Biosyst. Eng.* **2019**, *181*, 140–156. [CrossRef]
89. Zion, B.; Mann, M.; Levin, D.; Shilo, A.; Rubinstein, D.; Shmulevich, I. Harvest-order planning for a multiarm robotic harvester. *Comput. Electron. Agric.* **2014**, *103*, 75–81. [CrossRef]
90. Xiong, Y.; Ge, Y.; Grimstad, L.; From, P.J. An autonomous strawberry-harvesting robot: Design, development, integration, and field evaluation. *J. Field Robot.* **2020**, *37*, 202–224. [CrossRef]

91. Sarabu, H.; Ahlin, K.; Hu, A.-P. Graph-Based Cooperative Robot Path Planning in Agricultural Environments. In Proceedings of the 2019 IEEE/ASME International Conference on Advanced Intelligent Mechatronics (AIM), Hong Kong, China, 8–12 July 2019; pp. 519–525.
92. Ahlin, K.J.; Hu, A.-P.; Sadegh, N. Apple Picking Using Dual Robot Arms Operating Within an Unknown Tree. In Proceedings of the 2017 ASABE Annual International Meeting, Spokane, WA, USA, 16–19 July 2017; American Society of Agricultural and Biological Engineers: St. Joseph, MI, USA, 2017; pp. 1–11.
93. Sepulveda, D.; Fernandez, R.; Navas, E.; Armada, M.; Gonzalez-De-Santos, P. Robotic Aubergine Harvesting Using Dual-Arm Manipulation. *IEEE Access* **2020**, *8*, 121889–121904. [CrossRef]
94. Davidson, J.R.; Hohimer, C.J.; Mo, C.; Karkee, M. Dual robot coordination for apple harvesting. In Proceedings of the 2017 ASABE Annual International Meeting, Spokane, WA, USA, 16–19 July 2017; pp. 1–9. [CrossRef]
95. Ling, X.; Zhao, Y.; Gong, L.; Liu, C.; Wang, T. Dual-arm cooperation and implementing for robotic harvesting tomato using binocular vision. *Rob. Auton. Syst.* **2019**, *114*, 134–143. [CrossRef]
96. Pramod, A.S.; Jithinmon, T.V. Development of mobile dual PR arm agricultural robot. *J. Phys. Conf. Ser.* **2019**, *1240*, 012034. [CrossRef]
97. BACCHUS—Mobile Robotic Platforms for Active Inspection & Harvesting in Agricultural Areas. European Union's Horizon 2020 research and innovation programme under grant agreement No 871704. Available online: https://bacchus-project.eu/ (accessed on 10 May 2021).
98. Kaburlasos, V.G. The Lattice Computing (LC) Paradigm. In Proceedings of the Fifthteenth International Conference on Concept Lattices and Their Applications (CLA 2020), Tallinn, Estonia, 29 June–1 July 2020; Volume 2668, pp. 1–7.
99. Bazinas, C.; Vrochidou, E.; Lytridis, C.; Kaburlasos, V.G. Time-Series of Distributions Forecasting in Agricultural Applications: An Intervals' Numbers Approach. *Eng. Proc.* **2021**, *5*, 12. [CrossRef]

MDPI
St. Alban-Anlage 66
4052 Basel
Switzerland
Tel. +41 61 683 77 34
Fax +41 61 302 89 18
www.mdpi.com

Agronomy Editorial Office
E-mail: agronomy@mdpi.com
www.mdpi.com/journal/agronomy

www.ingramcontent.com/pod-product-compliance
Lightning Source LLC
LaVergne TN
LVHW070712100526
838202LV00013B/1080